Carbohydrate Chemistry

---

Volume 19
Part I

A Specialist Periodical Report

# Carbohydrate Chemistry

Volume 19

Part I
Monosaccharides, Disaccharides, and Specific Oligosaccharides

A Review of the Recent Literature Published during 1985

Senior Reporter
**N. R. Williams,** *Birkbeck College, University of London*

Reporters
**B. E. Davison,** *Polytechnic of North London*
**R. J. Ferrier,** *Victoria University of Wellington, New Zealand*
**R. H. Furneaux,** *D.S.I.R., Petone, New Zealand*
**R. H. Wightman,** *Heriot-Watt University, Edinburgh*

ROYAL SOCIETY OF CHEMISTRY

ISBN 0-85186-222-5
ISSN 0951-8428



Published by The Royal Society of Chemistry
Burlington House, Piccadilly, London W1V 0BN

Printed in Great Britain by
Whitstable Litho Ltd., Whitstable, Kent

# *Preface*

This report summarizes the literature for 1985 available to us by March 1986. We regret that we have not yet been able to improve the speed of publication, but we live in hopes!

Despite the value of this report expressed by some readers, the financial viability of it remains a cause for concern. In line with other titles in the Specialist Periodical Reports series, there has been a steady decline in sales from over 1000 in the initial period to between 500 - 600 in recent years, and of these the great majority are to libraries rather than individuals. Unfortunately this situation forces an unattractively high price for the volume, even at a concessionary rate for members of the Royal Society of Chemistry, and minimizes the financial encouragement of reporters. It is not clear to the Publications Committee of the RSC how this vicious circle can best be broken, and constructive advice would be welcome. At the present time the publication of these reports is under review, and future volumes cannot be guaranteed. We are sorry that in view of this situation we are unable to supply reprints for any part of this report, as this obviously adds to costs and detracts from sales.

I would like to thank my willing band of reporters for all their efforts, and also Dr.P.G.Gardam and his staff at the Royal Society of Chemistry for the production of this volume in its final form.

April 1987 | Neil R.Williams

# R E P R I N T S

In response to several queries, the situation regarding reprints of chapters of *Specialist Periodical Reports* titles is that they are not made available because even a relatively small consequent decrease in sales would have a disproportionately large adverse effect on the precarious finances of this specialist series of books.

# Contents

# Abbreviations

The following abbreviations have been used:

| | |
|---|---|
| Ac | acetyl |
| Ad | adenin-9-yl |
| Bn | benzyl |
| Boc | t-butoxycarbonyl |
| Bz | benzoyl |
| Cbz | benzyloxycarbonyl |
| c.d. | circular dichroism |
| CI | chemical ionization |
| DAST | diethylaminosulphur trifluoride |
| DBU | 1,5-diazabicyclo[5,4,0]undec-5-ene |
| DCC | dicyclohexylcarbodi-imide |
| DEAD | diethyl azodicarboxylate |
| DIBAL | di-isobutylaluminium hydride |
| DMAP | 4-dimethylaminopyridine |
| DMF | N,N-dimethylformamide |
| DMSO | dimethyl sulphoxide |
| EE | 1-ethoxyethyl |
| e.s.r. | electron spin resonance |
| FAB | fast-atom bombardment |
| GC | gas chromatography |
| HMPT | hexamethylphosphorous triamide |
| i.r. | infrared |
| LAH | lithium aluminium hydride |
| LDA | lithium di-isopropylamide |
| MCPBA | m-chloroperbenzoic acid |
| MEM | methoxyethoxymethyl |
| MOM | methoxymethyl |
| m.s. | mass spectrometry |
| Ms | methanesulphonyl |
| NBS | N-bromosuccinimide |
| NIS | N-iodosuccinimide |
| n.m.r. | nuclear magnetic resonance |
| o.r.d. | optical rotatory dispersion |
| PCC | pyridinium chlorochromate |
| PDC | pyridinium dichromate |
| Py | pyridine |
| SIMS | secondary-ion mass spectrometry |
| TASF | tris(dimethylamino)sulphonium difluorotrimethyl silicate |
| Tf | trifluoromethanesulphonyl |
| THF | tetrahydrofuran |
| Thp | tetrahydropyranyl |
| TBDMS | t-butyldimethylsilyl |
| TMS | trimethylsilyl |
| TPP | triphenylphosphine |
| TPS | tri-isopropylbenzenesulphonyl |
| Tr | triphenylmethyl |
| Ts | toluene p-sulphonyl |
| U | uracil-1-yl |

# 1
# Introduction and General Aspects

The past year has seen a continued development of the research area covered in this report towards specific biological applications of mono- and di-saccharides, especially the synthesis of disaccharides and oligosaccharides which are increasingly sought as probes for investigating immunological interactions, and the synthesis of analogues of antibiotics and nucleosides in the search for more effective antibiotics and antimetabolites. Nevertheless, the contents of the other chapters clearly indicate that wider aspects of monosaccharide chemistry continue to excite the interest of carbohydrate chemists, and also others who find carbohydrates convenient chiral sources for a wide range of biological compounds. Modern techniques of n.m.r. spectroscopy and mass spectrometry are also proving of great benefit for the analysis of complex carbohydrate structures. The fact that we record over 1400 references contributed by nearly 3000 workers speaks for itself!

Fringe areas have continued to pose a problem as to where to draw a line between carbohydrates and non-carbohydrate materials. In Chapter 4 we have concentrated on the synthesis of specific oligosaccharides, and the title of the report has been modified to reflect the distinction we wish to make between oligosaccharides occurring as such or attached to aglycones, which includes many antibiotics, and oligosaccharides recognized as structural units of polysaccharides. Likewise the coverage of other materials which are only part carbohydrate has been generally selective for those papers which discuss the chemistry of the carbohydrate moiety; obviously, distinctions are not always clear, and we hope that not too many papers get omitted which would be of interest to readers of this report.

Papers of more general carbohydrate interest published this year have included a description of a computer programme for selecting protecting groups of alcohols, ketones, and alkenes,[1] details of a new ball-stick building set for sugars,[2] and reviews (in Japanese) on the reaction of sugar derivatives with Grignard reagents ( asymmetric reactions, acetal cleavage, deoxygenation of vic-diol monosulphonates, anomerization and ring-opening of furanosides)[3] and on the application of electrochemical oxidation and reduction of carbohydrates.[4] A relationship has been established for several carbohydrates between the partial molar heat capacity and the number of equatorial hydroxy groups in aqueous solution; exceptionally large values are caused by specific interaction of equatorial hydroxy groups with water molecules.[5]

## References


1 E.J.Corey, A.K.Long, T.W.Greene, and J.W.Miller, J. Org. Chem., 1985, 50, 1920.

2 E.Menzel, Prax. Naturwiss., Chem., 1984, 33, 289 (Chem. Abstr., 1985, 102, 5121.

3 M.Kawana, Yuki Gosei Kagaku Kyokaishi, 1985, 43, 226 (Chem. Abstr., 1985, 102, 204 178).

4 I.Kitagawa and M.Yoshikawa, Yuki Gosei Kagaku Kyokaishi, 1985, 43, 594 (Chem. Abstr., 1985, 103, 88 129).

5 H.Uedaira and H.Uedaira, J. Chem. Thermodyn., 1985, 17, 901 (Chem. Abstr., 1985, 103, 215 658).

# 2
# Free Sugars

The processing of hardwoods to produce xylitol and other polyols, monosaccharides and furfural is the subject of a Russian review.[1] The methods for the preparation of lactulose from lactose have been reviewed.[2]

## 1 Theoretical Aspects

A theoretical study of solvent effects on the free energies of cellobiose conformers, taking into account electrostatic, dispersion and cavity terms in ten solvents, has been carried out using the PCILO quantum chemical method: the main conclusion was that the structure observed in aqueous solution is similar to that found in the crystal. The solvent-induced conformational changes for β-cellobiose were compared with those for β-maltose and the differences attributed to the different solvation. Comparisons between calculated and observed n.m.r. spectra were made.[3] *Ab initio* calculations on water dimers with ten model geometries and seventeen which simulate the hydrogen bond contacts encountered in crystals of β-D-fructose, -arabinose, and -turanose examined the relationship between variations in geometry and hydrogen bond energy. The distance between the donor and recipient was varied and the calculated frequency shift was examined in relation to the hydrogen bond energy and the absorption intensity. The absence of a one-to-one relationship led to the conclusion that the observed variations are due to differences in the separation distances. The results were used to determine which hydroxy-oxygen contacts give rise to the narrow hydroxy i.r. bands observed in sugar crystals.[4]

## 2 Synthesis

In a study of the formose reaction carried out in the presence of chitosan-lanthanum hydroxide, the sorption-desorption processes of the reaction and the stereochemical effectiveness were shown to depend mainly on the pH of the medium.[5]

Conversion of D-glucose to D-fructose proceeded in a maximum yield of 67.7% when catalyzed by disodium pentasilicate in aqueous methanol. The reaction was second order with an activation energy of 129 kJ $mol^{-1}$ and proceeded faster with increasing concentration of methanol.[6]

Hydrothermal equilibration of 1,3-dihydroxy-2-propanone and glyceraldehyde with their dehydration product methylglyoxal, a process related to the degradation of biomass, has been studied in the temperature range 180-240 °C.[7]

A further paper on the use of deuterium displacement of the products of photobromination of anhydrosugars (see Vol.17, p.6) describes the synthesis of the two optical antipodes of D-[5-$^{2}H_{1}$]-ribose. The key reaction shown in Scheme 1 for the (5-R)-isomer proceeded with 100% optical purity in 80% chemical yield; the (5R)- and (5S)-isomers were obtained using the previously described procedures.[8]

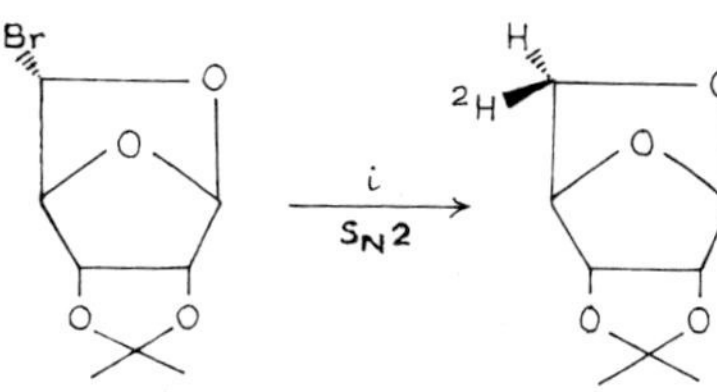


Reagents: i, $LiEt_3B^2H$

Scheme 1

Chain-extension reactions via aldehyde additions have been used to synthesize higher aldoses, enuloses, branched-chain enuloses and 1-deoxy-alduloses. Addition of 2-lithio-1,3-dithiane to partially blocked carbohydrate derivatives proceeds in certain cases with high stereoselectivity; the authors suggest that the reaction is controlled by chelation of the oxygen functions at C-1, C-2, and C-4 with the lithium cation as shown in formula (1) of Scheme 2. The same paper[9] describes the reaction with 2,3;5,6-di-O-isopropylidene D-allofuranose to yield the heptose derivative (2) and its C-2 epimer in a ratio of greater than 30:1. Addition of 2-lithio-1,3-dithiane to 2,3:5,6-di-O-isopropylidene-D-mannofuranose gave rise to the D-glycero-D-galacto-isomer (3) which was converted via a sequence involving oxidation at C-7 and reduction at C-1 into L-glycero-D-mannoheptose (4).[10] Stereocontrolled homologation of α-hydroxy aldehydes has also been achieved by addition of silazoles. Thus the protected derivative (5) of D-erythrose was preparable from

2,3-$\underline{O}$-isopropylidene-D-glyceraldehyde as shown in Scheme 3; using the same sequence on (5), the D-ribose derivative (6) was prepared.[11] The Wittig reaction using the reagents (7) and (8) on 2,3-iso-

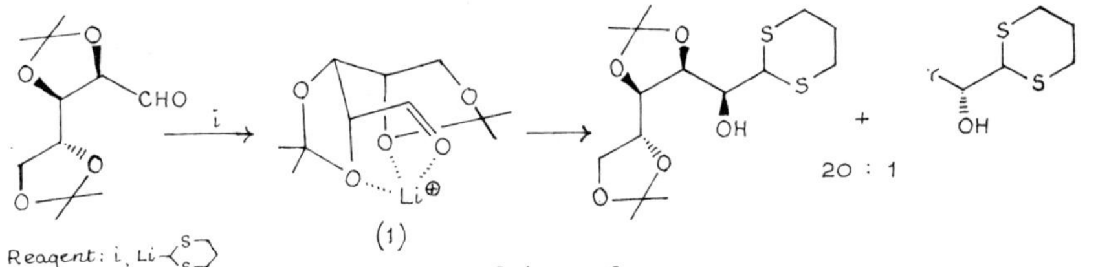


Scheme 2

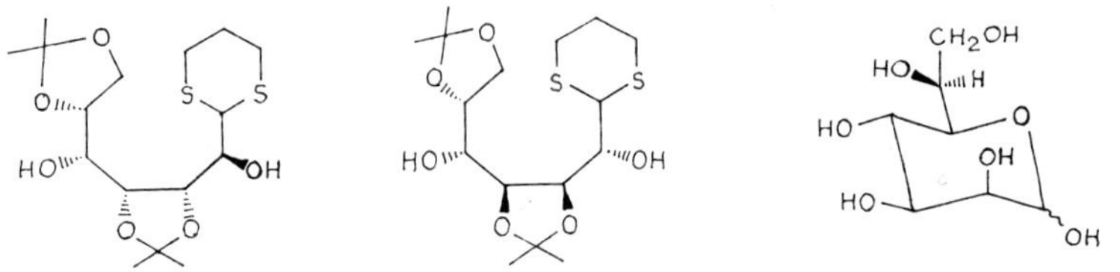

(2) (3) (4)

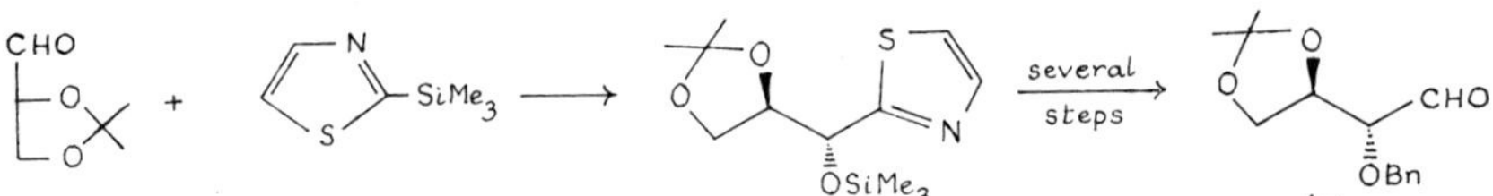


Scheme 3

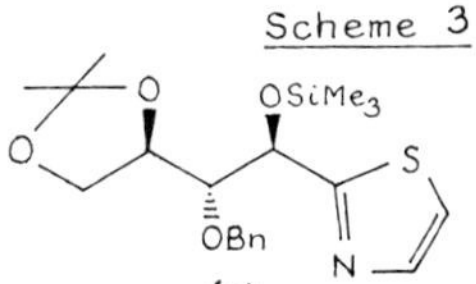

(6)

propylidene-D-glyceraldehyde also proceeded stereoselectively; in the case of (7), the ($\underline{E}$)-isomer (9) predominated over the ($\underline{Z}$)-isomer (10) by 64% to 25% isolated yield, while (8) gave exclusively the ($\underline{E}$)-product (11).[12] Hydroxylation of (9) with osmium tetroxide-potassium chlorate was used to synthesize 1-deoxy-D-tagatose and 1-deoxy-D-psicose as shown in Scheme 4.[13]

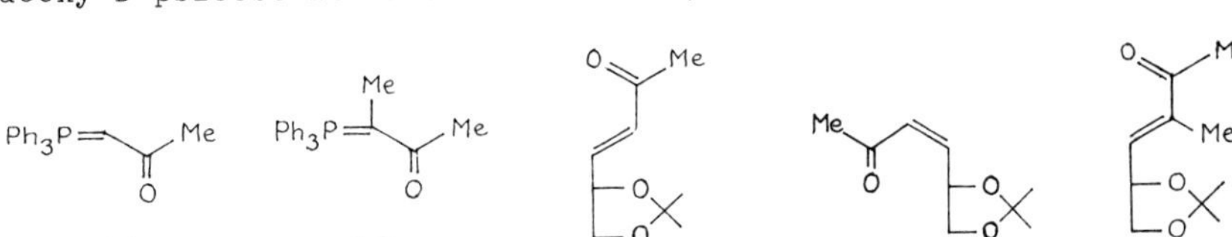

(7) (8) (9) (10) (11)

Catalytic osmylation of the allylic ethers and alcohols (12) led to the octose derivative (13) as the predominant isomer; the (Z)-alkene (14) gave the 7-epimer (15), in agreement with the empirical rule of Kishi that there will be an *erythro*-relationship between the oxygens at C-5 and C-6.[14] The reaction sequence was also extended

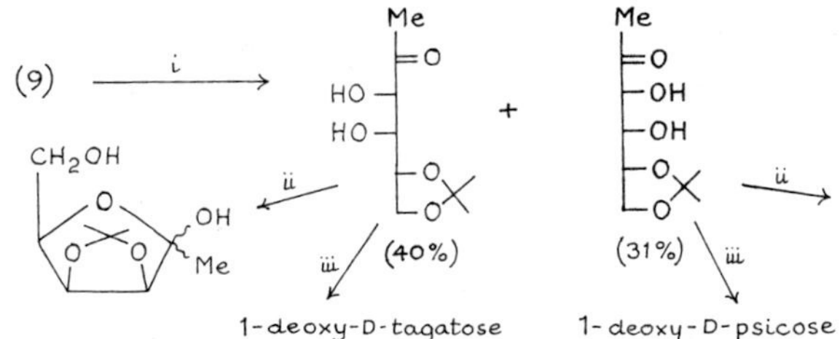


Reagents: i, $OsO_4$-$KClO_3$; ii, $Me_2CO$-$H^+$; iii, $H_3O^+$

Scheme 4

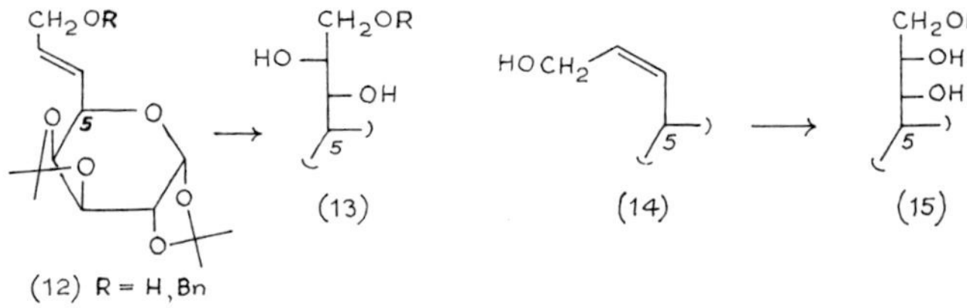


to the synthesis of decose derivatives.[15] The hydroxylation of the $\alpha,\beta$-unsaturated ester derivative (16) yielded the *erythro*-acetal (17). Reduction of the diacetate of (17) with lithium aluminium hydride followed by acid hydrolysis and deacetylation gave DL-erythrose. A similar sequence starting with the (E)-isomer of (16) led to DL-threose.[16] (+)-Di-isopropyl tartrate catalysed t-butylperoxide

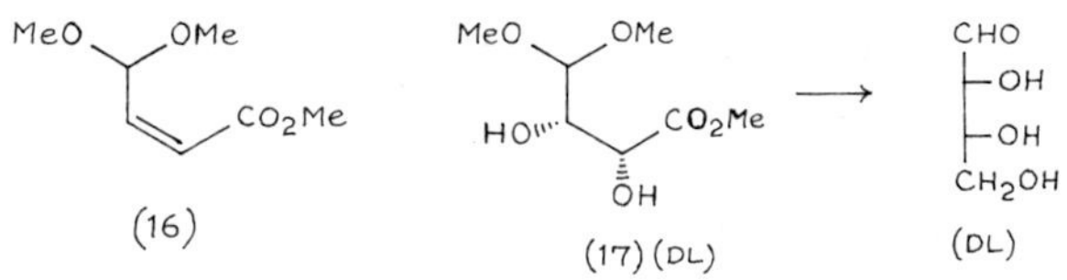


epoxidation of the unsaturated acetal (18), itself prepared by Grignard addition to aldehyde (19), gave a mixture of the *ribo*- and *lyxo*-isomers (20) and (21) in a ratio of 24:1. Separation and base-catalyzed ring opening of (20) yielded the *ribo*-hexose (22), acid hydrolysis of which gave 2,6-dideoxy-DL-*ribo*-hexose (23) (Scheme 5).[17] The total synthesis of methyl 3,4,5-tri-O-acetyl-1,7-di-O-benzyl-α-DL-gluco-hept-2-ulopyranoside (24) has been achieved (Scheme 6).[18]

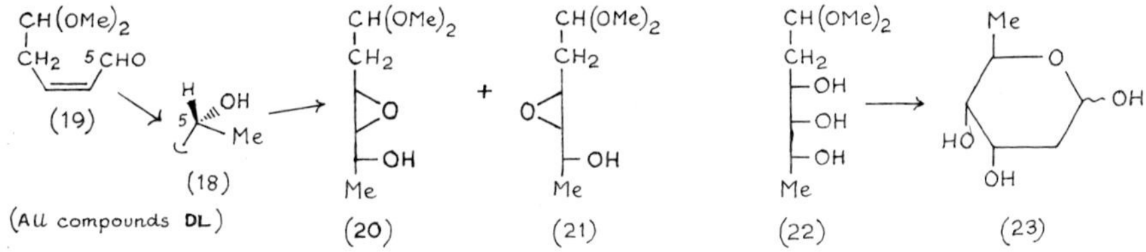

Scheme 5

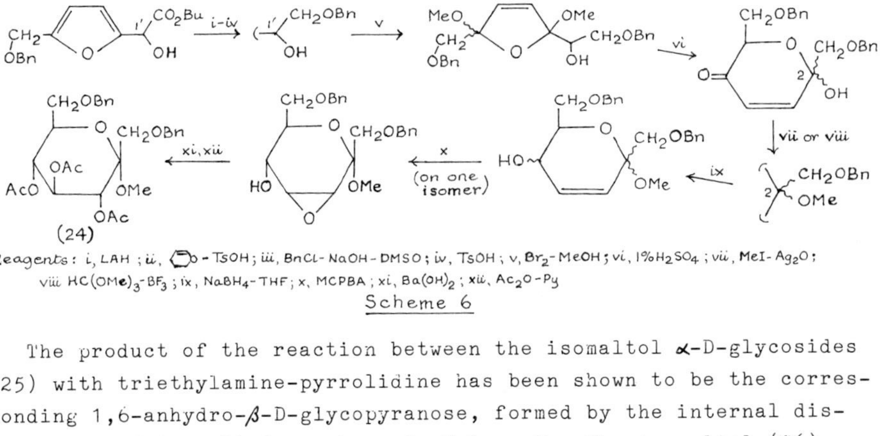

Reagents: i, LAH ; ii, (dihydropyran)-TsOH ; iii, BnCl-NaOH-DMSO ; iv, TsOH ; v, $Br_2$-MeOH ; vi, 1% $H_2SO_4$ ; vii, MeI-$Ag_2O$ ; viii $HC(OMe)_3$-$BF_3$ ; ix, $NaBH_4$-THF ; x, MCPBA ; xi, $Ba(OH)_2$ ; xii, $Ac_2O$-Py

Scheme 6

The product of the reaction between the isomaltol α-D-glycosides (25) with triethylamine-pyrrolidine has been shown to be the corresponding 1,6-anhydro-β-D-glycopyranose, formed by the internal displacement of isomaltol as shown in Scheme 7. The isomaltol (26) gives a polymer, which is also produced in the browning reaction.[19]

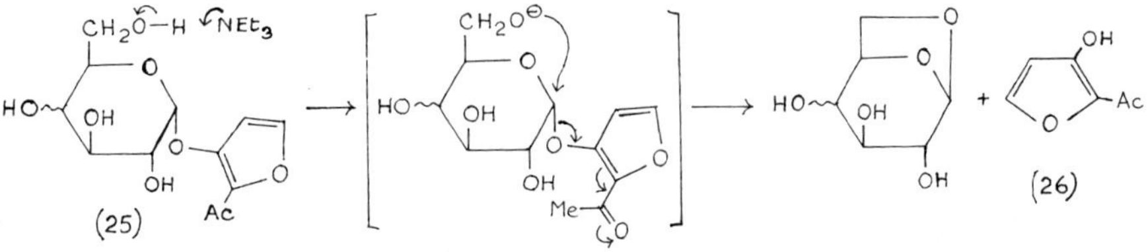

Scheme 7

1-O-Acetyl groups in substituted hexopyranoses may be hydrolyzed by aqueous tin(IV) chloride. The reaction is complete within 1h; trans-1,2-acetoxy groups are hydrolyzed at RT, whereas a temperature of 40 °C is required for 1,2-cis compounds. The method is satisfactory in the presence of 2-phthalimido and 2-O-tosyl groups, while for perbenzyl 1-acetates it is virtually quantitative. 1,2-Trans-di-acetoxy compounds do not give complete formation of the glycose due to a migration of the 2-O-acetyl group to C-1.[20] Regioselective 1-O-deacylation of peracylated glycopyranoses has also been achieved using ammonia in an aprotic solvent such as acetonitrile, dimethoxyethane, or THF, sometimes with the addition of a little methanol. The reaction proceeded in high yield at room temperature. Two examples are shown in Scheme 8.[21]

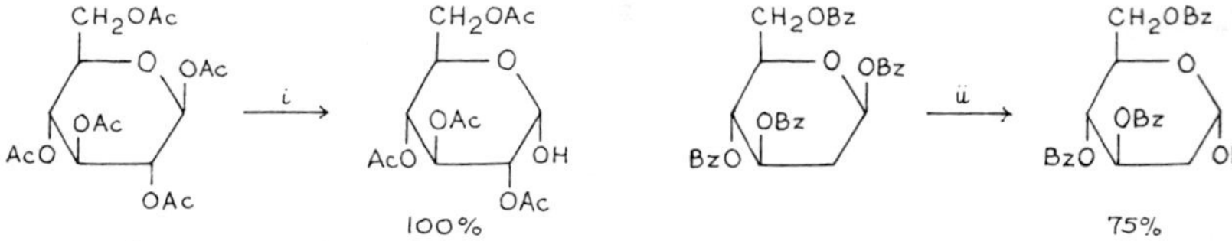


Reagents: i, $NH_3$-MeCN, 6h; ii, $NH_3$-THF-MeOH, 5h

Scheme 8

Galactose oxidase has been used to prepare L-sugars from alditols; the conversions are incomplete due to product inhibition of the enzymes. The general arrangement shown in (27) is necessary although not all examples are good substrates; it was found that addition of ferricyanide ion increased the reaction rate. The enzyme was used to prepare L-xylose, L-galactose, L-glucose, and D-threose, following Scheme 9.[22]

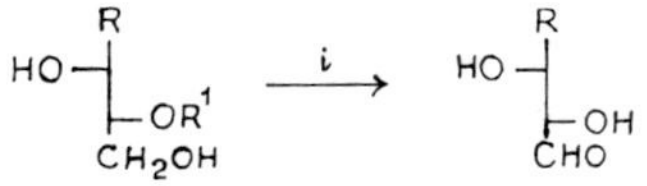


(27)

Reagent: i, Galactose oxidase

Scheme 9

## 3 Physical Measurements

The n.m.r. line width and saturation transfer method has been used to determine the thermodynamic and kinetic parameters for ring-opening and closing of aldo- and ketofuranoses and their phosphate esters.[23] The kinetics of the mutarotation of sugars have been determined using an h.p.l.c. technique, in which the change in the chromatogram with time following dissolution of a pure anomer in water was determined. For the α-L-rhamnopyranose⇌β-L-rhamnopyranose equilibrium the rate constant was determined to be $6.8 \times 10^{-4}\,s^{-1}$ at 25° in good agreement with values from optical rotation methods. The equilibrium of anomers in different solvents was also studied.[24]

Excess enthalpies of aqueous solutions of aldopyranosides have been measured at 25° and the self- and cross-interaction coefficients determined.[25] The same microcalorimetric methods were used to measure the excess enthalpies of eight ternary aqueous solutions of four aldopentoses and either glycine or N-acetylglycinamide at 25°C. The results showed that these amino-acids do not provide good models for the peptide-sugar interactions (see also Vol.18, p.7).[26] A rigorous method of analysis has been applied to determine the temperature dependent activation parameters for sucrose hydrolysis. It was concluded that the inclusion of a temperature dependent activation enthalpy was unwarranted.[27]

Sugar-water interactions have been evaluated from diffusion measurements. The Stokes-Einstein relation for mono- and trisaccharides was discussed and the diffusion coefficients for D-ribose and 2-deoxy-D-ribose were measured. The latter has the larger coefficient, showing that 2-deoxy-ribose breaks the local water structure, whereas D-ribose hardly affects the structure. It was suggested that the mean number of equatorial hydroxy groups is a good parameter for describing sugar hydration properties.[28] It has been demonstrated that, over a wide range of concentration, the ideal and non-ideal solution models of D-glucose are inadequate unless hydration is taken into account. A hydration number of 3.5 was found for molecular D-glucose, while dissociated D-glucose ions are not hydrated: this result was discussed in relation to intramolecular hydrogen bonding within D-glucose ions produced in alkaline isomerization reactions.[29] A study by $^{13}$C n.m.r. spectroscopy of D-idose using partially deuterated hydroxy groups has shown that intramolecular hydrogen bonding stabilizes the α-pyranose in the $^{4}C_{1}$ conformation in non-aqueous solvents (see also Chapter 21).[30] The constants for H-bridging complex formation between D-glucose, cellobiose, xylose, and phenol as models for cellulose and different O-basic dipolar molecules such as N-methyl-caprolactam, HMPA, and DMSO in chloroform and ethanol have been determined.[31] Evidence of hydrogen bonding in D-fructose as shown in (28) has been obtained from variable temperature high field n.m.r. spectroscopy. The preponderance of the β-furanose form in DMSO is attributed to this feature. The kinetics of the tautomeric equilibria were studied.[32]

The concentration and structure of free radicals generated in D-glucose, lactose and cellulose by γ-irradiation have been determined by the chemiluminescence produced on contact with distilled water or aqueous solutions of various substances. It was possible to

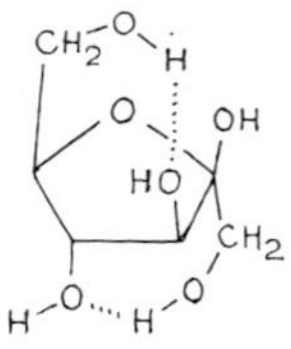

(28)

distinguish between singlet oxygen and excited carbonyl species by the wavelength of the chemiluminescence (630 nm and 500 nm respectively).[33] $\gamma$-Irradiation of cellobiose gave a paramagnetic intermediate formed by cleavage of the C-5-O bond as shown by e.s.r. spectroscopy. A mechanism for radiolysis of cellobiose was proposed.[34] E.s.r. spectroscopy has also been used to show that the source of lyoluminescence in irradiated L-rhamnose is the recombination of generated peroxide radicals to yield carbonyl-containing species.[35]

The density, viscosity, and ultrasonic velocity of aqueous D-fructose solutions in the presence of metal ions have been determined at 30$^{o}$C. Partial molal volumes, partial molal compressibility, free energy change, and viscosity **B**-coefficients of these solutions were calculated from the data. It was concluded that solvent-solute interactions are suppressed in the presence of metal ions.[36] "Hydrophobic indices" calculated from the hydrophobic and hydrophilic surface areas of seven monosaccharides were calculated and found to correlate with the partition coefficients between polystyrene gel and water and with the free energy change for transfer from water to butanol. The hydrophobic surface area was defined as the surface area occupied by methine and methylene groups and the hydrophilic surface as that occupied by hydroxy and ether oxygen functions. Different conformations of anomers of pyranose and furanose forms were considered where appropriate.[37]

When manganese(III) ions are used to induce reaction between hypobromous acid and sugars, oscillations appear in the potential of a platinum electrode in the solution. Nine common mono- and disacccharides were studied by such polarography and sustained oscillations of the Belousov-Zhabotinsky type were noted when a nitrogen flow was used.[38] N.m.r. spectroscopy of D-[2-$^{13}$C]fructose has been used to study its mutarotation in alkaline solution. Thermodynamic parameters for the interconversion of the $\alpha$- and $\beta$-furanose forms, the $\beta$-pyranose and the _keto_-form were determined.[39]

## 4 Oxidation

A mechanism for the oxidation of D-ribose with cerium(IV) in sulphuric acid has been proposed on the basis of its kinetics.[40] Activation parameters for the oxidation of D-glucose, D-galactose, D-mannose and D-gulose by silver(I) and mercury(II) ions at pH 1 have been determined. Rate constants were linearly related to the free energy of aldose equilibrium forms as well as to the redox potentials of the oxidants.[41] The kinetics and mechanisms of oxidation of D-glucopyranose 6-phosphate and D-ribofuranose 5-phosphate by chromium (VI) in perchloric acid media have been determined using u.v. spectroscopy. The reactions were acid catalyzed and accelerated by the addition of sodium perchlorate.[42] There is a variable dependence of rate on the concentration of oxidant and hydrogen ion concentration in the oxidation of D-fructose by vanadium(V) ions. The kinetic data and activation parameters were compared with those for simpler mono- and polyhydric alcohols, and a three-step mechanism involving C-H fission yielding glucosones as primary products has been suggested.[43] The oxidation of diols as models for starch and cellulose using manganese(III), cerium(IV), and vanadium(V) ions has been studied using u.v. and e.s.r. spectroscopy. The reaction was thought to proceed <u>via</u> a stable complex and showed a first order dependence on hydrogen ion concentration. Acyl radical spin adducts were detected as intermediates from aldehydes and diols by spin trapping techniques. <u>Cis</u>-1,2-diols were oxidized four times as fast as <u>cis</u>-3,4-diols, while <u>trans</u>-diols and isolated primary hydroxy groups showed negligible reactivity. It was concluded that C-1 - C-2 and C-2 - C-3 diols are the predominant sites on the polysaccharides for the initiation of graft co-polymers.[44] In the oxidation of melibiose and cellobiose by tetramminecopper(II) in ammoniacal and buffered media the rate of reaction is first order in disaccharide concentration, order one half in ammonia concentration, and independent of the concentration of copper(II). On addition of ammonium chloride, the rate is decreased by the common ion effect. A mechanism involving an intermediate enediolate ion with the rate of reaction being equal to the rate of enolization was proposed.[45] The kinetics and mechanism of thallium(III) oxidation of cellobiose in acid medium have been determined: the acid-catalyzed reaction was first order in each reactant, and the active oxidant was thallium(III) diacetate. The sugar products were identified as D-gluconic acid and glucose.[46]

A stopped-flow technique has been used to study the kinetics of oxidation of L-ascorbic acid by copper(II) at different pH values. The reaction was first order in the carbohydrate and copper(II) concentrations, but was retarded by increasing hydrogen ion concentration. The calculated Arrhenius parameters were comparable to those found in oxidations of ascorbic acid by other metal ions.[47] Similar negative dependence on hydrogen ion concentration was found for oxidation of L-ascorbic acid by hexacyanoferrate(III) in perchloric acid media at 18 - 26 $^{\circ}$C when hydrogen ion concentrations were less than 0.3 mol dm$^{-3}$; when this value was greater than 0.7 mol dm$^{-3}$, the reaction became first order in this concentration.[48] The oxidation of D-[1-$^{14}$C]glucose and D-[6-$^{14}$C]-glucose with oxygen in the presence of water at 100 $^{\circ}$C to yield formic acid, acetic acid, carbon dioxide and glycolic acid has been studied in connection with conversion of plant materials into organic chemicals. The carbon dioxide is mainly produced from C-2 to C-5, the formic acid from C-1 and the acetic acid from C-6. Addition of aluminium(III) chloride greatly increased the yield of carbon dioxide.[49] The electro-oxidation of D-glucose at platinum is markedly catalyzed by sub-monolayers of heavy metals such as thallium, palladium, or bismuth deposited at underpotentials; the suggested reason is that these metals reduce electrode poisoning.[50] The rates of oxidation of several sugars in DMF using $RhH(PPh_3)_4$ have been determined by measuring the quantity of the (E)-enone (29) reduced. The level of oxidation of the sugar varies with structure: at 40$^{\circ}$, D-galactose was completely oxidized within two hours, whereas sucrose was less than 2% oxidized during 24 hours. One mol D-glucose or 2,3,4,6-tetra-O-methyl-D-glucose reduced only one mol of (29) even when (29) was in excess, showing that the reaction is anomeric oxidation: a separate experiment showed that the α-anomer is oxidized twice as fast as the β-anomer.[51] Oxidation of D-glucose and D-fructose by oxygen has also been studied in relation to the

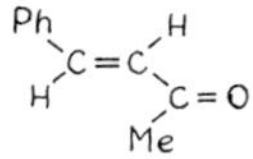


(29)

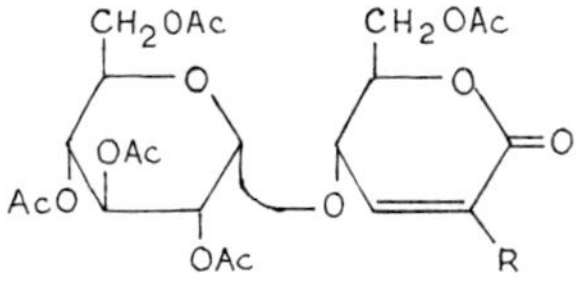


(30) R=H
(31) R=OAc

cleavage of the intermediate hydroperoxides. The use of $^{18}$O-enriched oxygen showed that D-glucose decomposes *via* C-1 and C-2 hydroperoxides, with the latter predominating, the products from both sugars being D-erythronic acid.[52] The kinetics of ozone oxidation of sucrose, D-glucose, and D-fructose in 0.01 M borax buffer have been reported.[53] Radicals formed by persulphate oxidation of maltose and isomaltose have been studied by e.s.r. spectroscopy, and their degradation products determined. The radicals were produced by proton abstraction from the reducing units, in maltose predominantly at C-1 and C-4, and in isomaltose at C-5. Two intermediate products, the enones (30) and (31), were analyzed by FAB m.s. and 400 MHz $^{1}$H-n.m.r. spectroscopy.

## 5 Other Reactions

Equimolar mixtures of D-glucose and D-fructose underwent quantitative transfer hydrogenation in the presence of platinum-carbon or rhodium-carbon in alkaline medium at room temperature under nitrogen to yield equal amounts of D-gluconic acid and a hexitol mixture consisting of mannitol and sorbitol. Other Group VIII metals were less efficient. The influence of several reaction variables was studied, and it was found that the rate determining step is the dehydrogenation reaction of D-glucose to form D-gluconic acid and chemisorbed hydrogen. Subsequently the latter is rapidly consumed by co-adsorbed D-fructose to yield mannitol and sorbitol in the ratio of 1.5-1.9:1. The alkaline degradation products were of minor importance at temperatures below 25$^{\circ}$C. A stirred batch and a continuous flow column process were devised.[55] Acidic products from the alkaline degradation of sugars have been efficiently separated by isotachophoresis. Formic, glycolic, glyceric, lactic, 2-hydroxybutyric, 2-hydroxyvaleric, α- and β-glucometasaccharinic, and α-D-glucoisosaccharinic acids were identified as products of the degradation of D-glucose by g.l.c.-m.s. of their per-O-trimethylsilyl ethers.[56] The kinetics of the reaction of lysine and other amino acids[57] and of human serum albumin[58] with aldoses have been studied by means of c.d., and it was concluded that pyrrole rings are formed. Other aspects of the Maillard reaction are dealt with in Chapter 10.

## References

1 N.V.Lebedev, V.I.Sarzhe, V.E.Kononov, V.M.Kuznetsov, Kompleksn. Ispol'z. Drev. Syr'ya, 1984, 92 (Chem. Abstr., 1985, 102, 115 324).
2 M.Saito, Chiryogaku, 1985, 14, 629 (Chem. Abstr., 1985, 103, 196 327).

3 I.Tvaroska, Biopolymers, 1984, 23, 1951 (Chem. Abstr., 1985, 102, 113 821).
4 J.G.C.M.Van Duijneveldt-Van de Rijdt, F.B.Van Duijneveldt, J.A.Kanters, and D.R.Williams, Theochem., 1984, 18, 351 (Chem. Abstr., 1985, 102 95 946).
5 A.G.Osinovskii and B.V.Erofeev, Dokl. Akad. Nauk BSSR, 1984, 28, 1006 (Chem. Abstr., 1985, 102, 95 926).
6 H.Asaoka, Carbohydr. Res., 1985, 137, 99.
7 G.Bonn, M.Rinderer, and O.Bobleter, J. Carbohydr. Chem., 1985, 4, 67.
8 H.Ohrui, T.Misawa, and H.Meguro, Agric. Biol. Chem., 1985, 49, 239.
9 H.Redlich and S.Thormahlen, Tetrahedron Lett., 1985, 26, 3685.
10 H.Paulsen, M.Schuller, M.A.Nashed, A.Heitmann, and H.Redlich, Tetrahedron Lett., 1985, 26, 3689.
11 A.Dondoni, M.Fogagnolo, A.Medici, and P.Pedrini, Tetrahedron Lett., 1985, 26, 5477.
12 I.Izquierdo Cubero, M.D.Portal Olea, and D.Garcia Posa, Carbohydr. Res., 1985, 138, 135.
13 I.Izquierdo Cubero and D.Garcia Posa, Carbohydr. Res., 1985, 138, 139.
14 J.S.Brimacombe, R.Hanna, and F.Bennett, Carbohydr. Res., 1985, 135, C17.
15 J.S.Brimacombe, R.Hanna, and F.Bennett, Carbohydr. Res., 1985, 140, C9.
16 F.Farina and M.Delores Parellada, An. Quim., Ser. C, 1983, 79, 165 (Chem. Abstr., 1985, 102, 62 523).
17 Yu.E.Raifel'd, L.L.Zil'berg, B.M.Arshava, and S.M.Makin, Zh. Org. Khim., 1984, 20, 1330 (Chem. Abstr., 1985, 102, 24 928).
18 O.Achmatowicz and M.H.Burzynska, Carbohydr. Res., 1985, 141, 67.
19 J.C.Goodwin, Carbohydr. Res., 1985, 143, 61.
20 A.Banaszek, X.Bordas Cornet, and A.Zamojski, Carbohydr. Res., 1985, 144, 342.
21 J.Fiandor, M.T.Garcia-Lopez, F.G.de las Heras, and P.P.Mendez-Castrillon, Synthesis, 1985, 1121.
22 R.L.Root, J.R.Durrwachter, and C.-H.Wong, J. Am. Chem. Soc., 1985, 107, 2997.
23 J.Pierce, A.S.Serianni, and R.Barker, J. Am. Chem. Soc., 1985, 107, 2448.
24 M.Moriyasu, A.Kato, M.Okada, and Y.Hashimoto, Anal. Lett., 1984, 17, 1533 (Chem. Abstr., 1985, 102, 79 215).
25 G.Barone, P.Cacace, G.Castronuovo, and V.Elia, J. Solution Chem., 1984, 13, 625 (Chem. Abstr., 1985, 102, 113 816).
26 G.Barone, G.Castronuovo, V.Elia, K.Stassinopoulou, and G.Della Gatta, J. Chem. Soc., Faraday Trans. 1, 1984, 80, 3095 (Chem. Abstr., 1985, 102, 79 219).
27 J.R.Ward, Int. J. Chem. Kinet., 1985, 17, 11 (Chem. Abstr., 1985, 103, 88 140).
28 H.Uedaira and H.Uedaira, J. Solution Chem., 1985, 14, 27 (Chem. Abstr., 1985, 102, 204 181).
29 J.A.W.M.Beenackers, B.F.M.Kuster, and H.S.Van der Baan, Carbohydr. Res., 1985, 140, 169.
30 J.Reuben, J. Am. Chem. Soc., 1985, 107, 5867.
31 W.Berger, M.Keck, D.V.Sang, and B.Philipp, Z. Phys. Chem. (Leipzig), 1985, 266, 421 (Chem. Abstr., 1985, 102, 186 873).
32 P.Dais and A.S.Perlin, Carbohydr. Res., 1985, 136, 215.
33 A.V.Lymarev, I.V.Yudin, S.L.Panasyuk, and V.V.Matyushkov, Vysokomol. Soedin., Ser. A, 1985, 27, 58 (Chem. Abstr., 1985, 102, 115 351).
34 B.G.Ershov and O.V.Isakova, Izv. Akad. Nauk SSSR., Ser. Khim., 1984, 2234 (Chem. Abstr., 1985, 102, 113 824).
35 S.L.Panasyuk and I.V.Yudin, Khim. Vys. Energ., 1984, 18, 418 (Chem. Abstr., 1985, 102, 79 216).
36 A.Shukla, M.Aradhana, and R.K.Shukla, Acoust. Lett., 1984, 7, 161 (Chem. Abstr., 1985, 102, 24 939).
37 K.Miyajima, K.Machida, and M.Nakagaki, Bull. Chem. Soc. Jpn., 1985, 58, 2595.

38 P.Sevcik and L.Adamcikova, J. Phys. Chem., 1985, 89, 5178.
39 W.J.Goux, J. Am. Chem. Soc., 1985, 107, 4320.
40 M.Verma and K.C.Nand, Middle East Tech. Univ. J. Pure Appl. Sci., 1984, 17, 171 (Chem. Abstr., 1985, 102, 167 043).
41 V.I.Krupenskii, Zh. Obshch. Khim., 1984, 54, 2579 (Chem. Abstr., 1985, 102, 79 223).
42 K.K.Sen Gupta, S.Sen Gupta, S.K.Mandal, and S.N.Basu, Carbohydr. Res., 1985, 139, 185.
43 P.Olavi, I.Virtanen, and S.Kurkisuo, Carbohydr. Res., 1985, 138, 215.
44 T.Doba, C.Rodehed, and B.Ranby, Macromolecules, 1984, 17, 2512 (Chem. Abstr., 1985, 102, 24 958).
45 K.C.Gupta and S.P.Panday, Z. Phys. Chem. (Leipzig), 1985, 266, 787 (Chem. Abstr., 1985, 103, 178 538).
46 S.N.Shukla and R.N.Keserwani, React. Kinet. Catal. Lett., 1984, 26, 409 (Chem. Abstr., 1985, 103, 6635).
47 P.Martinez, J.Zuluaga, and C.Sieiro, Z. Phys. Chem. (Leipzig), 1984, 265, 1225 (Chem. Abstr., 1985, 102, 113 847).
48 L.Ferrari and A.Alonso, An. Quim., Ser. A, 1983, 79, 531 (Chem. Abstr., 1985, 102, 7012).
49 C.J.Biermann, G.D.McGinnis, L.L.Ingram, and T.F.Kellogg, Carbohydr. Res., 1985, 143, 256.
50 N.Xonoglou and G.Kokkinidis, Bioelectrochem. Bioeng., 1984, 12, 485 (Chem. Abstr., 1985, 102, 149 654).
51 M.Massoui, D.Beapere, G.Goethals, and R.Uzan, J. Mol. Catal., 1985, 29, 7 (Chem. Abstr., 1985, 102, 113 810).
52 T.Vuorinen, Carbohydr. Res., 1985, 141, 319.
53 R.Perez, M.Gomez, and R.Ramos, Rev. Cienc. Quim., 1984, 15, 105 (Chem. Abstr., 1985, 103, 37 666).
54 Y.-Y.Chen, R.Oshima, T.Uryu, J.Kumanotani, and J.Tsuchiya, Carbohydr. Res., 1985, 141, 77.
55 A.J.Van Hengstum, A.P.G.Kieboom, and H.Van Bekkum, Starch/Staerke, 1984, 36, 317 (Chem. Abstr., 1985, 102, 24 931).
56 J.Polonsky, L.Repasova, A.Pekarovicova, and M.Kosik, J. Chromatogr., 1985, 320, 111.
57 G.Candiano, G.M.Ghiggeri, G.Delfino, C.Quierolo, C.Cuniberti, E.Gianazza, and P.G.Righetti, Carbohydr. Res., 1985, 145, 99.
58 G.M.Ghiggeri, G.Candiano, G.Delfino, C.Querolo, G.Vecchio, E.Gianazza, and P.G.Righetti, Carbohydr. Res., 1985, 145, 113.

# 3
# Glycosides and Disaccharides

## 1 O-Glycosides

1.1 Synthesis of Monosaccharide Glycosides.- A monograph on the stereo- and regio-control of synthesis in carbohydrate chemistry has covered several glycosidation procedures,[1] while a second review has dealt with the use of organotin derivatives in glycoside synthesis.[2]

Careful methanolysis of oligosaccharides causes glycosidation at greater rates than inter-unit cleavage and, in consequence, various di- and tri-saccharide glycosides have been obtained (see Section 1.2).[3] Applied to L-rhamnose, methanolysis gives access to the crystalline β-furanoside (flash chromatographic separation) together with the other previously reported isomers. The $^{1}H$ and $^{13}C$ n.m.r. spectra were recorded for the four glycosides and were compared with those for the mannosides.[4]

Treatment of the diethyl dithioacetal of D-fucose with mercury(II) salts in methanol at 0° gave the β-D-furanoside in 93% yield, whereas at 65° 20% of the α-anomer was produced. Similar proportions of this compound were also obtained following anomerization of methyl tri-O-benzyl-β-D-fucofuranoside.[5]

Acid-catalysed alcoholysis of D-fructose with 2-chloroethanol affords the crystalline β-pyranoside (1) in 90% yield, and an extensive range of further derivatives have been reported. For example, treatment of (1) with ethoxide anion gave the spiro-product (2), and the disulphonate (3) afforded the tricyclic epoxide

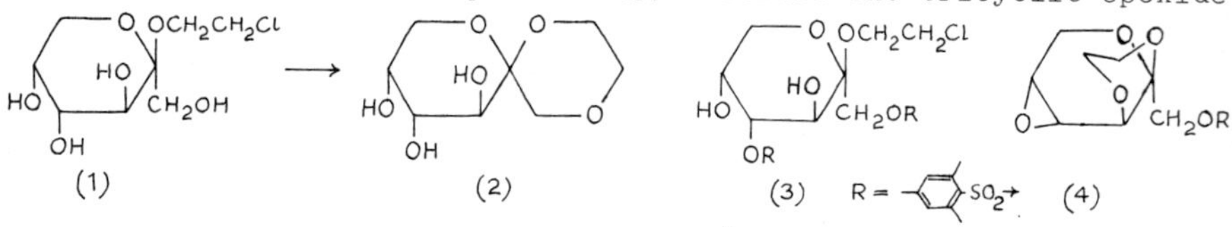


(1) (2) (3) (4)

(4) when caused to react with methoxide.[6]

A short synthesis of the natural cerebroside (5) involved the use of the trichloroacetimidate method applied to either the racemic (6) (in which case the diastereoisomeric products could be

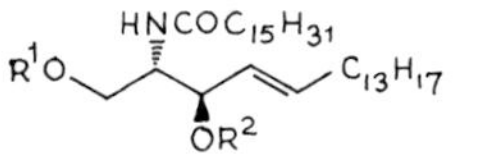

(5) $R^1 = \beta$-D-Glc-p, $R^2 = H$

(6) $R^1 = H$, $R^2 = Bz$

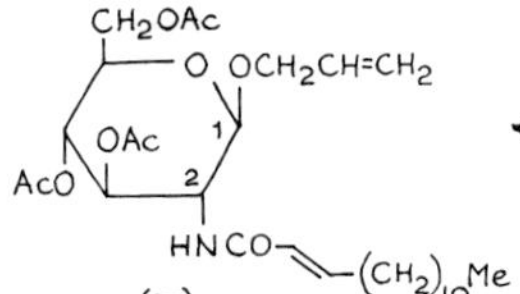

(7)

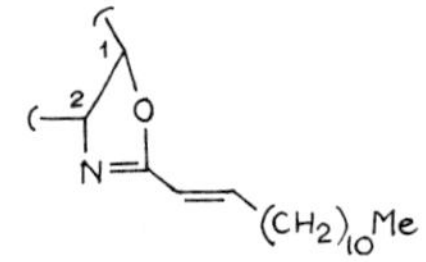

(8)

separated) or the enantiomerically pure form,[7] and in studies related to the synthesis of lipid A the allyl glycoside (7) has been made by ring opening of the oxazoline (8).[8]

A novel approach to glycoside synthesis involves the use of the dimethylphosphinothiolate (9) with alcohols in benzene solution in the presence of silver perchlorate (Scheme 1). Cholesterol, protected L-serine and partially protected sugars were used, and α,β-

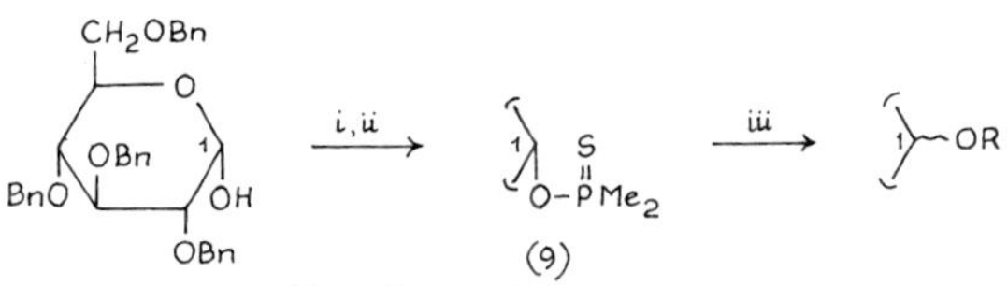

Reagents: i, BuLi; ii, $Me_2P(S)Cl$; iii, ROH, $AgClO_4$

Scheme 1

ratios varying from 3:1 to 9:1 were observed.[9] A further novel way of making specific α-glycosides with moderate selectivity involves the reaction of glycosyloxytin derivatives with allylic acetates in the presence of palladium derivatives (e.g. Scheme 2).[10]

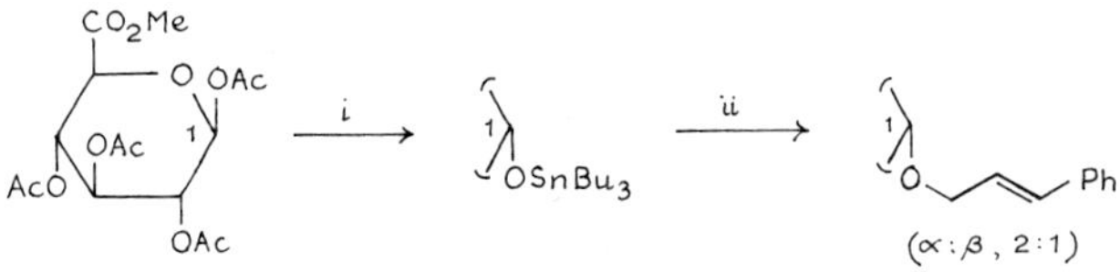

Reagents: i, $Bu_3SnOMe$ or $(Bu_3Sn)_2O$; ii, $Pd(PPh_3)_4$, Ph⁀⁀OAc

Scheme 2

β-Glucuronides are formed with high selectivity when methyl 2,3,4-tri-O-acetyl-D-glucuronate is treated with most alcohols in the presence of trimethylsilyl triflate. However, with alcohols which readily form stable carbonium ions, such as diphenylmethanol, α-glycosides are produced almost exclusively - presumably by alkylation of the α-form of the starting sugar.[11] This same glycosidation reaction, but with 2,3,4,6-tetra-O-acetyl-β-D-glucose, has been used to determine the configurations of asymmetric secondary alcohols. Chemical shift changes, particularly those of protons at β-positions relative to the alcoholic group, correlate with the configuration; the method proved applicable with steroidal and terpenoid alcohols.[12] Likewise the method has given 2-azido-2-deoxy-α-D-galactopyranosyl-L-serine and -L-threonine derivatives which have been peptide coupled to give the glycosylated pentapeptides (10) and (11) which are the antigenic aminoterminal portions of human glycophorin $A^N$ and $A^{MC}$, respectively.[13]

$H_2N$-Leu-Ser*-Thr*-Thr*-Glu-OH (10)

$H_2N$-Ser-Ser*-Thr*-Thr*-Glu-OH (11)

*These amino acids carry 2-acetamido-2-deoxy-α-D-galactopyranosyl groups.

Glycal derivatives continue to be used for the preparation of glycosides, and the N-iodosuccinimide procedure has afforded a route to steroidal glycosides of deoxy sugars. In some cases, 2,3-unsaturated glycosidic products were obtained by allylic rearrangement and these led to 2,3-dideoxy saturated products. Several cardenolide glycosides were made in these ways and, in the course of the same work, dideoxyhexose peracetates and the derived glycosyl bromides were used to give analogous products.[14] An unusual, alternative use of an unsaturated compound has led to the synthesis of 1,2-*cis*-related methyl hexopyranosides from a pentose derivative (Scheme 3).[15]

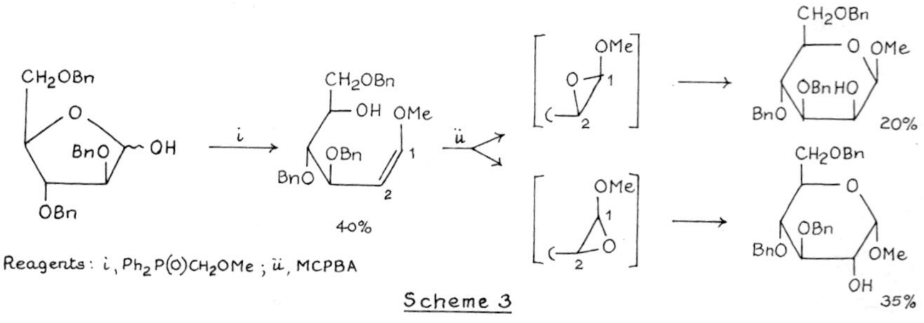


Scheme 3

Glycosyl fluorides continue to attract attention as glycosylating agents. Reaction of 1,2-*trans*-related acetylated glycosyl fluorides with phenols in the presence of boron trifluoride and a hindered base gave aryl glycosides in 32-80% yield. For example, tetra-O-acetyl-α-D-mannopyranosyl fluoride treated with 2,4-dinitro-phenol in benzene in the presence of the catalyst and 2,2,6,6-tetramethyl-4-piperidone gave the 2,4-dinitrophenyl α-mannopyranoside tetra-acetate in 64% yield.[16] In related fashion, various substituted glycosyl fluorides, treated with alcohols or their trimethylsilyl derivatives in the presence of boron trifluoride, afforded glycosylated products. In some cases mixtures of anomers were obtained with the ratios depending on the nature of the alcohols; with tetra-O-pivaloyl-α-D-glucopyranosyl fluoride β-products were obtained exclusively, and with 2,3:5,6-di-O-isopropylidene-α-D-mannofuranosyl fluoride only α-anomers were formed.[17] Similar observations have been made previously (K.C. Nicolaou et al., J. Chem. Soc., Chem. Commun., 1984, 1155).

Glycosyl bromide derivatives retain their importance as glycosylating agents; in the furanosyl series, compound (12) has been used in the preparation of the α- and β-anomers of 7-O-(3-amino-3,5-dideoxy-D-ribofuranosyl)adriamycinone,[18] and the interesting and unusual boronate ester (13) gives access to β-D-mannofuranosides.[19]

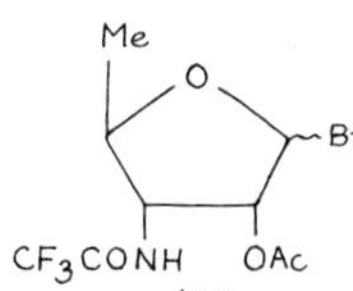


(12)

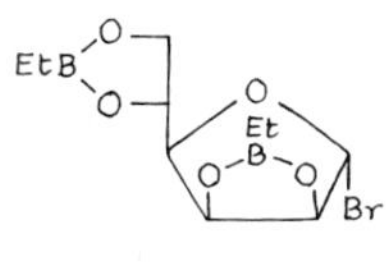


(13)

A review has been prepared on the synthesis of glycopyranosyl halides and the silver imidazolate-assisted glycosylation reaction applied to oligosaccharides.[20] A specific study of the reactions of 2,3,4,6-tetra-O-acetyl-α-D-glucopyranosyl bromide and the corresponding 2-O-benzoate with mono-, di- and tri-chloroethanol in the presence of mercury(II) salts or silver triflate found that more α-linked glycosides were formed with increasing halogenation of the alcohols. A mechanism involving orthoester intermediates was proposed to account for the results.[21] The effects of crown ethers, cryptands, and supported ligands, on the reaction of acetobromoglucose with alcohols in the presence of silver nitrate have been examined. With tert-butanol, glycosides and glycosyl nitrates were produced, and polymer-bound macrocycles increased the proportions of glycoside formed.[22]

Koenigs Knorr β-D-glucopyranosylation has been carried out to give the glycosides (14)[23] and (15),[24] the latter being the fruiting

$CH_2{=}CHCH{=}CH(CH_2)_{10}OR$

(14) R = β-D-Glc-p

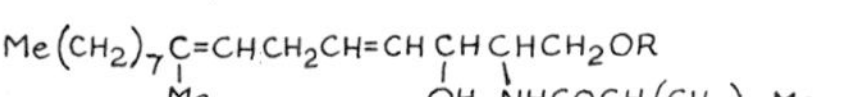


(15)

-inducing cerebroside of the basidiomycete Schizophyllum commune. Galactosylation of a partially protected serine followed by peptide synthesis gave the tripeptide derivative alanyl-(O-β-D-galactopyranosylserine)-alanine.[25]

A wide range of aryl β-D-galactopyranosides have been produced by a phase transfer method using tetra-O-acetyl-α-D-galactopyranosyl bromide with phenols.[26]

Koenigs Knorr glucosylation of the sodium enolate of malondialdehyde gave the enal glycoside (16) which, by Wittig

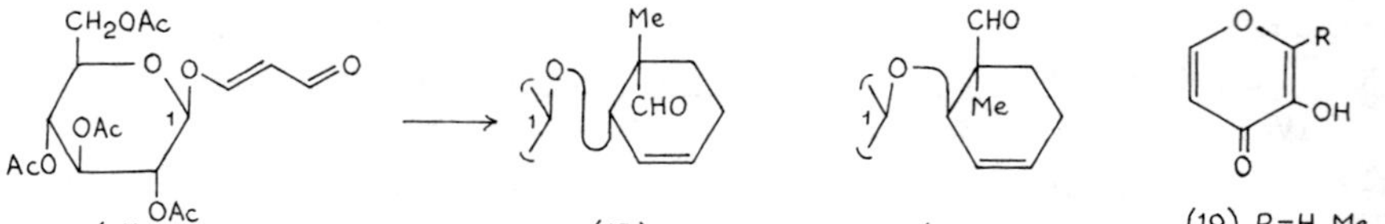


(16) (17) (18) (19) R=H, Me, $CH_2OH$

methylenation, followed by deacetylation and hetero-Diels Alder condensation with methacrolein in water and reacetylation, gave the cyclohexenyl glycosides (17) and (18).[26A] The advantage of using water is illustrated by the observation that the condensation occurred at 25°C in 3.5 h in this solvent whereas the corresponding tetra-acetate required 15 days at 60° in toluene.

β-D-Glucopyranosides have been prepared from the hydroxypyrones (19)[27] and the terpenoids(±)α-terpineol[28] and (S)-β-citronellol, (S)-*cis*-verbinol, 3R-linalool, dihydromyrcenol, terpinen-4-ol, α-ionol, (+)-cedrol, sclareol and *cis*-abienol.[29]

The Koenigs Knorr glycosylation of hydroxycardenolides has been reviewed (in Russian)[30] and a specific report has been published of the reaction applied to betulin,[31] β-D-glucuronides of allo-tetrahydro-11-deoxycortisol (3- and 21-derivatives),[32] and various 11-deoxycortisol metabolites (3-derivatives).[33]

Several references to syntheses related to glycophospholipid A which involve the production of glyceryl α-D-glucopyranosides and phosphorylated derivatives are given in Chapter 7.

By use of phenyl β-D-glucoside or -galactoside as donors and the transferases from *Aspergillus oryzae* which are active in aqueous dioxane or acetonitrile, various water insoluble cardiac genins have been glycosylated,[34] and 2,2,2-trichloroethyl α- and β-D-galactopyranosides have been prepared for use as galactosidase substrates.[35]

## 1.2 Synthesis of Disaccharides and Their Derivatives

.- A review has been published on silver imidazolate-assisted glycosidation reactions in the synthesis of disaccharides.[20]

Methanolysis of oligosaccharides gives the corresponding glycosides faster than it causes interunit cleavage.[3]

Several n.m.r. techniques have been developed for use with oligosaccharides (see Chapter 21); in particular they can be applied to sequence determination and to conformational analysis.

In the area of non-reducing disaccharides various 2-deoxy, 3-deoxy and 2,3-dideoxy derivatives of α,α-trehalose have been made by reductive desulphonylation of tosylates with lithium triethylborohydride,[36] and a bis-*allo*-2,3-epoxide has afforded a route to 3-amino-3-deoxy-α-D-mannopyranosyl 3-amino-3-deoxy-α-D-mannopyranoside by way of a D-altrose disaccharide intermediate which was sulphonylated at *O*-3 and 3'.[37]

D-Galactopyranosyl D-glucopyranosides with α,β-, β,β- and β,α-configurations and α-D-galactopyranosyl β-D-galactopyranoside have been prepared by Koenigs Knorr procedures, and their utilization by *Bifidobacterium* was examined.[38]

A review has appeared on the chemistry of the mycobacterial cord factor and related natural and synthetic trehalose esters,[39] and an enzymic synthesis of trehalose from maltose (60% efficiency) has been reported.[40]

The following reducing disaccharides are categorized according to their non-reducing sugars.

Tin(II) triflate has been used to catalyse the condensation between tetra-*O*-acetyl-α-D-glucopyranosyl bromide and primary and secondary alcohol groups on other monosaccharide derivatives. The products had the β-configuration exclusively but the yields were modest (30-65%).[41] In a related study the influences of catalyst, solvent, and temperature on the reaction of *O*-(tetra-*O*-benzyl-α-D-glucopyranosyl)trichloroacetimidate with primary and secondary sugar alcohols were examined. With boron trifluoride in dichloromethane at -18--40°C good yields (*e.g.* 80% for a secondary alcohol) and anomeric ratios (α,β, *ca.* 1:4) were obtained.[42]

Glycosylation of 2-*O*-benzyl-3-*O*-alkyl-*sn*-glycerols with acetobromomaltose gave the β-maltosyl glycosides which are model glyceryl ether lysoglycolipids.[43] In related work selective 2-*O*-α-glucosylation of the 4,6-TIPS-protected compound (20) was used to prepare the major streptococcal glycophospholipid (21) and

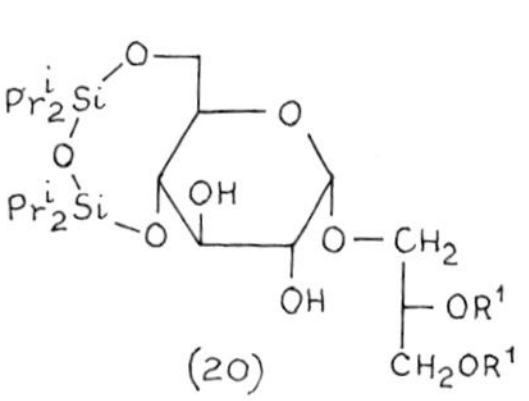

(20)

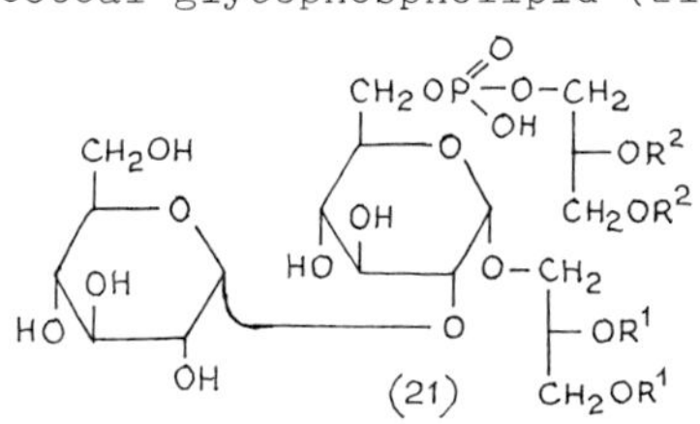

(21)

$R^1$ = Palmitoyl
$R^2$ = Oleoyl

similar compounds.

Further examples of the use of the TIPS protecting group allow selective substitutions at the 2,3- and 2',3'-hydroxy groups of maltose and cellobiose, respectively, as indicated by the structures (22) and (23) of the main mono-TIPS derivatives of the 1,6-anhydro-

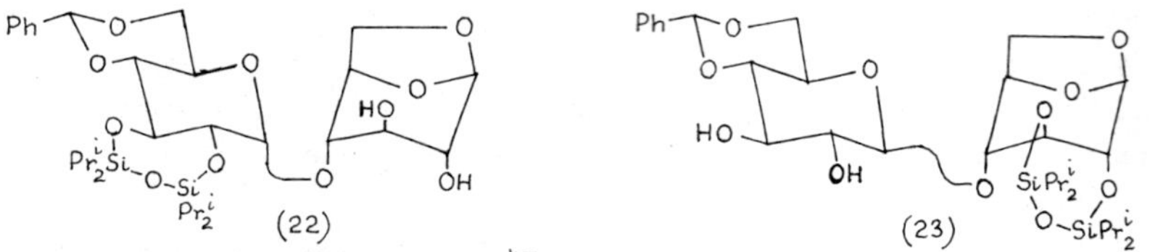

(22) (23)

4',6'-O-benzylidene compounds.[45] Various long chain alkyl β-maltosides and maltotriosides can be made *via* the acetobromo compounds;[46] methanolysis gives methyl glycosides in preference to products of interunit cleavage.[3]

The influences of O-allyl, benzyl, acetyl, trichloroacetyl and levulinoyl substituents at O-3, 4 or 6 on the α,β ratio in the products of silver silicate or silver zeolite activated coupling of 2-O-benzyl-α-D-gluco- and manno-pyranosyl bromide derivatives with the anhydrides (24) and (25) have been examined. Highest β:α-ratios were obtained for 4-O-acylated compounds.[47]

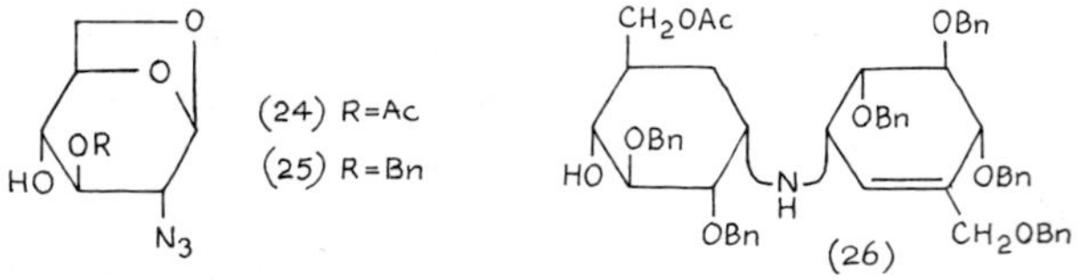

(24) R=Ac (25) R=Bn (26)

β-D-Glucosylation of compound (26) has led to a synthesis of validamycin A,[48] and the condensation between hepta-O-acetyl-α-cellobiosyl bromide and allyl alcohol gave the β-glycoside from which the aldobiuronic acid glycoside (27) was obtained by chromic acid oxidation of a specifically protected derivative. In the same work the isomeric compound (28) was prepared following condensation between 3,4,6-tri-O-acetyl-1,2-O-[1-(*exo*-cyano)ethylidene]-α-D-glucopyranose and 5-O-acetyl-1,2-O-isopropylidene-3-O-trityl-α-D-glucofuranuronate. These disaccharides of (27) and (28) are overlapping fragments of the linear chain of the capsular polysaccharide from *streptococcus pneumoniae* type 3 and the allyl glycosides were each copolymerized with acrylamide to give synthetic antigens.[49]

Other glucopyranosyl-based disaccharides and derivatives to have been reported are the α-kojibioside of methyl 9-hydroxynonanoate,[50] methyl 5-O-α-D-glucopyranosyl-β-D-arabinofuranoside and

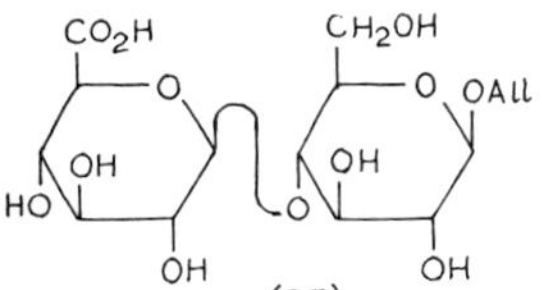
(27)

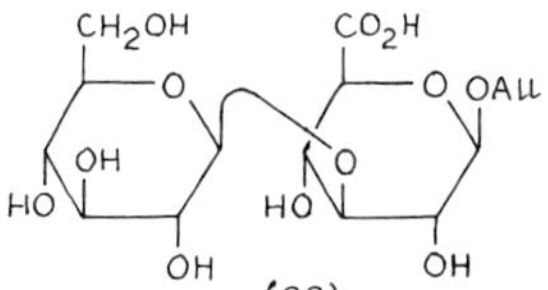
(28)

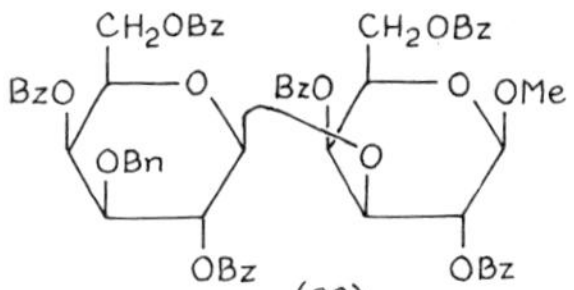
(29)

related glucosylpentose derivatives made by enzymic procedures,[51] disaccharides produced using 2-O-acetyl-3-O-allyl-4-O-benzyl-6-O-tert-butyldiphenylsilyl-D-glucosylchloride which is a multi-functional glycosylating agent giving products which can be deblocked in several ways,[52] and 3-O-methyl and 3,6-di-O-methyl-D-glucopyranosyl disaccharides produced by use of the corresponding methylated glycosyl halides.[53,54] The 1,2- and 1,3-β-linked glucosylgalactoses were also synthesized.[55]

In the area of galactose disaccharides, methyl 2,4,6-tri-O-benzoyl β-D-galactopyranoside and the corresponding glycosyl acetate have been made for use in the synthesis of (1 → 3)-β-D-galactose linked oligosaccharides, and the dimeric compound (29), from which a trimer can be synthesized, has been reported.[56,57] In related work the 6- and 6'-deoxy derivatives of methyl 4-O-α-D-galactopyranosyl-β-D-galactopyranoside have been made in connection with studies of the binding of firmbriated E. coli to urinary tract epithelium.[58] 3-O-Methyllactose has been made by galactosylation of a 3-O-methyl-D-glucose derivative for the determination of intestinal lactase.[59]

The glycosylating disaccharide derivative (30) was of limited value when used with the diserine compound (31). A better method for making the disubstituted dipeptide involved condensation of two disaccharide serine derivatives.[60] Two deoxy derivatives of

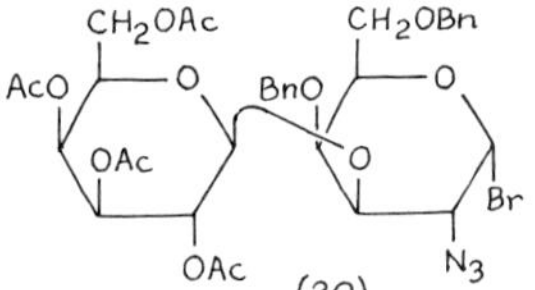
(30)

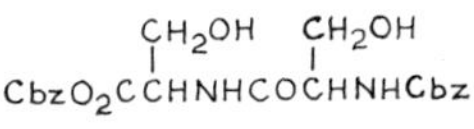
(31)

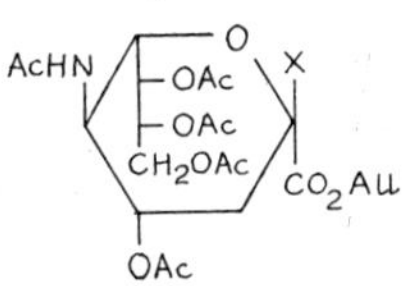
(32) X=Cl
(33) X=F

N-acetyllactosamine have been prepared with deoxy groups at C-3 in the glucosamine unit and at C-2 in the galactose unit (both as β-methyl glycosides and as glycosides of methyl 9-hydroxynonanoate) as have 4- and 2'-deoxy analogues of the analogous 1 → 3 linked disaccharide glycosides. They were tested as acceptors of β-D-galactosyl- and 2-acetamido-2-deoxy-β-D-glucosyl-α-L-fucosyl transferases.[61]

Moraprenyl 3-O-α- and β-D-mannopyranosyl-α-D-galactopyranosyl pyrophosphate, intermediates in the biosynthesis of Salmonella O-antigenic polysaccharide serotypes $C_2$ and $C_3$, have been synthesized by phosphorylation of the disaccharides followed by reaction with moraprenylphosphoimidazole. The synthesis of the two free disaccharides was also described.[62] A new approach to the synthesis of β-linked 2-acetamido-2-deoxy-D-mannosyl disaccharides uses 2-O-acetyl-3,4,6-tri-O-benzyl-α-D-glucopyranosyl bromide followed by deacetylation of the products, then oxidation and oximation.[63]

The disaccharides formed by α- and β-linkage of N-acetylneuraminic acid to O-6 of D-glactose have been made by use of the glycosylating agents (32) and (33),[64] and the α-linked analogue involving 2-acetamido-2-deoxy-D-galactose, which occurs as a branch point in O-glycoproteins, has been made by use of the acetylated glycosyl chloride methyl ester.[65] An ingenious approach to the

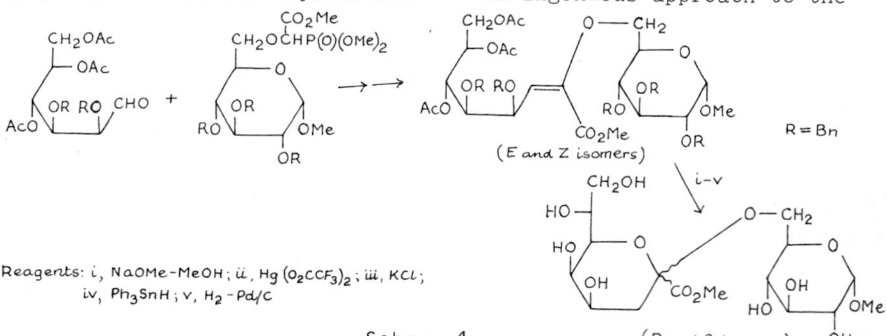


Scheme 4

synthesis of 3-deoxy-D-manno-2-octulosonic acid disaccharides depends upon Wittig-like condensation between acyclic aldehydo D-mannose derivatives and phosphonates developed at substitutent groups at the position of linkage to the reducing enol sugar (Scheme 4). In analogous fashion compound (34), the repeating unit of K antigen of Neisseria meningitidis, was made using a D-galactose derivative bearing the phosphonate-containing functional

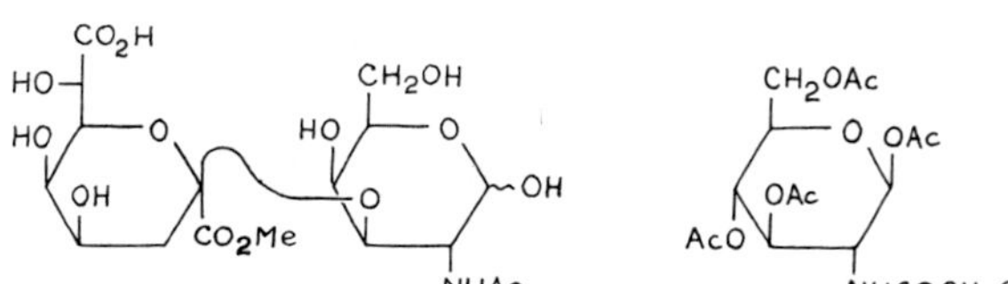


(34) (35)

group at O-3.[66]

Disaccharides of 2-amino-2-deoxy-D-glucose can be prepared by use of the glycosylating agent (35) and iron(III) chloride-catalysed condensations. With 1,2:3,4-di-O-isopropylidene-D-galactose, 80% of the 1 → 6 linked product was obtained.[67] Tri-deuteriomethyl 3-O-(2-acetamido-2-deoxy-β-D-glucopyranosyl)-β-D-galactopyranoside was made using the acetylated N-phthalimido-glycosyl bromide for n.O.e. studies,[68] and during condensation between the same bromide and 4-O-benzyl-1,2-O-benzylidene-β-L-rhamnose the expected β-1 → 3-linked disaccharide was obtained as major product, but smaller proportions of the 1,1'-linked compound with the benzylidene acetal at the 2,3-positions were formed, showing that, in the presence of the silver triflate catalyst, some acetal migration had occurred.[69]

The glycosyl azide derivative of fully acetylated chitobiose has been reduced to the glycosylamine and coupled to aspartic acid linked peptides to produce the linkage regions of two naturally occurring glycopeptides.[70]

Two reviews of the total synthesis and biological activity of lipid A analogues have appeared. These and other references to glucosamine-based disaccharides are discussed in Chapter 7.

An α-L-rhamnopyranosyl compound has been prepared by standard methods and tested as a substrate for an Aspergillus niger enzyme.[71,72]

In the pentose series, condensation of glycosyl fluorides with trimethylsilyl ethers in the presence of boron trifluoride affords a good glycosylation method. Thus tri-O-acetyl-β-D-xylopyranosyl fluoride and methyl 2,3-O-isopropylidene-4-O-trimethylsilyl-α-L-rhamnopyranoside gave the β-1 → 4 linked disaccharide in 70% yield.[73] The "cyanoalkylidene method" using trityl tetrafluoroborate has been used to give α-1 → 5 linked D-arabinofuranose di- and tri-saccharides,[74] and Kochetkov's group have used the procedure with trityl perchlorate to obtain the α-1 → 2 linked L-arabinofuranose disaccharide. Similarly, derivatives of D-galactofuranose and D-glucofuranose can be made by this approach.[75] An interesting and unexpected displacement took place (Scheme 5)

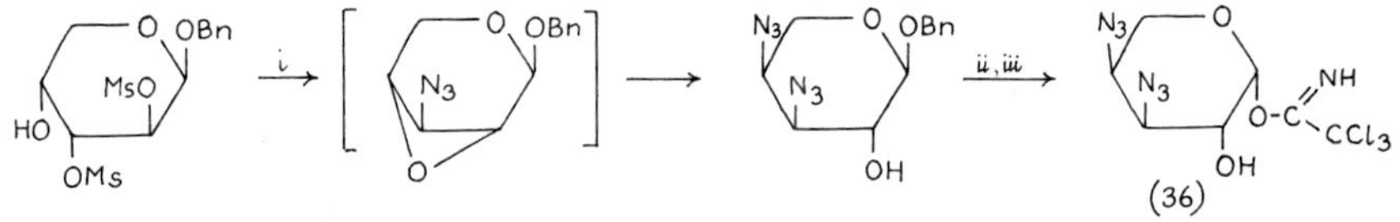


Reagents: i, $NaN_3$; ii, $H^+$; iii, NaH-$CCl_3CN$

Scheme 5

during the preparation of the trichloroacetimidate (36) which was used in the synthesis of the triaminodisaccharide derivative (37).[76] The preparation of 4-thioxylobiose, which was used as a non-metabolized inducer of xylan-degrading enzymes, was based on reaction of 1-thio-D-xylose with 1,2,3-tri-O-benzoyl-β-L-arabinopyranose 4-triflate.[77]

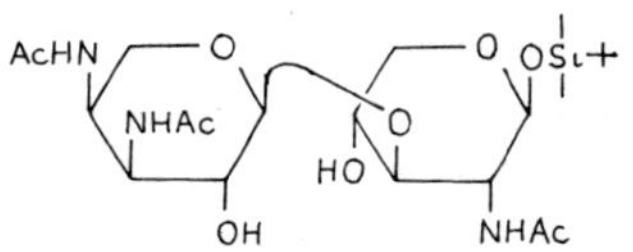

(37)

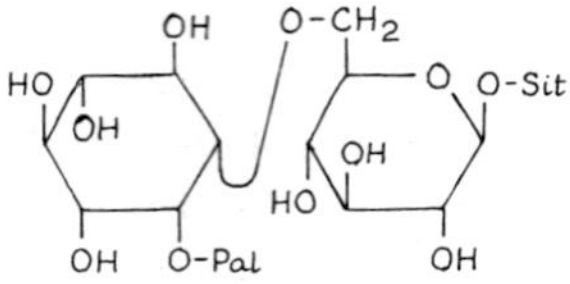

(38)

Pal = Palmitoyl
Sit = 3-Sitosteryl

1.3 O-Glycosides Isolated from Natural Products.- Allose has been identified as a constituent of a flavonoid glycoside from Sideritis leucantha,[78] and D-altrose as an aromatic glycosidic product of a marine gorgonian.[79] This, it is claimed, is the first finding of this sugar in a natural product. 3,4-Di-O-(2-methylcrotonoyl)-α-L-arabinose occurs in a steroidal glycoside of Majidea fosteri,[80] and sitosterol 3-[2"-O-palmitoyl-myo-inosityl (1" → 6')]-β-D-glucopyranoside (38) and a related disaccharide derivative have been claimed to be new anti-ulcerogenic components of bananas. However structure (38) may be incorrect, since inositol and glucose are obtained on acidic hydrolysis.[81] Some highly deoxygenated pyranosyl moieties incorporated in antibiotics are referred to in Chapter 19.

The insect antifeedant compound (39), present in fresh apple leaves, has been shown to have the illustrated conformation by $^{13}C$

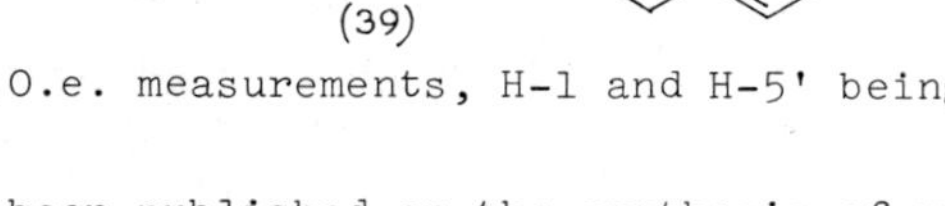

(39)

and $^{1}H$ n.m.r. n.O.e. measurements, H-1 and H-5' being in close proximity.[82]

A review has been published on the synthesis of modified digitalis cardiotonics by saturation of the lactone ring, polyformylation, methylation, acetylation, fluorination or by modification of the sugars of the natural products.[83]

<u>1.4 Hydrolysis and Other Reactions and Features</u>.- The kinetics of hydrolysis of sucrose using sulphonated polystyrene-divinylbenzene at 40-70°C have been investigated,[84] and the $^{18}O$ kinetic isotope effects for the perchloric acid catalysed hydrolysis of isopropyl and p-nitrophenyl[1-$^{18}$O]-α-arabinofuranoside have been reported. The latter reacts with exocyclic C-O bond cleavage, the former by endocyclic cleavage.[85] Addition of dioxane can increase the rate of the acid-catalysed hydrolysis of various glycosides by 5 to 10 fold.[86] Application of molecular orbital theory to the protonation of methyl β-D-glucopyranoside allowed an effort to determine the site of initial protonation during acid hydrolysis of the glycoside.[87]

The instability towards alkali of certain β-D-galactopyranosides has been detailed, and the particular sensitivity of β-D-(1 → 4) linked galactopyranosides in oligosaccharides was noted.[88] A study has been undertaken of the isomerization and subsequent hydrolysis of cellobiose, maltose and lactose catalysed by various minerals and zeolites. Results suggested that ketose formation was an important part of the processes.[89]

Differential pulse polarography cleaves the glycosidic bond of daunorubicin and several derivatives as illustrated in Scheme 6, the reaction rate depending on the orientation of the sugar to the π-system.[90,91]

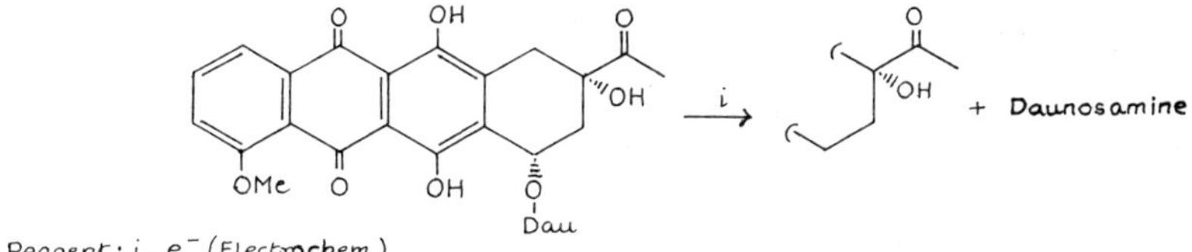


Scheme 6

A specific method for cleaving the sugar-aglycone bond of saponins involves the use of diazomethane (Scheme 7).[92]

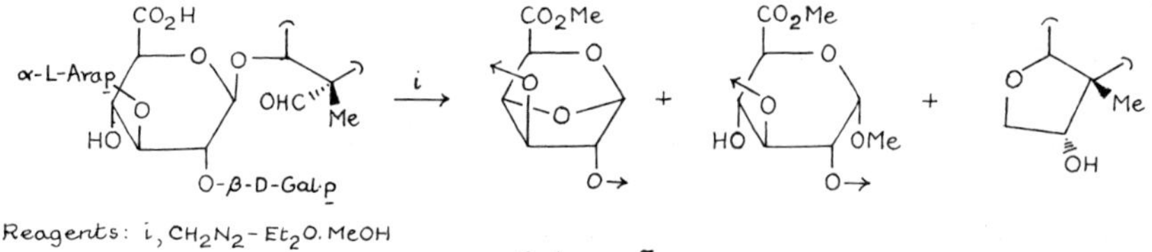


Scheme 7

Allyl ether protecting groups are isomerized to prop-1-enyl ethers by trans-[$Pd(NH_3)_2Cl_2$]; the latter can then be cleaved by heating in <u>tert</u>-butanol under reflux for several hours. Prop-1-

enyl glycosides derived from allyl analogues are less readily cleaved; in the presence of neighbouring hydroxy groups they afford propylidene acetals (Scheme 8).[93]

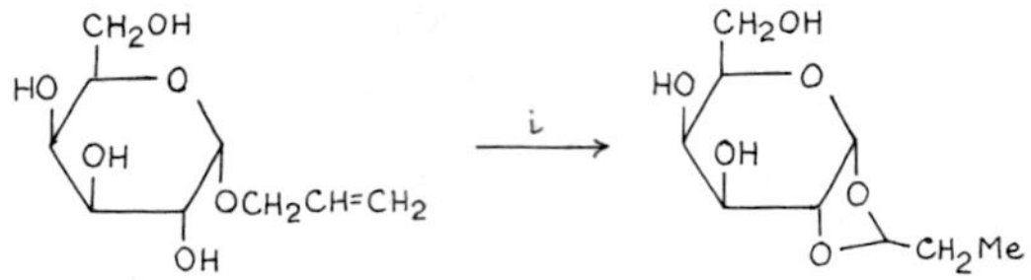

Reagent: i, *trans*-[$Pd(NH_3)_2Cl_2$]

Scheme 8

The hydrogen-bond energy in crystalline sucrose has been calculated from i.r. spectral data to be 134.4 kJ $mol^{-1}$,[94] and sucrose in DMSO-$d_6$ solution has an intramolecular hydrogen-bond network that is sufficiently stable to be manifested by SIMPLE $^1H$ n.m.r. measurements of isotope-shifted resonances of hydroxy groups in samples in which 50% of the exchangeable protons have been replaced by deuterons. There are two inter-residue hydrogen bonds in competition: OH-1'....O-2 and OH-3'....O-2 and weaker networks of bonding "nucleated" by these.[95]

Consistent force field methods have been used to calculate the energy of the conformations of methyl β-D-ribofuranoside taking into account the orientations of the ring substituents; the results were correlated with conformations determined by observed $^1H$ n.m.r. coupling constants.[96]

The potentials of stearoyl glucoside or galactoside to form liposome-like vesicles, and thereby to act as carriers which encapsulate drugs, have been evaluated.[97]

## 2 S-Glycosides

The homologous series of n-hexyl to n-nonyl 1-thio-β-D-glucopyranosides have been synthesized in high yield from glucose by alkylation of the tetra-*O*-acetylated 1-thiol, and found to have potential as detergents in biological systems, *e.g.* for the solubilization of membrane proteins.[98]

The use of thioglycosides as precursors of glycosyl halides used in oligosaccharide synthesis is noted in Chapter 4.

The use of S-phenyl-1-thioglycosides to prepare glycosyl fluorides and hence complex glycosidic products has been further developed.[99,100] A similar type of phenyl thioglycoside of 2-deoxysugars on oxidation to the sulphones offers access to 1-lithiated products from which *C*-glycosides can be derived.

Ingenious control has allowed specific synthesis of α- and β-products (Scheme 9[101] and Scheme 10[102]). The third paper in the series[103] described a further route to α-linked <u>C</u>-glycosides (<u>c.f.</u> Vol.18, p.30, 32).

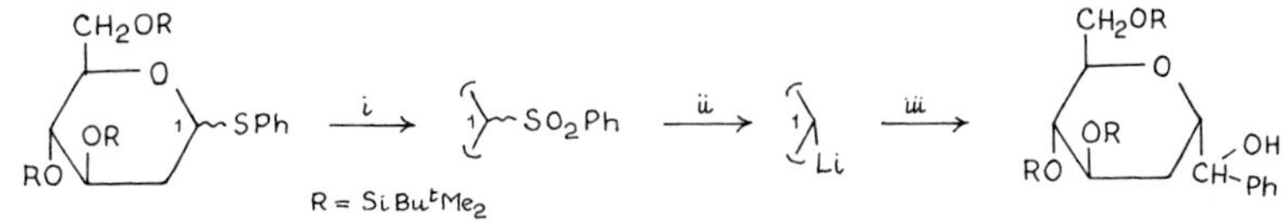

Reagents: i, MCPBA; ii, Li naphthalenide; iii, PhCHO

<u>Scheme 9</u>

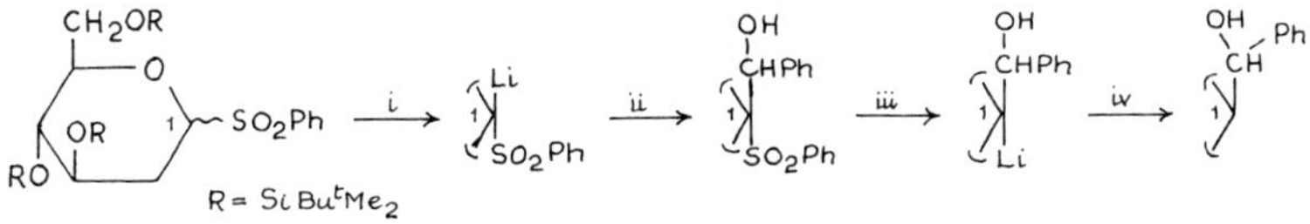

Reagents: i, LDA; ii, PhCHO; iii, Li naphthalenide; iv, $H^+$

<u>Scheme 10</u>

Pyridyl 1-thioglycosides also afford routes to C-glycosides (Scheme 11; <u>c.f.</u> R.M. Williams and A.O. Stewart, <u>Tetrahedron Lett.</u>, 1983, <u>24</u>, 2715).[104]

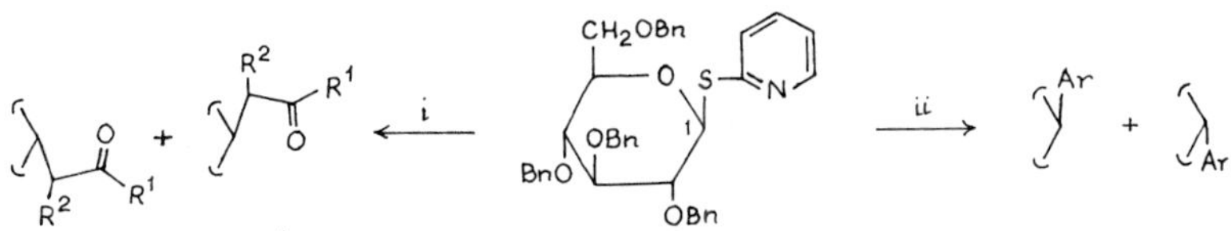

Reagents: i, $R^2CH{=}C(R^1)OTMS$; ii, AgOTf-ArH

<u>Scheme 11</u>

## <u>3 C-Glycosides</u>

<u>3.1 Pyranoid Compounds</u>.- A review on the formation of C-C bonds by radical reactions involving organotin or organomercury compounds includes instances of the synthesis of <u>C</u>-glycosides by this approach.[105]

Reaction of tetra-<u>O</u>-acetyl-α-D-glucopyranosyl bromide[106] or chloride[107] with aryl or alkylmagnesium bromides affords the corresponding β-<u>C</u>-glycosides in 85-90% yield, and corresponding benzylated or methylated glycosyl bromides with sodium pentacarbonylmanganate afford glycosyl manganese compounds which can be converted into C-glycosidic products (Scheme 12).[108] Glycosyl fluorides, <u>e.g.</u> (40), (41), react with trialkyl- or dialkylalkenyl-

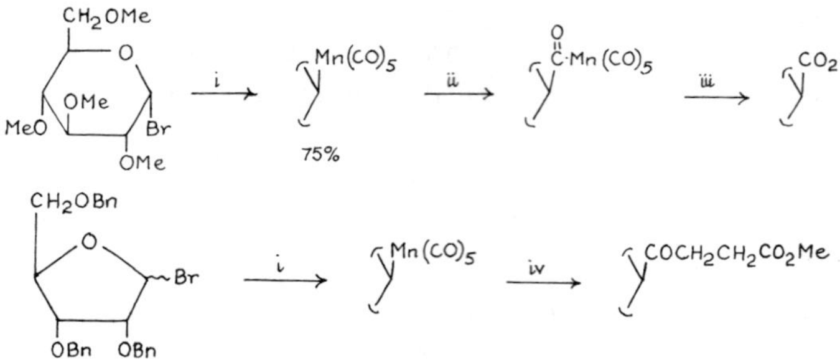

Reagents: i, $Mn(CO)_5Na$; ii, CO; iii, $MeOH-Na_2CO_3$; iv, $CH_2=CHCO_2Me$ - hν

Scheme 12

or dialkylalkynyl-aluminiums to give alkyl or unsaturated C-glycosides, as indicated in Scheme 13.[109] The reaction with triethylaluminium was also used to introduce an ethyl group at C-6 of 1,6-anhydrotribenzyl-D-glucose via the 6-fluoro derivative.

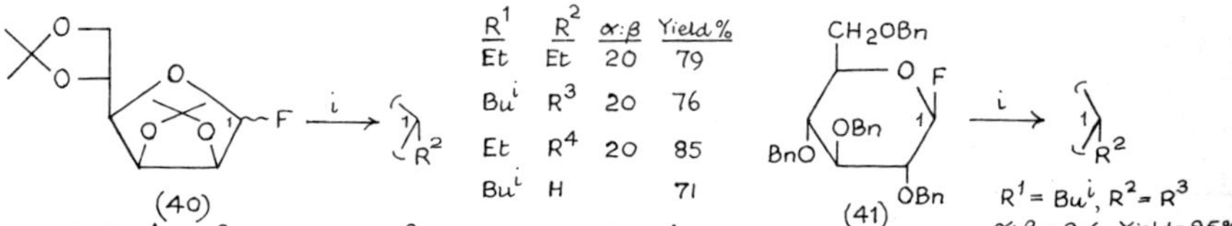


| $R^1$ | $R^2$ | α:β | Yield % |
|---|---|---|---|
| Et | Et | 20 | 79 |
| $Bu^i$ | $R^3$ | 20 | 76 |
| Et | $R^4$ | 20 | 85 |
| $Bu^i$ | H | | 71 |

Reagent: i, $R^1_2AlR^2$ $R^3 = -CH=CHC_6H_{13}$ $R^4 = -C{\equiv}CC_6H_{13}$

Scheme 13

Penta-O-acetyl-β-D-glucopyranose with allyltrimethylsilane and boron trifluoride gave the corresponding α-C-allyl glycoside in 81% yield and high selectivity when the reaction was carried out in acetonitrile; in 1,2-dichloroethane, however, the α- and β-anomers were produced in equal proportions.[110]

Ring closures of appropriate hydroxyalkenes also provide access

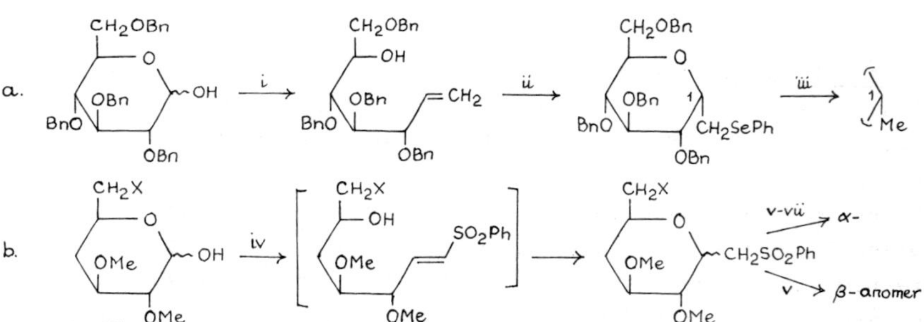

X = Cl, CN

Reagents: i, $Ph_3P=CH_2$; ii, N-SePh (phthalimide); iii, $Bu_3SnH$; iv, $(EtO)_2P(O)\overset{\ominus}{C}HSO_2Ph\ \overset{\oplus}{Na}$; v, BuLi (-78°C); vi, $H^+$; vii, NaH

Scheme 14

to C-glycosides (Scheme 14a[111] and 14b[112]).

The radical addition reactions illustrated in Scheme 15 give access to C-glycosides of ketoses from the corresponding tertiary nitro-compounds.[113]

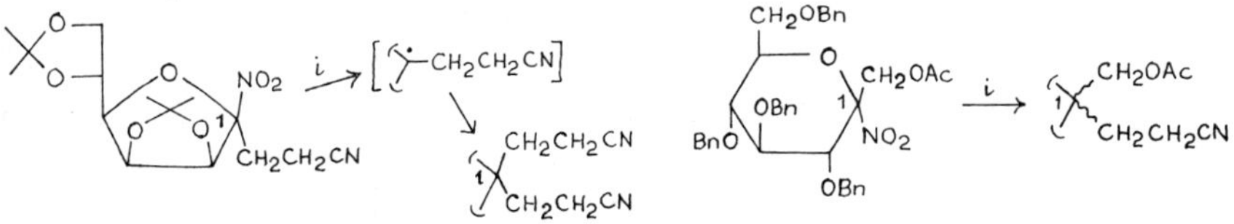

Reagents: i, $Bu_3SnH$ - $CH_2$=CHCN

Scheme 15

Glycosyl lithium compounds (c.f. Schemes 9 and 10) can also be prepared from benzylated glycal derivatives and hence converted selectively into α- and β-C-glycosides (Scheme 16).[114] Reaction

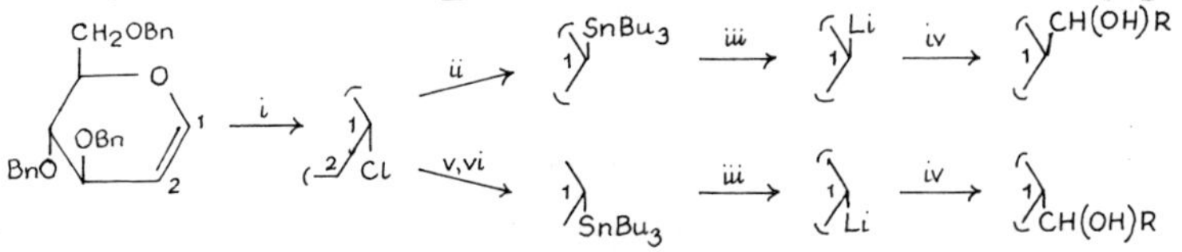

Reagents: i, HCl; ii, $Bu_3SnLi$; iii, BuLi; iv, RCHO; v, Li naphthalenide; vi, $Bu_3SnCl$

Scheme 16

of the allylic trifluoroacetate (42) with diethyl malonate anion in the presence of palladium dibenzylideneacetone bis(diphenyl-phosphino)ethane gave the unsaturated C-glycoside (43) in 56% yield and, similarly, the analogue (44) could be produced (Scheme 17).[115] The reaction did not proceed with tri-O-acetyl-D-glucal.

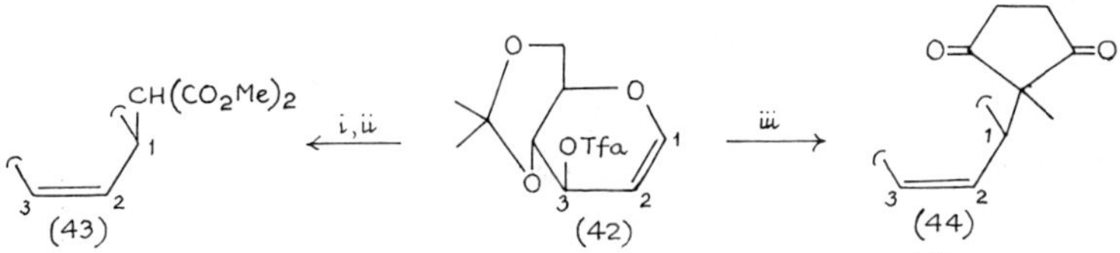

(43) (42) (44)

Reagents: i, $KCH(CO_2Me)_2$; ii, $(PhCH{=}CH)_2CO.Pd.Ph_2PCH_2CH_2PPh_2$; iii

Scheme 17

Reference to the use of the diene (45) in the synthesis of part of the complex marine toxin brevetoxin B is made in Chapter 24. Compounds (46)[116] and (47)[117] have been made following hetero-Diels

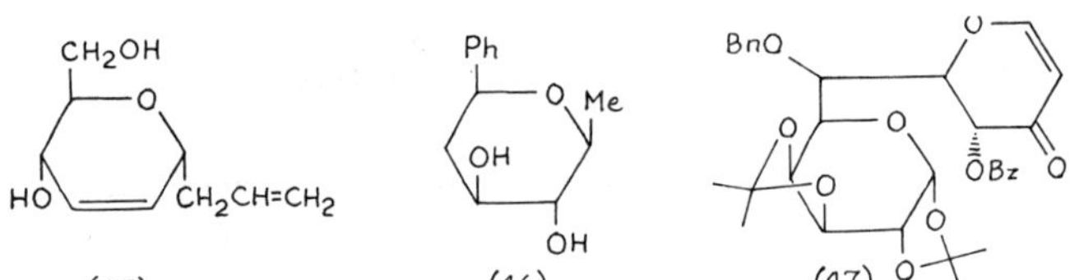

(45) (46) (47)

Alder procedures applied with oxygenated dienes to benzaldehyde and the appropriate heptodialdose derivative, respectively.

3.2 Furanoid Compounds.- Several such compounds have been reported in conjunction with pyranoid analogues which were noted above (see, *e.g.*, Schemes 12, 13 and 15).

Radicals generated thermally with photochemical AIBN initiation from sugars containing halogen, phenylthio, phenylseleno, thionocarbonate, or xanthate substituents can react with allyl- or methallyl-tributylstannane, leading to extended chain compounds, branched chain compounds and allyl C-glycosides (*e.g.* Scheme 18).[118]

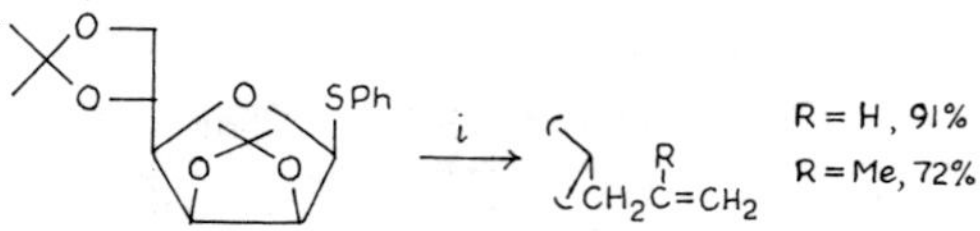


Reagents: i, $CH_2=C(R)CH_2SnBu_3$ - hν

Scheme 18

With tin(IV) chloride, benzylated glycofuranosyl acetates give anomeric carbonium ions which cause Friedel Crafts alkylation of O-2 benzyl substituents. Scheme 19 illustrates the reaction in the D-ribose series; the corresponding L-arabinose-based product was produced in 60% yield.[119]

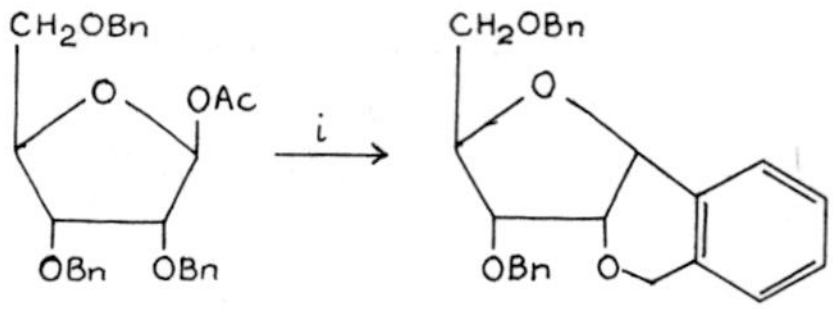


Reagent: i, $SnCl_4$ 46% (isolated)

Scheme 19

As in the pyranoid series (Scheme 14), furanoid C-glycosides can be made by ring closure procedures. Compounds (48),[120] (49),[121] (50)[122] have been made from alkenes derived from the corresponding

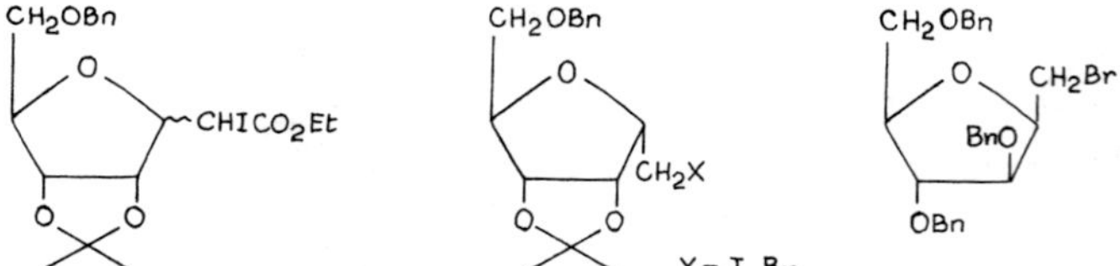


(48) (49) X = I, Br, PhSe, HgOAc (50)

furanoses by treatment with Wittig reagents. In the first of these the Z-alkene gave a β-product with specific stereochemistry at the iodine-carrying atom whereas the E-isomer afforded a specific α-product. An interesting preferential pyranoid-forming reaction, however, occurred with the E-alkene (51) (Scheme 20).[120]

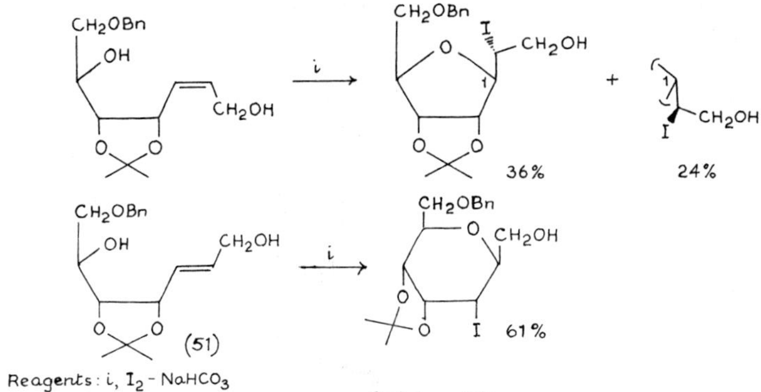

Scheme 20

The Wittig reaction-derived (52) and (53) have been configurationally characterized by $^1$H n.m.r. methods.[123] Lactonization of

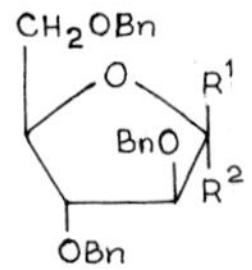

| | $R^1$ | $R^2$ |
|---|---|---|
| (52) | H | $CH{=}CHCO_2Et$ |
| (53) | $CH{=}CHCO_2Et$ | H |

the ester (54) was effected by way of the methyl selenoester (55) (Scheme 21).[124]

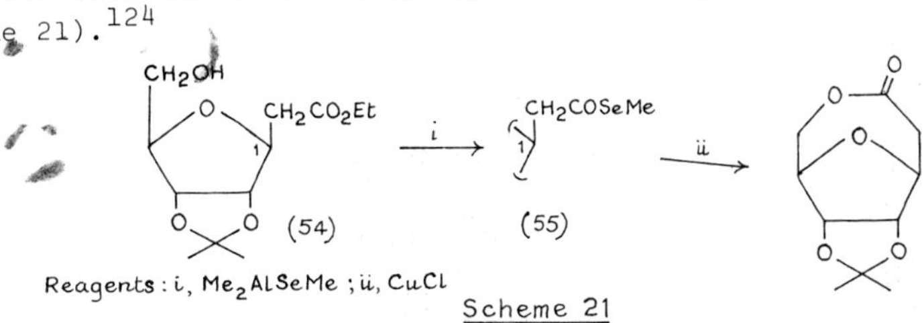

Scheme 21

1-Deoxyalditol-1-yl pyrazoles and C-glycofuranosyl pyrazoles have been prepared from sugar aldehydes (Scheme 22).[125] 2,3-O-Isopropylidene-5-O-trityl-D-ribofuranose has similarly been converted

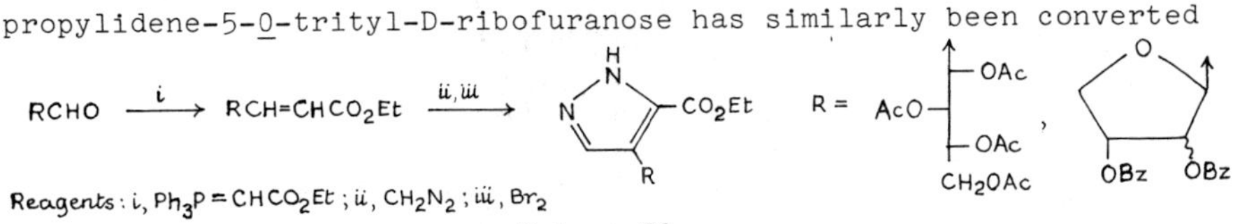

Scheme 22

into the anomeric pyrazofurins (56),[126] and the glycofuranosyl-pyrroles (57) and (58) have been made by dehydration of corresponding 4(D-arabino- and D-xylo-tetrosyl)pyrrole derivatives.[127]

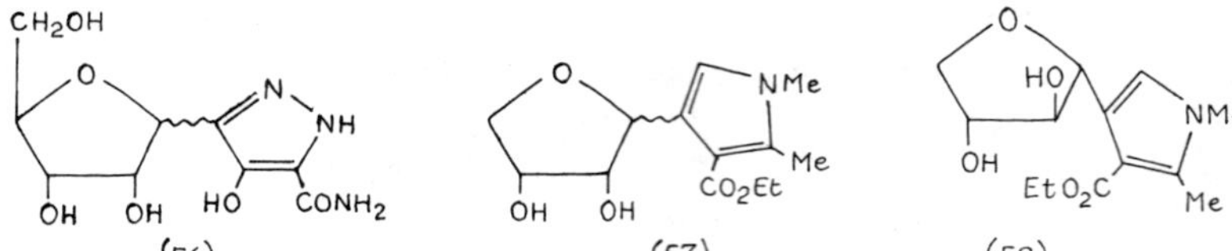

(56) (57) (58)

3.3 C-Glycosides Isolated from Natural Sources.- $^1H$ and $^{13}C$ n.m.r. analyses of carminic acid have confirmed the structure (59),[128]

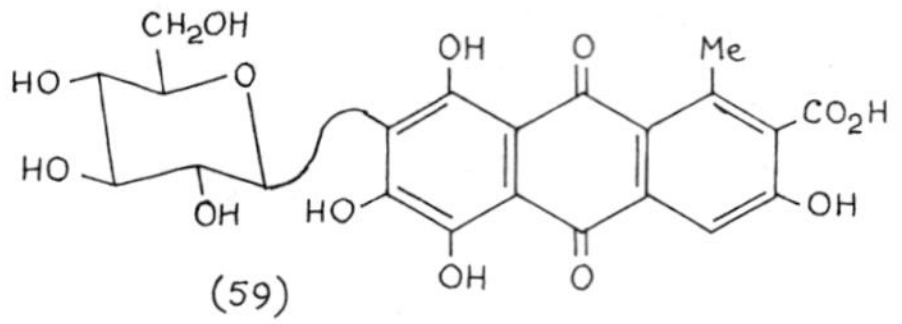

(59)

and 2,6-dideoxy-D-arabino-hexose has been found C-linked to a quinonoid tetracycle as part of new antitumour compounds (see Chapter 19).

The syntheses of 6-C-β-D-galactopyranosyl, 6-C-β-D-xylopyranosyl and 6-C-β-D-arabinopyranosyl quercitin have been effected by use of the corresponding glycosyl halides and in the presence of lithium methoxide. Yields were only in the region of 3%.[129]

## References


1 R.Johansson, Chem. Commun., Univ. Stockholm, 1984, (Monograph) 29 pp (Chem. Abstr., 1985, 102, 95 909).

2 S.David and S.Hanessian, Tetrahedron, 1985, 41, 643.

3 N.W.H.Cheetham and G.Teng, Carbohydr. Res., 1985, 144, 169.

4 J.Stanek, Jr., J.Moravcova, and J.Jary, J. Carbohydr. Chem., 1985, 4, 79.

5 T.Kinoshita, T.Miwa, and J.Clardy, Carbohydr. Res., 1985, 143, 249.

6 J.Y.C.Chan, P.P.L.Cheong, L.Hough, and A.C.Richardson, J. Chem. Soc., Perkin Trans. 1, 1985, 1447.

7 R.R.Schmidt and R.Klager, Angew. Chem. Int. Ed. Engl., 1985, 24, 65.

8 C.Diolez, S.R.Sarfati, and L.Szabo, J. Chem. Soc., Perkin Trans. 1, 1985, 57.

9 T.Inazu, H.Hosokawa, and Y.Satoh, Chem. Lett., 1985, 297.

10 E.Keinan, M.Sahai, Z.Roth, A.Nudelman, and J.Herzig, J. Org. Chem., 1985, 50, 3558.

11 B.Fischer, A.Nudelman, M.Ruse, J.Herzig, H.E. Gottlieb, and E.Keinan, J. Org. Chem., 1984, 49, 4988.

12 R.Faghih, C.Fontaine, I.Horibe, P.M.Imamura, G.Lukacs, A.Olesker, and S.Seo, J. Org. Chem., 1985, 50, 4918.

13 B.Ferrari and A.A.Pavia, Tetrahedron, 1985, 41, 1939.

14 J.Thiem, S.Kopper, and J.Schwentner, Liebigs Ann. Chem., 1985, 2135.

15 F.Nicotra, L.Panza, F.Ronchetti, G.Russo, and L.Toma, Tetrahedron Lett., 1985, 26, 807.

16 Ya.V.Voznyi, I.S.Kalicheva, and A.A.Galoyan, Bioorg. Khim., 1984, 10, 1256 (Chem. Abstr., 1985, 102, 95 930).

17 H.Kunz and W.Sager, Helv. Chim. Acta, 1985, 68, 283.
18 N.Kumar and M.Israel, Can. J. Chem., 1984, 62, 2585.
19 W.V.Dahlhoff, A.Geisheimer, G.Schroth, and R.Mynott, Z.Naturforsch., B; Anorg. Chem., Org. Chem., 1984, 39, 1004 (Chem. Abstr., 1985, 102, 6972).
20 C.Ortega, Chem. Commun., Univ. Stockholm, 1984, (3) (Monograph, 36 pp) (Chem. Abstr., 1985, 102, 95 911).
21 P.J.Garegg, P.Konradsson, I.Kvarnstrom, T.Norberg, S.C.T.Svensson, and B.Wigilius, Acta Chem. Scand. Ser. B., 1985, 39, 569.
22 A.R.Card, J.Capillon, and C.Quivoron, Polymer, 1984, 25, 1136 (Chem. Abstr., 1985, 102, 79 226).
23 E.E.Rusanova, L.V.Volkova, and R.P.Evstigneeva, Bioorg. Khim., 1984, 10, 957 (Chem. Abstr., 1985, 102, 46 196).
24 K.Mori and Y.Funaki, Tetrahedron, 1985, 41, 2379.
25 P.Hoogerhout, C.P.Guis, C.Erkelens, W.Bloemhoff, K.E.T.Kerling, and J.H.Boom, Recl. Trav. Chim. Pays-Bas, 1985, 104, 54 (Chem. Abstr., 1985, 102, 204 275).
26 H.P.Kleine, D.V.Weinberg, R.J.Kaufman, and R.S.Sidhu, Carbohydr. Res., 1985, 142, 333.
26a A.Lubineau and Y.Queneau, Tetrahedron Lett., 1985, 26, 2653.
27 J.H.Looker and M.S.Fisher, J. Carbohydr. Chem., 1985, 4, 381.
28 J.Yamahara, H.Kimura, M.Kobayashi, T.Okamoto, T.Sawada, H.Fujimura, and T.Chisaka, Chem. Pharm. Bull., 1985, 33, 1669.
29 H.Paulsen, B.Le-Nguyen, V.Sinnwell, V.Heemann, and F.Seehofer, Liebigs Ann. Chem., 1985, 1513.
30 N.Sh.Pal'yants and N.K.Abubakirov, Khim. Prir. Soedin., 1984, 547 (Chem. Abstr., 1985, 102, 95908).
31 L.E.Odinokova, G.I.Oshitok, V.A.Denisenko, V.F.Anufriev, A.M.Tolkach, and N.I.Uvarova, Khim. Prir. Soedin., 1984, 182 (Chem. Abstr., 1985, 102, 24 940).
32 H.Hosoda, W.Takasaki, H.Miura, M.Tohkin, Y.Maruyama, and T.Nambara, Chem. Pharm. Bull., 1985, 33, 4281.
33 H.Hosoda, H.Yokohama, H.Shioya, and T.Nambara, Chem. Pharm. Bull., 1985, 33, 1049.
34 Y.Ooi, T.Hashimoto, N.Mitsuo, and T.Satoh, Chem. Pharm. Bull., 1985, 33, 1808.
35 B.Koppen and L.Dalgaard, J. Chromatogr., 1985, 321, 385.
36 H.H.Baer, M.Mekarska, and F.Baucher, Carbohydr. Res., 1985, 136, 335.
37 H.H.Baer, B.Radatus, and J.Defaye, Can. J. Chem., 1985, 63, 440.
38 Y.Nimami, K.Yazawa, K.Nakamura, and Z.Tamura, Chem. Pharm. Bull., 1985, 33, 710.
39 E.Lederer, Microbiol. Ser., 1984, 15, 361 (Chem. Abstr., 1985, 102, 20 816).
40 S.Murao, H.Nagano, S.Ogura, and T.Nishino, Agric. Biol. Chem., 1985, 49, 2113.
41 A.Lubineau and A.Malleron, Tetrahedron Lett., 1985, 26, 1713.
42 R.R.Schmidt and J.Michel, J. Carbohydr. Chem., 1985, 4, 141.
43 H.Prinz, L.Six, K.-P.Ruess, and M.Lieflander, Liebigs Ann. Chem., 1985, 217.
44 C.A.A.van Boeckel and J.H.van Boom, Tetrahedron, 1985, 41, 4567.
45 Y.Ichikawa, A.Manaka, and H.Kuzuhara, Carbohydr. Res., 1985, 138, 55.
46 D.E.Koeltzow, J. Carbohydr. Chem., 1985, 4, 125.
47 C.A.A.van Boeckel, T.Beetz, and S.F.van Aelst, Tetrahedron, 1984, 40, 4097.
48 S.Ogawa, T.Nose, T.Ogawa, T.Toyokuni, Y.Iwasawa, and T.Suami, J. Chem. Soc., Perkin Trans. 1, 1985, 2369.
49 A.Ya.Chernyak, K.V.Antonov, N.K.Kochetkov, L.N.Padyukov, and N.V.Tsvetkova, Carbohydr. Res., 1985, 141, 199.
50 G.O.Aspinall and M.K.Gurjar, Carbohydr. Res., 1985, 143, 266.
51 H.Hashmoto, M.Sekiguchi, and J.Yoshimura, Carbohydr. Res., 1985, 144, C6.
52 C.-M.Liu, C.D.Warren, and R.W.Jeanloz, Carbohydr. Res., 1985, 136, 273.
53 J.Gigg, R.Gigg, S.Payne, and R.Conant, Carbohydr. Res., 1985, 141, 91.
54 T.Fujiwara, S.Izumi, and P.J.Brennan, Agric. Biol. Chem., 1985, 49, 2301.
55 A.Temeriusz, J.Radomski, J.Stepinski, and B.Piekarska, Carbohydr. Res., 1985, 142, 146.

56 P.Kovac and C.P.J.Glaudemans, Carbohydr. Res., 1985, 138, C10.
57 P.Kovac, C.P.J.Glaudemans, and R.B.Taylor, Carbohydr. Res., 1985, 142, 158.
58 P.J.Garegg and S.Oscarson, Carbohydr. Res., 1985, 137, 270.
59 J.Hoffmann, O.Theander, M.Lindberg, and T.Norberg, Carbohydr. Res., 1985, 137, 265.
60 H.Paulsen, M.Schultz, J.-D.Klamann, B.Waller, and M.Paal, Liebigs Ann. Chem., 1985, 2028.
61 D.P.Khare, O.Hindsgaul, and R.U.Lemieux, Carbohydr. Res., 1985, 136, 285.
62 V.I.Torgov, K.A.Panosyan, A.T.Smelyanskii, and V.N.Shibaev, Bioorg. Khim., 1985, 11, 83 (Chem. Abstr., 1985, 102, 185 377).
63 E.Micheli, F.Nicotra, L.Panza, F.Ronchetti, and L.Toma, Carbohydr. Res., 1985, 139, C1.
64 H.Kunz and H.Waldmann, J. Chem. Soc., Chem. Commun., 1985, 638.
65 H.Paulsen, V.von Deessen, and H.Tietz, Carbohydr. Res., 1985, 137, 63.
66 F.Paquet and P.Sinay, J. Am. Chem. Soc., 1984, 106, 8313.
67 M.Kiso and L.Anderson, Carbohydr. Res., 1985, 136, 309.
68 D.M.Whitfield, J.P.Carver, and J.J.Krepinsky, J. Carbohydr. Chem., 1985, 4, 369.
69 Y.E.Tsvetkov and L.V.Backinowsky, Carbohydr. Res., 1985, 135, 312.
70 H.Kunz and H.Waldmann, Angew. Chem. Int. Ed. Engl., 1985, 24, 883.
71 S.Kamiya, S.Esaki, and R.Ito-Tanaka, Agric. Biol. Chem., 1985, 49, 55.
72 S.Kamiya, S.Esaki, and R.Ito-Tanaka, Agric. Biol. Chem., 1985, 49, 2351.
73 Ya.V.Voznyi, A.A.Galoyan, and O.S.Chizhov, Bioorg. Khim., 1985, 11, 276 (Chem. Abstr., 1985, 102, 167 047).
74 K.Hatanaka and H.Kuzuhara, J. Carbohydr. Chem., 1985, 4, 333.
75 L.V.Backinowsky, S.A.Nepogod'ev, A.S.Shashkov, and N.K.Kochetkov, Carbohydr. Res., 1985, 138, 41.
76 R.R.Schmidt, M.Faas, and K.-H.Jung, Liebigs Ann. Chem., 1985, 1546.
77 J.Defaye, H.Driguez, M.John, J.Schmidt, and E.Ohleyer, Carbohydr. Res., 1985, 139, 123.
78 F.A.T.Barberan, F.Tomas, and F.Ferreres, J. Nat. Prod., 1985, 48, 28 (Chem. Abstr., 1985, 102, 218 369).
79 N.Fusetani, K.Yasukawa, S.Matsunaga, and K.Hashimoto, Tetrahedron Lett., 1985, 26, 6449.
80 M.Kapundu, N.Lami, M.Z.Dimbi, R.Warin, C.Delaude, and R.Huls, Bull. Soc. Chim. Belg., 1984, 93, 497 (Chem. Abstr., 1985, 102, 21 184).
81 S.Ghosal, Phytochemistry, 1985, 24, 1807.
82 I.Kubo and A.Matsumoto, Chem. Pharm. Bull., 1985, 33, 3817.
83 F.Rawdimbivololona, P.Pellegrin, and M.Lesne, J. Pharm. Belg., 1984, 39, 225 (Chem. Abstr., 1985, 102, 113 794).
84 G.Siegers and F.Martinola, Int. Sugar. J., 1985, 87, 23 (Chem. Abstr., 1985, 102, 151 184).
85 A.J.Bennett, M.L.Sinnott, and W.S.S.Wijesundera, J. Chem. Soc., Perkin Trans. 2, 1985, 1233.
86 T.S.Seliverstova, L.G.Matusevich, M.A.Kushner, V.M.Reznikov, Khim. Drev., 1985, 70 (Chem. Abstr., 1985, 102, 115 328).
87 T.J. Elder and S.D.Worley, Wood Fiber Sci., 1985, 17, 199 (Chem. Abstr., 1985, 103, 215 665).
88 H.A.El-Shenawy and C.Schuerch, J. Carbohydr. Chem., 1985, 4, 215.
89 R.Shukla, X.E.Verykios, and R.Mutharasan, Carbohydr. Res., 1985, 143, 97.
90 V.Malatesta, S.Penco, N.Sacchi, L.Valentini, A.Vigevani and F.Arcamone, Can. J. Chem., 1984, 62, 2845.
91 A.Anne, F.Bennani, J.-C.Florent, J.Moiroux, and C. Monneret, Tetrahedron Lett., 1985, 26, 2641.
92 R.Higuchi, Y.Tokimitsu, N.Hamada, T.Komori, and T.Kawasaki, Liebigs Ann. Chem., 1985, 1192.
93 T.Bieg and W.Szeja, J. Carbohydr. Chem., 1985, 4, 441.
94 L.G.Belostotskii, A.E.Arkhipets, R.Ts.Mishchuk, and L.P.Leva, Izv. Vyssh. Uchebn. Zaved., Pishch. Technol., 1984, 20 (Chem. Abstr., 1985, 102, 149 669).
95 J.C.Christofides and D.B.Davies, J. Chem. Soc., Chem. Commun., 1985, 1533.
96 J.Wiorkiewicz-Kuczera and A.Rabszenko, J. Chem. Soc., Perkin Trans. 2, 1985, 798.

97 H.Kiwada, H.Nimura, and Y.Kato, Chem. Pharm. Bull., 1985, 33, 2475.
98 S.Saito and T.Tsuchiya, Chem. Pharm. Bull., 1985, 33, 503.
99 R.E.Dolle and K.C.Nicolaou, J. Am. Chem. Soc., 1985, 107, 1691.
100 R.E.Dolle and K.C.Nicolaou, J. Am. Chem. Soc., 1985, 107, 1695.
101 J.-M.Beau and P.Sinay, Tetrahedron Lett., 1985, 26, 6185.
102 J.-M.Beau and P.Sinay, Tetrahedron Lett., 1985, 26, 6189.
103 J.-M.Beau and P.Sinay, Tetrahedron Lett., 1985, 26, 6193.
104 A.O.Stewart and R.M.Williams, J. Am. Chem. Soc., 1985, 107, 4289.
105 B.Giese, Angew. Chem. Int. Ed. Engl., 1985, 24, 553.
106 F.Marquez, M.V.Arriandiaga, and M.T.Urbieta, An. Quim., Ser. C., 1983, 79, 428 (Chem. Abstr., 1985, 102, 149 656).
107 S.Chen, Y.Gao, and M.Cai, Huaxue Tongbao, 1985, 19 (Chem. Abstr., 1985, 103, 215 661).
108 P.De Shong, G.A.Slough, V.Elango, and G.L.Trainor, J. Am. Chem. Soc., 1985, 107, 7788.
109 G.H.Posner and S.R.Haines, Tetrahedron Lett., 1985, 26, 1823.
110 A.Giannis and K.Sandhoff, Tetrahedron Lett., 1985, 26, 1479.
111 J.-M.Lancelin, J.-R.Pougny, and P.Sinay, Carbohydr. Res., 1985, 136, 369.
112 N.J.Barnes, A.H.Davidson, L.R.Hughes, and G.Procter, J. Chem. Soc., Chem. Commun., 1985, 1292.
113 J.Dupuis, B.Giese, J.Hartung, M.Leising, H.-G.Korth, and R.Sustmann, J. Am. Chem. Soc., 1985, 107, 4332.
114 P.Lesimple, J.-M.Beau, and P.Sinay, J. Chem. Soc., Chem. Commun., 1985, 894.
115 T.V.RajanBabu, J. Org. Chem., 1985, 50, 3642.
116 S.J.Danishefsky, B.J.Uang, and G.Quallich, J. Am. Chem. Soc., 1985, 107, 1285.
117 S.J.Danishefsky, W.H.Pearson, D.F.Harvey, C.J.Maring, and J.P.Springer, J. Am. Chem. Soc., 1985, 107, 1256.
118 G.E.Keck, E.J.Enholm, J.B.Yates, and M.R.Wiley, Tetrahedron, 1985, 41, 4079.
119 O.R.Martin, Tetrahedron Lett., 1985, 26, 2055.
120 F.Freeman and K.D.Robarge, Tetrahedron Lett., 1985, 26, 1943.
121 F.Freeman and K.D.Robarge, Carbohydr. Res., 1985, 137, 89.
122 A.B.Reitz, S.O.Nortey, and B.C.Maryanoff, Tetrahedron Lett., 1985, 26, 3915.
123 J.A.Lopez Sastre, J.Isac Garcia, J. Molina Molina, and C.Romero-Avila, An. Quim., Ser. C, 1983, 79, 283 (Chem. Abstr., 1985, 102, 62 524).
124 A.P.Kozikowski and A.Ames, Tetrahedron, 1985, 41, 4821.
125 J.Molina Molina, A.Sanz Tejedor, and J.A.Lopez Sastre, An. Quim., Ser. C., 1984, 80, 111 (Chem. Abstr., 1985, 102, 149 698).
126 F.J.Lopez Herrera and C.Uraga Baelo, Carbohydr. Res., 1985, 143, 161.
127 J.Fernandez-Bolanos, J.Fuentes Mota, and I.Robina Ramirez, An. Quim., Ser. C., 1984, 80, 123 (Chem. Abstr., 1985, 102, 149 699).
128 P.Schmitt, H.Guenther, G.Haegele, and R.Stilke, Org. Magn. Reson., 1984, 22, 446 (Chem. Abstr., 1985, 102, 6 960).
129 L.Rasolojaona and P.Mastagli, Carbohydr. Res., 1985, 143, 246.

# 4
# Oligosaccharides

## 1 General

As before, this Chapter deals with specific tri- and higher oligosaccharides; most references relate to their syntheses by specific chemical methods. It does not deal with compounds made by the oligomerization of monosaccharide derivatives, nor does it deal with the cyclodextrins (although this year, for the first time, mention is made of the chemical synthesis of such compounds). The synthesis of, *e.g.*, pentasaccharides is dealt with under that heading, and the required preparations of constituent parts are assumed and are not covered in their respective sections. Frequently, specific derivatives of the basic compounds are involved and this fact is often not recorded in the structural formulae used.

The syntheses, by newer procedures, of the complex glycan chains of glycoproteins and protoglycans present on cell surfaces have been illustrated in a published lecture.[1] Ogawa and colleagues, during their work on cell-surface glycans, have also used oligosaccharide trichloroacetamidates for production of corresponding gangliosides.[2]

A particular sensitivity to alkali of 1 → 4 linked β-D-galactopyranosyl units in oligosaccharides has been noted.[3]

Conformational calculations have been made on the determinant tri- and tetra-saccharides with $Le^a$, $Le^b$ and H group immunospecificity, and were predicted to be structurally rigid in aqueous solution.[4]

The chemical shifts and $^1J_{C,H}$ coupling constants in 3-deoxy-D-*manno*-octulosonic acid (KDO)-containing oligosaccharides can be used to determine the positions of substitution of the acid units.[5]

## 2 Trisaccharides

### 2.1 Linear (and Cyclic) Homotrisaccharides

2.1 Linear (and Cyclic) Homotrisaccharides.- The trisaccharide *O*-β-D-Glc*p*(1 → 4)-*O*-β-D-Glc*p*(1 → 3)-*O*-β-D-Glc*p* was prepared as its *N*-

methylazoxymethylglycoside, and several analogous disaccharide glycosides were obtained by use of transglycosylation enzymes,[6] and an enzymic synthesis of cellotriose from D-glucose and cellobiose has utilized a β-glucosidase covalently immobilized on polyacrylamide beads.[7] A series of β-(1 → 4) linked oligosaccharide derivatives have also been made by successive condensations between acetylated oligosaccharide glycosyl fluorides and phenyl 2,3-di-*O*-acetyl-6-*O*-benzyl-1-thio-β-D-glucopyranoside. The condensation products were then converted into the next highest glycosyl-fluorides and the process was repeated. In this way the tri- to penta-saccharides were produced as their 1-(propane-1,2-diol)-glycosides, which are the naturally occurring rhynchosporosides.[8]

2-*O*-Acetyl-3,4,6-tri-*O*-benzyl-D-glucose used together with *p*-nitrobenzenesulphonyl chloride, silver trifluoromethanesulphonate and triethylamine were the key reagents in the preparation of *O*-β-D-Glc*p*-(1 → 2)-*O*-β-D-Glc*p*-(1 → 6)-D-Glc,[9] and the interesting cyclic compounds cyclogentiotriose and cyclogentiotetraose (1) have been prepared as crown ethers with the potential of coordinating Group IA and IIA cations.[10]

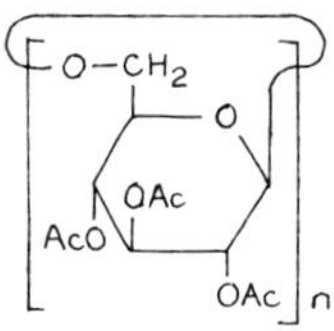


(1) n = 3,4

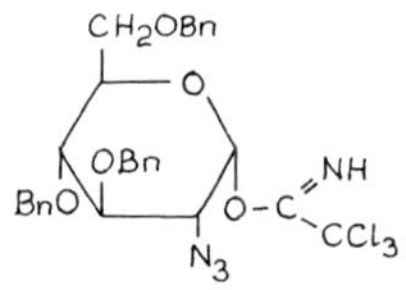


(2)

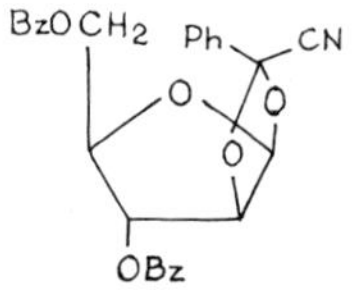


(3)

The trichloroacetimidate (2) has been used in the synthesis of the repeating unit of the capsular polysaccharide of *Neisseria meningitidis* [*O*-β-D-GlcNAc-(1 → 3)-*O*-β-D-GlcNAc-(1 → 3)-*O*-β-D-GlcNAc].[11] The β-(1 → 3)-linked D-galactopyranose trisaccharide has been prepared by a stepwise procedure,[12] and two groups of workers have reported preparations of the β-(1 → 6)-linked analogue in the form of various derivatives;[13,14] a set of papers has appeared on analogues having fluorine atoms in place of the C-3 hydroxy groups of different galactose units of this trimer.[15-17]

Condensation of the cyanoalkylidene compound (3) with 5-*O*-trityl compounds was used in the key step of the synthesis of the α-(1 → 5)-linked D-arabinofuranose trimer.[18]

2.2 Branched Homotrisaccharides.- *O*-α-D-Man*p*-(1 → 3)-*O*-[α-D-Man*p* (1 → 6)]-*O*-β-D-Man*p* was made as its *p*-trifluoroacetamidophenyl glycoside by use of *p*-nitrophenyl 2-*O*-benzoyl-4,6-*O*-benzylidene-β

D-mannopyranoside as the starting material.[19]

2.3 Linear Heterotrisaccharides.- The non-reducing α-D-galactopyranosyl *O*-β-D-fructofuranosyl-(2 → 6)-β-D-fructofuranoside has been made by an enzymic procedure.[20]

Reducing compounds of this class are now dealt with first according to their reducing termini and then according to the adjacent moieties.

*O*-α-L-Ara*f*-(1 → 6)-*O*-α-D-Glc*p*-(1 → 4)-D-Glc has been prepared in connection with studies of an anti-tumour plant polysaccharide,[21] and *O*-α-D-Gal*p*-(1 → 4)-*O*-β-D-Gal*p*-(1 → 4)-D-Glc, the human $P^k$ antigenic determinant, was synthesized.[22] Regioselective glycosylations at the unsubstituted 3' and 4' positions of benzyl penta-*O*-benzyl-β-D-lactoside, induced respectively by silver triflate and silver silicate, have been used to prepare derivatives of *O*-β-D-$GlcNH_2$*p*-(1 → 3)-*O*-β-D-Gal*p*-(1 → 4)-D-Glc and *O*-β-D-$GlcNH_2$*p*-(1 → 4)-*O*-β-D-Gal*p*-(1 → 4)-D-Glc.[23] A further synthesis of the first of these as its methyl glycoside has also been described,[24] as has the analogue of the second with *N*-acetylgalactosamine replacing the glucosamine moiety.[25] The synthesis of sialosyllactose, *O*-α-NeuAc*p*-(2 → 3)-O-β-D-Gal*p*-(1 → 4)-D-Glc, and the β-linked neuraminic acid isomer have been prepared[26] and the former was converted into $G_{M3}$ ganglioside by way of the trichloroacetimidate.[27]

Compounds having reducing D-mannose units have been synthesized as follows: *O*-β-D-GlcNAc*p*-(1 → 2)-*O*-α-D-Man*p*-(1 → 3)-D-Man (the trisaccharide component of a bivalve spermatoza),[28] *O*-α-L-Fuc*p*-(1 → 3)-*O*-β-D-GlcNAc*p*-(1 → 2)-D-Man[29] and *O*-β-D-Gal*p*-(1 → 4)-O-β-D-GlcNAc*p*-(1 → 2)-D-Man[30] (both being glycoprotein trisaccharides).

An interesting approach to the preparation of a variety of trisaccharides involves the use of the lactose-derived compound (4), and with it the following D-galactose reducing trisaccharides were made: *O*-β-D-Gal*p*-(1 → 4)-*O*-β-D-Man*p*-(1 → 6)-D-Gal, O-β-D-Gal*p*-(1 → 4)-*O*-β-D-ManNAc*p*-(1 → 6)-D-Gal and *O*-β-D-Gal*p*-(1 → 4)-*O*-α-D-GlcNAc*p*-(1 → 6)-D-Gal.[31] Related syntheses using maltose-

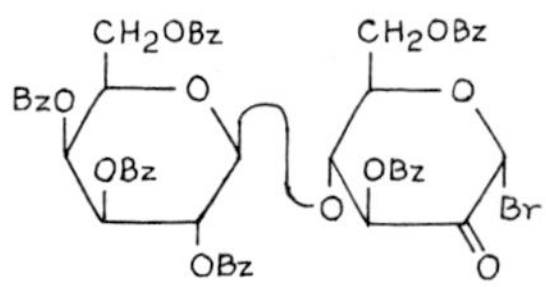


(4)

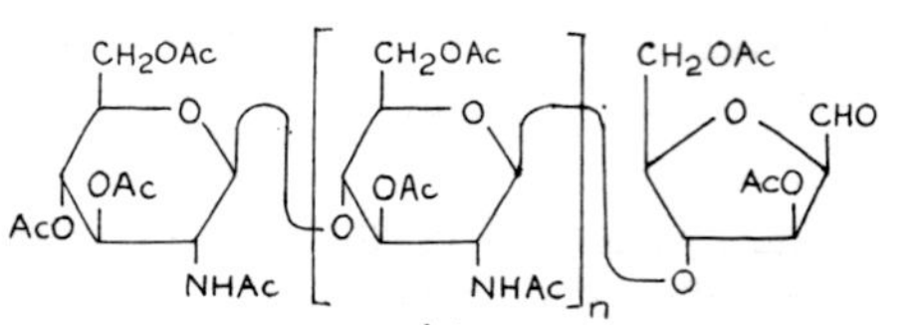


(5)

and cellobiose-derived analogues of compound (4) have also been reported.[32] Independent work described the preparation of O-β-D-Galp-(1 → 4)-O-β-D-GlcNAcp-(1 → 3)-D-Gal.[33]

Compounds reported with 2-amino-2-deoxy-D-glucose at the reducing termini are: O-α-D-Galp-(1 → 3)-O-β-D-Galp-(1 → 4)-D-GlcNAc (made as a glycoside with a spacer arm required for immunological studies),[34] O-α- and β-D-NeuAcp-(2 → 6)-O-β-D-Galp-(1 → 4)-D-GlcNPhth,[35,36] O-β-D-Manp-(1 → 4)-O-β-D-$GlcNH_2$-(1 → 6)-D-$GlcNH_2$ (prepared as a long-chain fatty acid diamide for studies of Lipid A),[37] O-α-D-$GlcNH_2$p-(1 → 4)-O-β-D-GlcAp-(1 → 4)-$GlcNH_2$ (made as a sulphate derivative for studies of heparin).[38] Compounds obtained by nitrous acid deaminative cleavage of partially N-acetylated chitosan followed by acetylation included the 2,5-anhydro-D-mannose-containing dimer (5, n = 0). Higher members of this series (5; n = 1, 2, 3) were also isolated.[39]

Trisaccharides with 2-amino-2-deoxy-D-mannose and -galactose termini to have been synthesized are: O-α-D-Glcp-(1 → 3)-O-α-L-Rhap(1 → 3)-β-D-ManNAcp (the repeating unit of the O-specific chain of Aeromonas lipopolysaccharides),[40] O-β-D-Galp-(1 → 3)-O-β-D-Galp-(1 → 3)-GalNAc[41] and O-β-D-Galp-(1 → 4)-O-β-D-GlcNAcp-(1 → 3)-GalNAc.[42] Some of these were prepared as 8-methoxycarbonyl-octyl glycosides for immunological studies.

Pentose-based trisaccharides to have been reported are O-α-L-Araf-(1 → 3)-O-β-D-Xylp-(1 → 4)-D-Xyl (synthesis[43] and its isolation as a 5-(hydroxymethoxy)cinnamate ester from a plant source),[44] and O-α-D-KDO-(2 → 2)-O-β-D-Ribf-(1 → 2)-D-Rib (the repeating unit of an E. coli capsular polysaccharide).[45]

Deoxysugar-containing compounds to have been synthesized are O-α-D-Glcp-(1 → 3)-O-α-L-Rhap-(1 → 2)-L-Rha (Shigella polysaccharide component)[46] and compound (6), which is a derivative of the "lower" oligosaccharide chain of aureolic acids.[47]

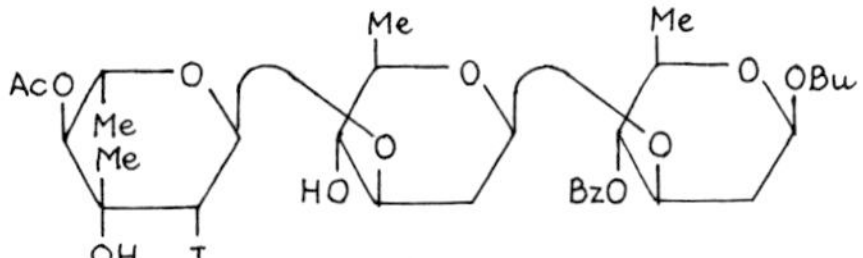


(6)

2.4 Branched Heterotrisaccharides.- Compounds of this category to have been synthesized are: O-α-D-Glcp-(1 → 4)-O-[α-L-Araf-(1 → 6)]-O-β-D-Glcp-OMe,[21] O-β-D-GlcNAcp-(1 → 4)-O-[α-L-Fucp-(1 → 6)]-O-β-

D-GlcNAc-OMe,[48] O-β-D-Gal*p*-(1 → 3)-*O*-[β-D-GlcA*p*-(1 → 4)]-O-α-L-Rha*p*-OMe,[49] O-β-D-GlcNAc*p*-(1 → 2)-*O*-[α-D-Glc*p*-(1 → 3)]-*O*-α-L-Rha*p*-OMe,[50] and O-β-D-Rib*f*-(1 → 2)-O-[α-D-KDO-(2 → 3)]-D-Rib.[45]

## 3 Tetrasaccharides

As with the trisaccharides, the following tetrasaccharides are classified according as to whether they have linear or branched structures, and then by the nature of the sugars at the reducing termini.

### 3.1 Linear and Cyclic Tetrasaccharides

.- Cyclogentiotetraose (the cyclic β-1→6 linked D-glucose compound) behaves like a crown ether and forms 1:1 complexes with Group I metal ions.[51]

D-Hexose termination compounds to have been reported, often as derivatives, are: O-β-D-GalNH$_2$*p*-(1 → 3)-*O*-α-D-Gal*p*-(1 → 4)-*O*-β-D-Gal*p*-(1 → 4)-D-Glc,[52] *O*-β-D-Gal*p*-(1 → 3)-*O*-β-D-GalNAc*p*-(1 → 4)-*O*-β-D-Gal*p*-(1 → 4)-D-Glc,[25] *O*-β-D-Api*f*-(1 → 2)-*O*-β-D-Glc*p*-(1 → 2)-*O*-α-L-Ara*p*-(1 → 6)-D-Glc,[53] and *O*-α- and *O*-β-D-NeuAc*p*-(2 → 6)-*O*-β-D-Gal*p*-(1 → 4)-*O*-β-D-GlcNAc*p*-(1 → 2)-D-Man.[35,36]

2-Amino-2-deoxyhexose compounds to have been noted are: O-α-D-KDO-(2 → 4)-*O*-α-D-KDO-(2 → 6)-*O*-β-D-GlcNH$_2$*p*-(1 → 6)-D-GlcNH$_2$ (oligosaccharide portion of a lipopolysaccharide characterized),[54] *N*-acetylchitotetraose and higher oligomers, which have been shown to enhance host-mediated microbiocidal effects in mice,[55] and the repeating tetrasaccharide of hyaluronic acid with a 2-acetamido-2-deoxy-D-glucose reducing unit, which, on degradation with alkali, gave the expected trisaccharide as well as *O*-β-D-GlcA*p*-(1 → 3)-*O*-β-D-GlcNAc*p*-(1 → 3)-D-*arabino*-trihydroxyglutaric acid.[56] The mannosamine-containing tetramer *O*-α-D-Glc*p*-(1 → 4)-*O*-α-D-Glc*p*-(1 → 3)-*O*-α-L-Rha*p*-(1 → 3)-*O*-β-D-ManNAc*p*-O(CH$_2$)$_8$CO$_2$Me has also been reported.[40]

Interest has been shown in the L-rhamnose derivative *O*-β-D-GlcNH$_2$*p*-(1 → 2)-*O*-α-L-Rha*p*-(1 → 2)-O-α-L-Rha*p*-(1 → 3)-L-Rha.[57,58]

### 3.2 Branched Tetrasaccharides

.- The following branched compounds or derivatives thereof have been reported: *O*-α-L-Fuc*p*-(1 → 2)-*O*-β-D-Gal*p*-(1 → 4)-*O*-[α-L-Fuc*p*-(1 → 3)]-D-Glc,[59] *O*-β-D-Gal*p*-(1 → 4)-*O*-β-D-GlcNAc*p*-(1 → 6)-*O*-[β-D-GlcNAc*p*-(1 → 3)]-D-GalNAc,[42] *O*-α-D-Glc*p*-(1 → 3)-α-L-Rha*p*-(1 → 2)-*O*-[α-D-Glc*p*-(1 → 3)]-*O*-α-Rha*p*-OMe,[60] and *O*-β-D-GlcNA*p*-(1 → 2)-*O*-[α-D-Glc*p*-(1 → 3)]-α-L-Rha*p*-(1 → 2)-*O*-α-L-

Rhap-OMe.[46]

## 4 Pentasaccharides

Syntheses of the following linear members of this category have been reported; the first is an N-glycoprotein component and was examined conformationally by theoretical HSEA and sophisticated n.m.r. methods: O-β-D-Galp-(1 → 4)-O-β-D-GlcNAcp-(1 → 2)-O-α-D-Manp-(1 → 6)-O-β-D-Manp-(1 → 4)-D-GlcNAc,[30] the isomer with a 1 → 3 linkage instead of the 1 → 6 bond,[30] O-α-D-Manp-(1 → 3)-O-α-D-Manp-(1 → 2)-O-α-D-Manp-(1 → 2)-O-α-D-Manp-(1 → 2)-D-Man,[61] and compound (7), which corresponds to the antithrombin binding region of heparin.

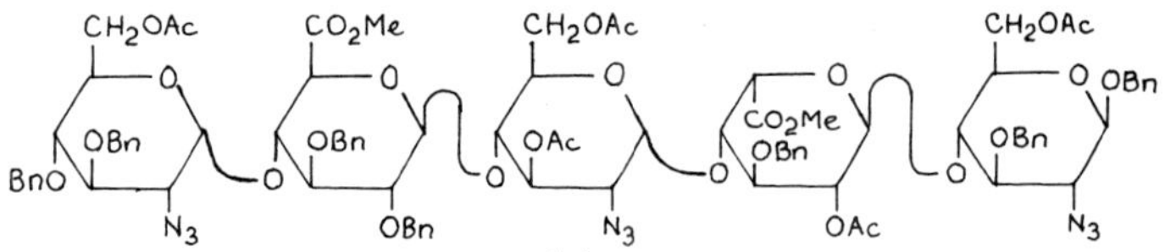


(7)

The acetyl groups indicate positions of sulphate esters in the natural material, the benzyl groups indicate hydroxy groups, and the azido groups are all precursors of amino functions.[62]

Branched-chain compounds to have been prepared are: O-β-D-GlcNAcp-(1 → 2)-O-α-D-Manp-(1 → 3)-O-[α-D-Manp-(1 → 6)]-O-β-D-Manp-(1 → 4)-D-GlcNAc,[63] O-β-D-GlcNAcp-(1 → 2)-O-α-D-Manp-(1 → 6)-O-[α-D-Manp-(1 → 3)]-O-β-D-Manp-(1 → 4)-D-GlcNAc,[63] O-α-D-Manp-(1 → 3)-O-[β-D-GlcNAcp-(1 → 4)]-O-[α-D-Manp-(1 → 6)]-O-β-D-Manp-(1 → 4)-GlcNAc[64] and O-β-D-GlcNAcp-(1 → 2)-O-[α-D-Glcp-(1 → 3)]-O-α-L-Rhap-(1 → 2)-O-[α-D-Glcp-(1 → 3)]-O-α-L-Rhap-OMe.[65]

## 5 Hexasaccharides

Two impressive reports from T. Ogawa's group describe the rational synthesis of α-cyclodextrin, i.e. the cyclic α-(1 → 4) linked D-glucose hexamer. β-Glycosyl fluoride derivatives in the presence of tin(II) chloride and silver triflate were employed for the glycoside bond formations.[66] The same group has described the preparation of the α-(1 → 3) linked D-mannopyranose oligomers up to the hexasaccharide, which are models for the fruit body polysaccharides of *Tremella fuciformis*.[68]

The further, branched D-hexose-based hexasaccharides (8),[69] (9; R = β-D-GlcNAcp)[70] and (10; R = H)[71] have been prepared.

```
α-D-Manp-(1 → 2)-α-D-Manp-(1 → 6)-α-D-Manp-(1 → 6)-α-D-Manp-OPr
                                       3
                                       ↑
                                       1
                α-D-Manp-(1 → 2)-α-D-Manp                    (8)
```

```
       R-(1 → 2)-α-D-Manp
                     1
                     ↓
                     6
β-D-GlcNAcp-(1 → 4)-D-Man
                     3
                     ↑
                     1
       R-(1 → 2)-α-D-Manp                                    (9)
```

```
                        R
                        1
                        ↓
                        3
β-D-Galp-(1 → 4)-β-D-GlcNAcp-(1 → 6)-β-D-Galp-(1 → 4)-D-Glc
                                         3
                                         ↑
                                         1
                  β-D-Galp-(1 → 3)-β-D-GlcNAcp
                                         3
                                         ↑
                                         1
                                         R                   (10)
```

## 6 Higher Saccharides

The hexaglucosylhexitol (11) is a plant regulatory substance - the first specific complex carbohydrate to be identified as such.[72] The heptasaccharide (12),[29] a glycoprotein component, has been prepared, and the octasaccharide (13)[30] was conformationally characterized. Octasaccharides (9; R = β-D-Gal*p*-(1 → 4)-β-D-GlcNAc*p*)[70] and (10; R = α-L-Fuc*p*)[71] have also been synthesized, as has the nonasaccharide (14).[73]

```
β-D-Glcp-(1 → 6)-β-D-Glcp-(1 → 6)-β-D-Glcp-(1 → 6)-β-D-Glcp-(1 → 6)-D-glucitol
                  3                                  3
                  ↑                                  ↑
                  1                                  1
               β-D-Glcp                           β-D-Glcp                              (11)

α-L-Fucp-(1 → 3)-β-D-GlcNAcp-(1 → 2)-α-D-Manp-(1 → 3)-D-Man
                                                      6
                                                      ↑
                                                      1
              α-L-Fucp-(1 → 3)-β-D-GlcNAcp-(1 → 2)-α-D-Manp                         (12)

β-D-Galp-(1 → 4)-β-D-GlcNAcp-(1 → 2)-α-D-Manp-(1 → 6)-β-D-Manp-(1 → 4)-D-GlcNAc
                                                        3
                                                        ↑
                                                        1
                β-D-Gal-(1 → 4)-β-D-GlcNAcp-(1 → 2)-α-D-Manp                        (13)

β-D-Galp-(1 → 4)-β-D-GlcNAcp-(1 → 2)-α-D-Manp-(1 → 3)-D-Man
                   3                                  6
                   ↑                                  ↑
                   1                                  |
                α-L-Fucp                              |
                              α-L-Fucp                |
                                 1                    |
                                 ↓                    |
                                 3                    1
              β-D-Galp-(1 → 4)-β-D-GlcNAcp-(1 → 2)-α-D-Manp                         (14)
```

## References


1 T.Ogawa, H.Yamomoto, T.Nukada, T.Kitajima, and M.Sugimoto, Pure Appl. Chem., 1984, 56, 779 (Chem. Abstr., 1985, 102, 6 984).
2 M.Sugimoto, T.Horisaki, and T.Ogawa, Glycoconjugate J., 1985, 2, 11 (Chem. Abstr., 1985, 103, 123 839).
3 H.A.El-Shenawy and C.Schuerch, J. Carbohydr. Chem., 1985, 4, 215.
4 V.E.Verovskii, G.M.Lipkind, and N.K.Kochetkov, Bioorg. Khim., 1984, 10, 1680 (Chem. Abstr., 1985, 102, 204 193).
5 R.Christian, G.Schulz, and F.M.Unger, Tetrahedron Lett., 1985, 26, 3951.
6 F.Yagi, K.Tadera, and A.Kobayashi, Agric. Biol. Chem., 1985, 49, 2985.
7 T.Tanaka and S.Ol, Agric. Biol. Chem., 1985, 49, 1267.
8 K.C.Nicolaou, J.L.Randall, and G.T.Furst, J. Am. Chem. Soc., 1985, 107, 5556.
9 S.Koto, N.Morishima, H.Sato, Y.Sato, and S.Zen, Bull. Chem. Soc. Jpn., 1985, 58, 120.
10 G.Excoffier, M.Paillet, and M.Vignon, Carbohydr. Res., 1985, 135, C10.
11 W.Kinzy and R.R.Schmidt, Liebigs Ann. Chem., 1985, 1537.
12 P.Kovac and C.P.J.Glaudemans, Carbohydr. Res., 1985, 138, C10.
13 P.Kovac, Carbohydr. Res., 1985, 144, C12.
14 G.Ekborg, B.Vranesic, A.K.Bhattacharjee, P.Kovac, and C.P.J.Glaudemans, Carbohydr. Res., 1985, 142, 203.
15 P.Kovac, H.J.C.Yeh, and C.P.J.Glaudemans, Carbohydr. Res., 1985, 140, 277.
16 P.Kovac and C.P.J.Glaudemans, Carbohydr. Res., 1985, 140, 289.
17 P.Kovac, C.P.J.Glaudemans, W.Guo, and T.C.Wong, Carbohydr. Res., 1985, 140, 299.
18 K.Hatanaka and H.Kuzuhara, J. Carbohydr. Chem., 1985, 4, 333.
19 G.Ekborg and C.P.J.Glaudemans, Carbohydr. Res., 1985, 142, 213.
20 Y.Nimami, K.Yazawa, K.Nakamura, and Z.Tamura, Chem. Pharm. Bull., 1985, 33, 710.
21 T.Fujiwara, T.Takeda, and Y.Ogihara, Carbohydr. Res., 1985, 141, 168.
22 J.-C.Jacquinet and P.Sinaÿ, Carbohydr. Res., 1985, 143, 143.
23 H.Paulsen, D.Hadamczyk, W.Kutschker, and A.Bunsch, Liebigs Ann. Chem., 1985, 129.
24 A.Maranduba and A.Veyrières, Carbohydr. Res., 1985, 135, 330.
25 H.Paulsen and M.Paal, Carbohydr. Res., 1985, 137, 39.
26 T.Ogawa and M.Sugimoto, Carbohydr. Res., 1985, 135, C5.
27 M.Sugimoto and T.Ogawa, Glycoconjugate J., 1985, 2, 5 (Chem. Abstr., 1985, 103, 142 296).
28 T.Takeda, S.Fujisawa, Y.Ogihara, and T.Hori, Chem. Pharm. Bull., 1985, 33, 540.
29 H.Lönn, Carbohydr. Res., 1985, 139, 105.
30 H.Paulsen, T.Peters, V.Sinnwell, R.Lebuhn, and B.Meyer, Liebigs Ann. Chem., 1985, 489.
31 F.W.Lichtenthaler and E.Kaji, Liebigs Ann. Chem., 1985, 1659.
32 F.W.Lichtenthaler, E.Kaji, and S.Weprek, J. Org. Chem., 1985, 50, 3505.
33 J.Dahmen, G.Gnosspelius, A.-C.Larsson, T.Lave, G.Noori, K.Palsson, T.Frejd, G.Magnusson, Carbohydr. Res., 1985, 138, 17.
34 P.J.Garegg and S.Oscarson, Carbohydr. Res., 1985, 136, 207.
35 H.Paulsen and H.Tietz, Carbohydr. Res., 1985, 144, 205.
36 H.Paulsen and H.Tietz, Angew. Chem. Int. Ed. Engl., 1985, 24, 128.
37 P.Westerduin, J.H.Van Boom, C.A.A.Van Boeckel, and T. Beetz, Carbohydr. Res., 1985, 137, C4.
38 Y.Ichikawa, R.Ichikawa, and H.Kuzuhara, Carbohydr. Res., 1985, 141, 273.
39 S.Hirano, Y.Kondo, and K.Fujii, Carbohydr. Res., 1985, 144, 338.
40 H.Paulsen and J.P.Lorentzen, Tetrahedron Lett., 1985, 26, 6043.
41 H.Paulsen, T.Hasenkamp and M.Paal, Carbohydr. Res., 1985, 144, 45.
42 G.Grundler and R.R.Schmidt, Carbohydr. Res., 1985, 135, 203.
43 S.Koto, N.Morishima, K.Takenaka, C.Uchida, and S.Zen, Bull. Chem. Soc. Jpn., 1985, 58, 1464.
44 Y.Kato and D.J.Nevins, Carbohydr. Res., 1985, 137, 139.

45 P.Kosma, R.Christian, G.Schulz, and F.M.Unger, Carbohydr. Res., 1985, 141, 239.
46 L.V.Bakinovskii, A.R.Gomtsyan, N.E.Bairamova, and N.K.Kochetkov, Bioorg. Khim., 1985, 11, 254 (Chem. Abstr., 1985, 103, 105 237).
47 J.Thiem and M. Gerken, J. Org. Chem., 1985, 50, 954.
48 D.A.Schwartz, H.-H.Lee, J.P.Carver, and J.J.Krepinsky, Can. J. Chem., 1985, 63, 1073.
49 F.C.Baines, P.M.Collins, and F.Farnia, Carbohydr. Res., 1985, 136, 27.
50 H.-P.Wessel and D.R.Bundle, J. Chem. Soc., Perkin Trans. 1, 1985, 2251.
51 C.Bosso, G.Excoffier, M.Vignon, and J.Ulrich, Spectros.: Int. J., 1984, 3, 72 (Chem. Abstr., 1985, 102, 149 665).
52 K.Leontein, M.Nilsson, and T.Norberg, Carbohydr. Res., 1985, 144, 231.
53 R.Lanzetta, G.Laonigro, M.Parrilli, and E.Breitmaier, Can. J. Chem., 1984, 62, 2874.
54 U.Zähringer, B.Lindner, U.Seydel, E.T.Rietschel, H.Naoki, F.M.Unger, M.Imoto, S.Kusamoto, and T.Shiba, Tetrahedron Lett., 1985, 26, 6321.
55 K.Suzuki, A.Tokoro, Y.Okawa, S.Suzuki, and M.Suzuki, Chem. Pharm. Bull., 1985, 33, 886.
56 Y.Inoue and K.Nagasawa, Carbohydr. Res., 1985, 141, 99.
57 N.K.Kochetkov, N.E.Byramova, Yu.E.Tsvetkov, and L.V.Backinowsky, Tetrahedron, 1985, 41, 3363.
58 N.E.Byramova, Y.E.Tsvetkov, L.V.Backinowsky, and N.K.Kochetkov, Carbohydr. Res., 1985, 137, C8.
59 K.Takeo and S.Tei, Carbohydr. Res., 1985, 141, 159.
60 A.R.Gomtsyan, N.É.Byramova, L.V.Backinowsky, and N.K.Kochetkov, Carbohydr. Res., 1985, 138, C1.
61 T.Ogawa and H.Yamamoto, Carbohydr. Res., 1985, 137, 79.
62 C.A.A.van Boeckel, T.Beetz, J.N.Vos, A.J.M.de Jong, S.F.van Aelst, R.H.van den Bosch, J.M.R.Mertens, and F.A.van der Vlugt, J. Carbohydr. Chem., 1985, 4, 293.
63 H.Paulsen, W.Rauwald, and R.Lebuhn, Carbohydr. Res., 1985, 138, 29.
64 H.Paulsen, M.Heume, Z.Gyorgydeak, and R.Lebuhn, Carbohydr. Res., 1985, 144, 57.
65 A.R.Gomitsyan, N.É.Byramova, L.V.Backinowsky, and N.K.Kochetkov, Carbohydr. Res., 1985, 138, C1.
66 T.Ogawa and Y.Takahashi, Carbohydr. Res., 1985, 138, C5.
68 T.Ogawa and H.Yamamoto, Agric. Biol. Chem., 1985, 49, 475.
69 T.Ogawa and T.Nukada, Carbohydr. Res., 1985, 136, 135.
70 J.Arnarp, M.Haraldsson, and J.Lönngren, J. Chem. Soc., Perkin Trans. 1, 1985, 535.
71 V.K.Dua, K.Goso, V.E.Dube, and C.A.Bush, J. Chromatogr., 1985, 328, 259.
72 J.K.Sharp, M.McNeil, and P.Albersheim, J. Biol. Chem., 1984, 259, 11321 (Chem. Abstr., 1985, 102, 2 243).
73 H.Lönn, Carbohydr. Res., 1985, 139, 115.

# 5
# Ethers and Anhydro-sugars

## 1 Ethers

Methyl Ethers.- Elloramycin, a new anthracycline antibiotic from *Streptomyces olivaceus*, is a glycoside of 2,3,4-tri-*O*-methyl-L-rhamnopyranoside.[1]

A new modification of the Hakamori permethylation procedure uses n-butyllithium to generate the sulphinyl carbanion from dimethyl-sulphoxide. It has been applied to methyl α-D-mannopyranoside, oligosaccharides, and polysaccharides, and is preferred to procedures using metal hydrides or amides because it is rapid, convenient, and results in fewer byproducts (especially important in the g.c. analysis of polysaccharide constituents *via* partially methylated alditol acetate derivatives).[2,3]

A review on the regioselective manipulation of hydroxy groups *via* organotin derivatives has covered regioselective alkylation, including methylation.[4] Regioselective monoalkylation (including methylation) of several pento- and hexo-pyranosides using the dibutyltin oxide method is covered in the next section.

2-*O*-Methyl-L-lyxose, a constituent of everninomycin, has been synthesized from D-galactose using standard protecting group methodology; C4-C5 periodate cleavage of 2,3-di-*O*-benzyl-4-*O*-methyl-D-galactitol generated the desired pentose.[5] 3- and 4-*O*-Methyl-α-D-galactopyranose tetraacetates have been synthesized conventionally from 1,6-anhydro-β-D-galactose.[6] Syntheses of the 3-*O*-methyl and 3-*O*-benzyl ethers of methyl 4-deoxy-β-D-*arabino*-hexopyranoside from levoglucosan *via* its 4-deoxy-4-iodo-2-*O*-tosyl derivative have been reported.[7] A practical synthesis of 6-*O*-methyl-D-glucose from 1,2-*O*-isopropylidene-D-glucofuranose without isolation of intermediates has been detailed.[8] New syntheses of 3-*O*-methyl-lactose, required as a substrate for a study of hydrolysis by intestinal lactase, have utilized partial benzylation of benzyl 3',4'-*O*-isopropylidene-β-lactoside and benzyl β-lactoside.[9]

Other Alkyl and Aryl Ethers.- Regioselective alkylation *via* organotin derivatives has been reviewed.[4] The regioselective mono-alkylation (benzylation, allylation, methylation, and methoxymethylation) of nine pento- and hexo-pyranosides using the dibutyltin oxide method (i, $Bu_2SnO$-MeOH; ii, alkylating agent, *e.g.* BnBr, -dioxan, reflux) has been effected without the need for protection of hydroxy groups. Equatorial hydroxy groups with an adjacent *cis*-oxygen function (OH or OMe) were selectively activated, even in the presence of a primary hydroxy group. Thus phenyl α-L-arabinopyranoside and methyl α- and β-D-galactopyranosides gave the corresponding 3-*O*-benzyl or -methoxymethyl ethers exclusively, while methyl β-L-arabinopyranoside gave in 92% yield a 85:15 mixture of the 3- and 4-*O*-benzyl ethers. Methyl α-D-mannopyranoside gave mainly 3-*O*-alkylated products, and methyl β-D-xylopyranoside gave the 4-*O*-benzyl ether exclusively. Other glycosides (Me α-D-xyl, Me α- and β-D-Glc) and other alkylating agents gave more complex mixtures, which nevertheless contained predominantly the products of monoalkylation. Products were identified by $^{13}C$-n.m.r. spectroscopy data on some forty-nine mono- and seven di-alkylated glycosides being reported.[10]

Allyl and benzyl trichloroacetimidates have been shown to be useful allylating and benzylating agents under mild acidic conditions compatible with benzoyl protecting groups. Using these reagents, the 2,4-di-*O*-allyl-D-galactoside (1) was obtained in 54% yield and the 3-*O*-benzyl-L-rhamnoside (2) in 71% yield.[11] Dimolar

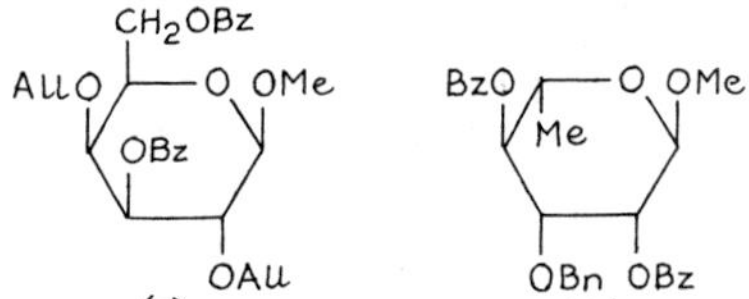


(1) (2)

allylation (allyl bromide -NaH-DMF) of L-ascorbic acid yielded the 2,3-diether as the major product in 63% yield. Allylation of 5,6-*O*-isopropylidene-L-ascorbic acid has also been investigated.[12] The perdeuterioallyl group has been shown to be useful for simplifying n.m.r. spectra of allyl-protected carbohydrates. The synthesis of perdeuterioallyl bromide from propynoic acid and the preparation and spectra of benzyl 3-*O*-allyl-2,6-dideoxy-α-L-*ribo*-hexopyranoside and its perdeuterioallyl analogue was described.[13] Allyl ether protecting groups can be isomerized by *trans*-$[Pd(NH_3)_2Cl_2]$ to prop-1-enyl ethers, which can then be cleaved on heating the reaction solution (in $Bu^tOH$) under reflux for several hours.[14] The relative stabilities of allyl ether, allyloxycarbonyl, and prop-2-enylidene acetal protecting groups to iridium, rhodium, and palladium catalysts have been examined. Conditions for selective cleavage of allyloxycarbonyl groups [with $Pd(PPh_3)_4$ or Wilkinson's catalyst] and selective isomerization of allyl ethers to prop-1-enyl ethers [with $Ir(COD)(PMePh_2)_2.PF_6$] were found.[15] Palladium derivatives have been used to effect allylation of tin alkoxides. Thus the 2,3-O-stannylidene derivative (3) gave a mixture of the 2- and 3-ethers (4) and (5) in 47 and 34% yield, respectively (Scheme 1).[16]

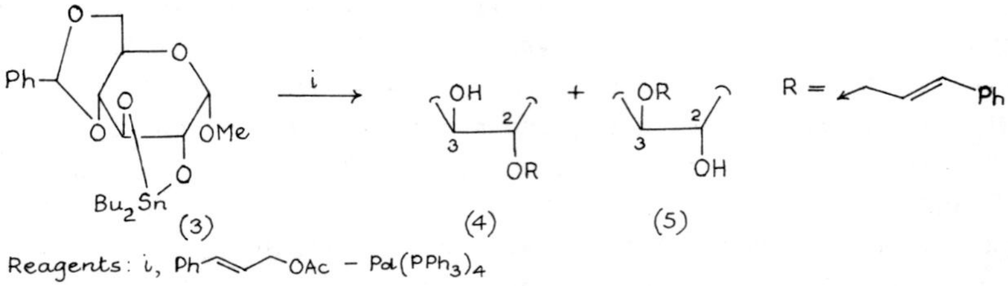

Scheme 1

A large number of α,α-dideuteriobenzyl (i.e. $PhCD_2$-) ethers of monosaccharides have been prepared so that their n.m.r. spectra could be compared with those of normal benzyl ethers.[17] Partial benzylation of methyl α-D-mannopyranoside yielded either the 2,3,6-triether (53%, using $PhCH_2Cl$-LiOH) or the 2,4,6-triether (41%, using $PhCH_2Cl$-KOH).[18] Several partially benzylated lactoses have been prepared and characterized.[9] A number of 1,3,4,6-tetra-O-substituted L-gulose derivatives, including benzyl 3,4,6-tri-O-benzyl-β-L-gulopyranoside, have been synthesized as part of a study of bleomycin, which has a 2-O-glycosylated L-gulopyranosyl moiety.[19]

New methods for cleavage of benzyl ethers have been reported. Catalytic transfer hydrogenolysis using ammonium formate as donor over palladium-on-charcoal has been found to work better than with formic acid or cyclohexene over palladium catalysts. Trityl ethers were also cleaved but not benzylidene acetals.[20] Ozone ($O_3/O_2$, $CH_2Cl_2$, <1 h, $0^oC$) has been shown to be a mild reagent for effecting debenzylation and compatible with glycosidic and acetal linkages. Methyl 2,3,4-tri-O-benzyl-α-D-xylopyranoside thus gave a mixture of the corresponding tribenzoate and partially deprotected derivatives, which on saponification yielded methyl α-D-xylopyranoside in 88% isolated yield.[21] Tri-O-benzyl protected arabinofuranosyl-nucleosides have been deblocked using ethanethiol-boron trifluoride diethyletherate.[22]

Reductive cleavage ($LiAlH_4$-$AlCl_3$, $NaCNBH_3$-HCl, or $AlCl_3$-$Me_3N$) of 2-substituted 1,6-anhydro-3,4-O-endo-(4-methoxybenzylidene)-β-D-galactopyranose derivatives gave predominantly the axial 3-O-(4-methoxybenzyl) ethers with a free 4-hydroxy group.[23] 4-Methoxy-benzyl ether groups survive acetolysis of a 1,6-anhydride bridge, but can be selectively removed with DDQ.[23,24]

Methyl 2,3,4-tri-O-butyl-α-D-glucopyranoside has been synthesized by selective protection at O-6 (>85%) with dihydropyran ($H^+$-DMSO) followed by phase-transfer catalyzed alkylation and hydrolysis. The 2,3,4,6-tetra-O-butyl ether was also synthesized.[25]

The novel ether-linked bis(1,6-anhydro-D-mannoses) (6) and (7) have been synthesized and then hydrolysed to the corresponding bis(D-mannose) compounds for testing as substrates and inhibitors

R-$(CH_2)_6$-R (6)

$RCH_2CH(NH_2)CH_2R$ (7)

R = 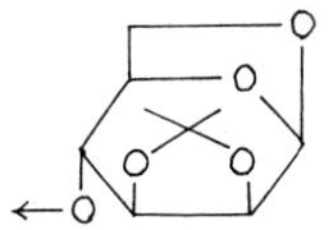

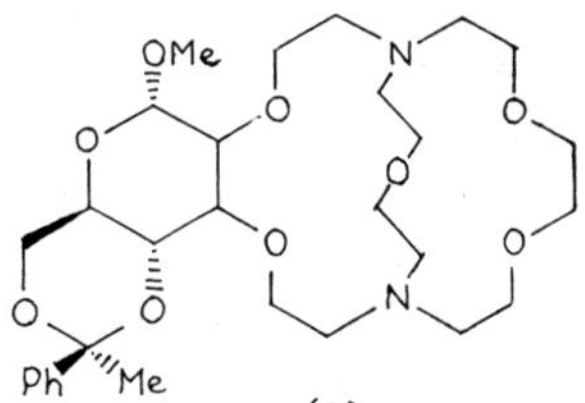

(8)

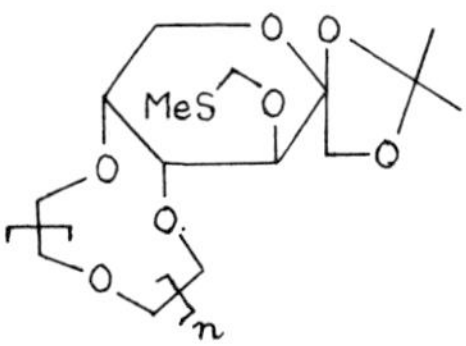

(9) n = 1–5

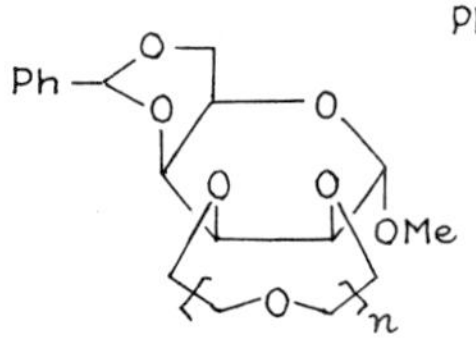

(10) n = 3 or 4

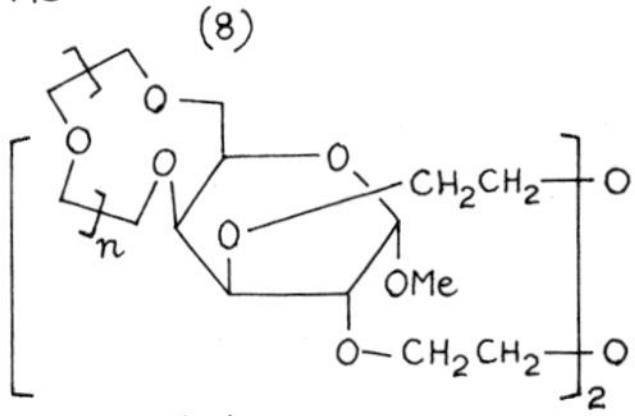

(11) n = 2 or 4

of cell-membrane transport systems.[26]

Several chiral crown ethers incorporating glycoside moieties have been reported. The [2.2.1]cryptand (8), with a 2,3-alkylated methyl 4,6-O-[(S)-phenylethylidene]-α-D-mannopyranoside moiety (having an axial phenyl group) has been synthesized following previously employed methodology (Vol.17, p.57).[27] Similarly, crown ethers based on 1,2-O-isopropylidene-3-O-(methylthiomethyl)-β-D-fructopyranose, i.e. (9),[28] methyl α-D-talopyranoside, e.g. its 2,3-alkylated derivatives (10) and analogues with macrocycles attached on both 2,3- and 4,6-positions,[29] and methyl α-D-galactopyranoside, e.g. compound (11) which has inter-residue ether linkages as well as large rings on the 4,6-positions, have been reported.[30]

Detritylation of 1,2,3,4-tetra-*O*-acetyl-6-*O*-trityl-β-D-galactopyranose with iodotrimethylsilane has been found to be better than other methods in minimizing side reactions.[31] The 4,4',4''-tris(levulinoyloxy)trityl group has been used for protection of primary hydroxy groups in dinucleotide synthesis. It is moderately stable to acids, but can be removed by hydrazinolysis followed by warming with pyridine-acetic acid , without removing acetyl and dimethoxytrityl groups.[32]

Silyl Ethers.- A review on the use of organosilicon protecting groups in organic synthesis has included carbohydrate and nucleoside examples, and the protection of 1,2- and 1,3-diols.[33]

Reaction of 1,6-anhydro-4',6'-*O*-benzylidene-β-maltose with 1,3-dichloro-1,1,3,3-tetraisopropyldisiloxane (*i.e.* TIPS-$Cl_2$) in pyridine yielded the 2',3'-*O*-TIPS ether, while the cellobiose analogue yielded the 2,3-*O*-TIPS ether (*i.e.* substituted on the anhydro-sugar moiety). Discrimination between the 2,3- and 2',3'-hydroxy groups of maltose and cellobiose can be effected in this way.[34] 3,4-*O*-TIPS-protected D-glucopyranose derivatives have been prepared in connection with the synthesis of glycophospholipids, by mild rearrangement (py.HCl-DMF) of the corresponding 4,6-*O*-TIPS derivatives.[35-37] The selective silylation of L-rhamnal and L-fucal is covered in Chapter 13, while n.m.r. studies of trimethylsilylated sugar derivatives are mentioned in Chapter 21.

## 2 Intramolecular Ethers (Anhydro-sugars)

Oxirans.- Synthesis of the 2,3-anhydro-4-deoxy-glycosides (12) from levoglucosan has been described.[7] Conformational studies of *cis*- and *trans*-2-methyl-3,4-epoxytetrahydropyrans are covered in Chapter 21.

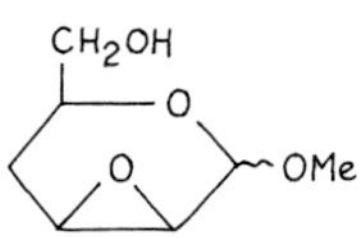


(12)

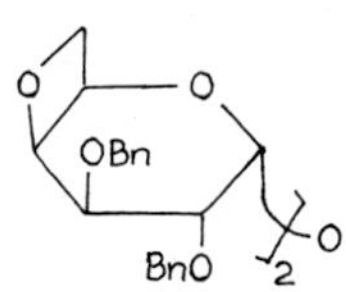


(13)

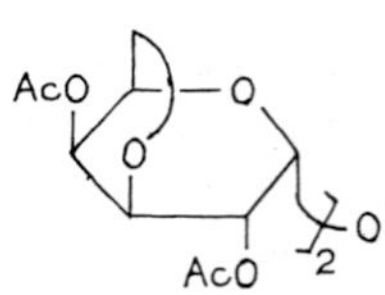


(14)

Other Anhydrides.- Ring-opening polymerizations of 1,3-anhydro-2,4,6-tri-*O*-(*p*-bromobenzyl)-β-D-glucopyranose,[38] 1,4-anhydro-2,3,6-tri-*O*-benzyl-α-D-glucopyranose,[39] 1,4-anhydro-2,3-*O*-cyclohexylidene

-α-D-ribopyranose,[40] and related derivatives have given stereoregular (1 → 3)-α-D-glucopyranan, (1 → 5)-α-D-glucofuranan, and (1 → 4)-β-D-ribopyranan respectively.

The 4,6;4',6'-dianhydride (13) and the 3,6;3',6'-dianhydride (14) have been prepared by intramolecular sulphonate displacements on α-D-galactosyl α-D-galactoside 6,6'-ditosylates, with and without 2,2',3,3'-tetrabenzyl ether protecting groups, respectively.[41] Synthesis of a 4,6-thioanhydride is covered in Chapter 11.

1,6-Anhydro-β-D-glucopyranose has been identified as a product from hydrothermolysis of cellulose and its degradation in aqueous solution at 200-240°C examined.[42] A mild procedure for the synthesis of 1,6-anhydrohexopyranoses involves treatment of the pentabromophenyl β-glycosides (obtained in two steps from the β-peracetates) with tetrabutylammonium hydroxide at 20°C, followed by acetylation for isolation. The 1,6-anhydrides of D-galactopyranose, D-glucopyranose, and cellobiose were obtained in this way.[43] The same authors have also been able to obtain 1,6-anhydro-β-D-glucopyranose (but not the galactose analogue) by the one-pot selective tosylation-alkaline treatment procedure developed by Georges and Fraser-Reid for the synthesis of the mannose analogue (Vol.18, Ch.5, Ref.37). 1,6-Anhydro-3,4-di-O-benzyl-β-L-gulopyranose (15) has been synthesized by two routes as a key intermediate in the synthesis of the carbohydrate moiety of bleomycin,[19] that starting from penta-O-acetyl-L-gulopyranose (16) (which is available from L-ascorbic acid in 3 steps) being shown in Scheme 2.

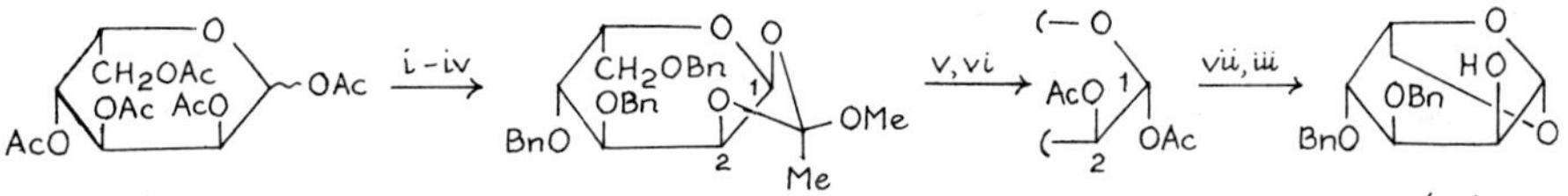


(16) (15)

Reagents: i, HBr-HOAc ; ii, $Et_4NBr$-MeOH ; iii, MeONa-MeOH ; iv, BnBr-NaH-DMF ; v, HOAc ; vi, $Ac_2O$-Py ; vii, $SnCl_4$

Scheme 2

D-Fructose, or inulin, is converted by anhydrous hydrogen fluoride into a mixture of difructose dianhydrides. Six were isolated, five being known compounds, but two of these were assigned the revised structures α-D-fructopyranose-β-D-fructopyranose 1,2':2,1'-dianhydride and α-D-fructofuranose-β-D-fructopyranose 1,2':2,1'-dianhydride. The new compound (17) was shown to be β-D-fructofuranose-β-D-fructopyranose 2,1':3,2'-dianhydride.[44]

Molecular mechanics and *ab initio* molecular orbital calculations have been performed on the neutral and anionic forms of methyl 3,6-

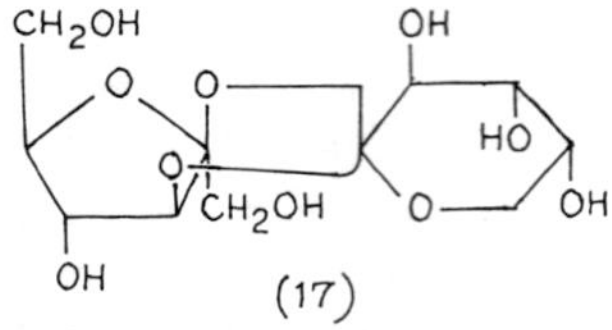

(17)

anhydro-α-D-galactopyranoside, and the results suggest that alkylation and esterification should occur at O-4 preferentially due to anion stabilization rather than steric factors. The 4-hydroxy group is of course equatorial whereas that at C-2 is axial.[45]

References


1 H.Drautz, P.Reuschenbach, and H.Zähner, J. Antibiot., 1985, 38, 1291.
2 A.B.Blakeney and B.A.Stone, Carbohydr. Res., 1985, 140, 319.
3 J.P.Parente, P.Cardon, Y.Leroy, J.Montreuil, B.Fournet, and G.Ricart, Carbohydr. Res., 1985, 141, 41.
4 S.David and S.Hanessian, Tetrahedron, 1985, 41, 643.
5 V.A.Oláh, A.Lipták, J.Harangi, and P.Nánási, Acta Chim. Hung., 1985, 118, 25.
6 A.Fernandez-Mayoralas and M.Martin-Lomas, An. Quim., Ser. C, 1984, 80, 184 (Chem. Abstr., 1985, 102, 132 371).
7 A.Banaszek, Bull. Pol. Acad. Sci., Chem., 1984, 32, 99 (Chem. Abstr., 1985, 102, 149 663).
8 J.W.Van Cleve, G.E.Inglett, and L.W.Tjarks, Carbohydr. Res., 1985, 137, 259.
9 A.Fernandez-Mayoralas, M.Martin-Lomas, and D.Villanueva, Carbohydr. Res., 1985, 140, 81.
10 M.E.Haque, T.Kikuchi, K.Yoshimoto, and Y.Tsuda, Chem. Pharm. Bull., 1985, 33, 2243.
11 H.-P.Wessel, T.Iverson, and D.R.Bundle, J. Chem. Soc., Perkin Trans. 1, 1985, 2247.
12 K.S.Kim, H.S.Lee, S.C.Kim, and C.S.Hahn, Bull. Korean Chem. Soc., 1984, 5, 264 (Chem. Abstr., 1985, 102, 149 684).
13 J.Theim, H.Mohn, and A.Heesing, Synthesis, 1985, 775.
14 T.Bieg and W.Szeja, J. Carbohydr. Chem., 1985, 4, 441.
15 M.Kloosterman, J.H.Van Boom, P.Chatelard, P.Boullanger and G.Descotes, Tetrahedron Lett., 1985, 26, 5045.
16 E.Keinan, M.Sahai, Z. Roth, A.Nudelman, and J.Herzig, J. Org. Chem., 1985, 50, 3558.
17 A.Zamojski, B.Grzeszczyk, A.Banaszek, J.Babinski, X.Bordas, K.Dziewiszek, and S.Jarosz, Carbohydr. Res., 1985, 142, 165.
18 S.Koto, K.Takenaka, N.Morishima, A.Sugimotc, and S.Zen, Bull. Chem. Soc. Jpn., 1984, 57, 3603.
19 K.Katano, P.-I.Chang, A.Millar, V.Pozsgay, D.K.Minster, T.Ohgi, and S.M.Hecht, J. Org. Chem., 1985, 50, 5807.
20 T.Bieg and W.Szeja, Synthesis, 1985, 76.
21 P.Angibeaud, J.Defaye, A.Gadelle, and J.-P.Utille, Synthesis, 1985, 1123.
22 T.Kulikowski, Z.Zawadzki, E.de Clercq, and D.Shugar, Acta Biochim. Pol., 1984, 31, 341 (Chem. Abstr., 1985, 103, 6656).
23 M.Kloosterman, T.Slaghek, J.P.G.Hermans, and J.H.Van Boom, Recl.: J.R. Neth. Chem. Soc., 1984, 103, 335 (Chem. Abstr., 1985, 102, 149 649).

24 Y.Oikawa, T.Nishi, and O.Yonemitsu, J. Chem. Soc., Perkin Trans. 1, 1985, 7.
25 R.Nouguier, C.Lambert, and O.Azria, Tetrahedron Lett., 1985, 26, 5769.
26 G.D.Holman and P.J.W.Midgley, Carbohydr. Res., 1985, 135, 337.
27 M.Pietraszkiewicz and J.Jurczak, J. Carbohydr. Chem., 1985, 4, 429.
28 R.Ellinghaus and G.Schröder, Liebigs Ann. Chem., 1985, 418.
29 H.Gruber and G.Schröder, Liebigs Ann. Chem., 1985, 421.
30 B.Walz and G.Schröder, Liebigs Ann. Chem., 1985, 426.
31 P.Kovac and C.P.J.Glaudemans, Carbohydr. Res., 1985, 140, 313.
32 M.Sekine and T.Hata, Bull. Chem. Soc. Jpn., 1985, 58, 336.
33 M.Lalonde and T.H.Chan, Synthesis, 1985, 817.
34 Y.Ichikawa, A.Manaka, and H.Kuzuhara, Carbohydr. Res., 1985, 138, 55.
35 C.A.A.van Boeckel and J.H.van Boom, Tetrahedron, 1985, 41, 4545.
36 C.A.A.van Boeckel, G.M.Visser, and J.H.van Boom, Tetrahedron, 1985, 41, 4557.
37 C.A.A.van Boeckel and J.H.van Boom, Tetrahedron, 1985, 41, 4567.
38 F.J.Good, Jr. and C.Schuerch, Macromolecules, 1985, 18, 595 (Chem. Abstr., 1985, 102, 167 264).
39 T.Uryu, C.Yamaguchi, K.Morikawa, K.Terui, T.Kanai, and K.Matsuzaki, Macromolecules, 1985, 18, 599 (Chem. Abstr., 1985, 102, 167 263).
40 T.Uryu, T.Kato, and S.Higuchi, Sen'i Gakkaishi, 1984, 40, T439 (Chem. Abstr., 1985, 102, 149 866).
41 A.Liav and M.B.Goren, Carbohydr. Res., 1985, 141, 323.
42 G.Bonn, J. Carbohydr. Chem., 1985, 4, 405.
43 M.Kloosterman, M.J.Dees, G.A.van der Marel, and J.H.van Boom, Recl. Trav. Chim. Pays-Bas, 1985, 104, 116 (Chem. Abstr., 1985, 103, 37 659).
44 J.Defaye, A.Gadelle, and C.Pedersen, Carbohydr. Res., 1985, 136, 53.
45 A.P.Mazurek and W.Szeja, J. Chem. Soc., Perkin Trans. 2, 1985, 57.

# 6
# Acetals

## 1 Isopropylidene Acetals

A useful method for preparing 2,3-O-isopropylidene-L-(S)-glyceraldehyde from L-arabinose has been described, involving the acetonation of its dibenzyldithioacetal derivative with copper sulphate catalysis and lead tetra-acetate oxidation of the resulting 4,5-O-isopropylidene derivative.[1] 1,2-O-Isopropylidene-L-threitol has been prepared from dimethyl tartrate in four steps, and converted into (R)-1,2-O-isopropylideneglycerol by periodate oxidation - borohydride reduction.[2] The direct isopropylidenation of L-arabinose N,N-dimethylhydrazone under various conditions has been studied. Four major products, the ratio of which varied with conditions, were identified as the 2,5-anhydropentitol derivatives (1) and (2), the diacetal (3), and the open-chain acetal (4). The structure of (4) was confirmed by hydrolysis with dilute sulphuric acid to 2,3:4,5-di-O-dispropylidene-L-arabinose, which was converted to the known crystalline branched-chain pentitol (5) by aldol-Cannizzaro reaction with formaldehyde.[3] Kinetic isopropylidenation of aldose diethyl-

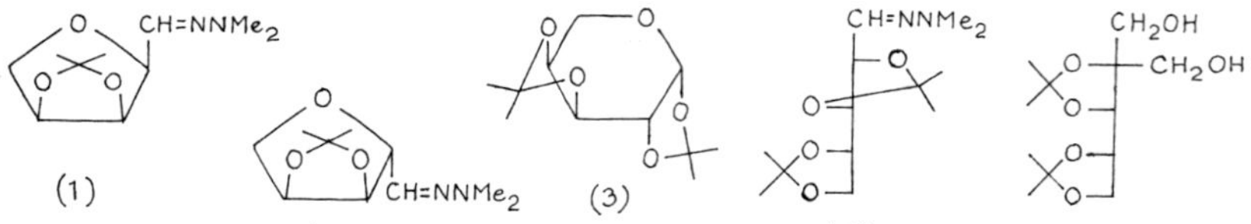


(1) (2) (3) (4) (5)

dithioacetals has been carried out using 1.2 equivalents of 2-methoxypropene in DMF at 0 °C with tosic acid catalysis. Five-membered ring acetals attached at the terminal diols were the major products in every case, with six-membered and seven-membered rings as minor products. $^{13}$C N.m.r. spectroscopy showed that C-2 of a 1,3-propandiol derivative, i.e. the six-membered acetal, was shifted up-field by 6-9 ppm compared to starting materials, and that shifts of the diol carbons also depend on the ring size. Similar observations were made on the cyclohexylidene derivatives prepared analagously from 1-ethoxycyclohexene, and on the methylene acetals.[4] D-Ribose diethyldithioacetal, on treatment with propanone-sulphuric acid in the presence of copper(II) sulphate gives, contrary to earlier claims, the 2,3:4,5- and the 2,5:3,4-diacetals (6) and (7) in equal quantities. De-thioacetalation with mercury(II) and reduction with

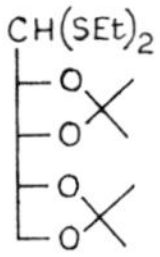
(6)
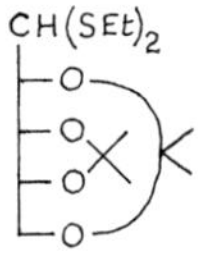
(7)
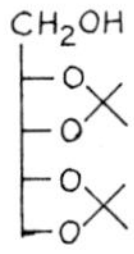
(8)
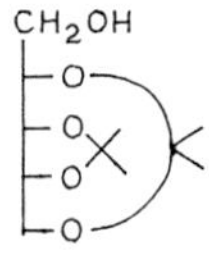
(9)
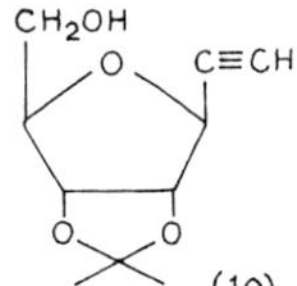
(10)

sodium borohydride gave the parent ribitol diacetals (8) and (9). Direct acetonation of ribitol gave (8) in 22% yield and (9) in 11% yield, but the major product was the 1,2:4,5-di-O-isopropylidene derivative (66%). The paper also describes the synthesis of the acetylenic derivative (10), used previously[5] in C-nucleoside synthesis from 2,3:4,5-di-O-isopropylidene-D-ribose. Reference to the n.m.r spectroscopy of acetals is made in Chapter 21. The dimer of 1,2-O-isopropylidene-α-D-xylodialdo-1,4-furanose has been confirmed to have the structure shown in (11) as previously postulated (see Vol. 1, p. 174), and its conformation[6] in $CDCl_3$ and $CD_3OD$ investigated by n.m.r. spectroscopy.

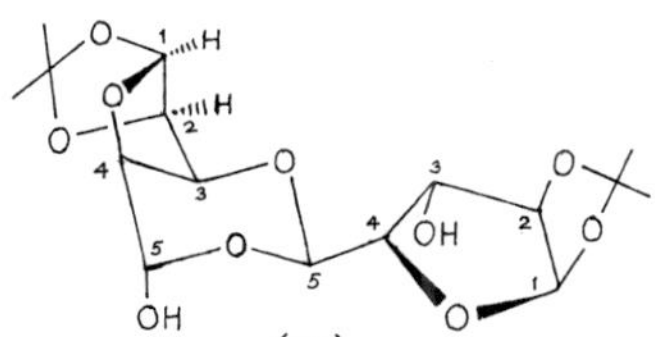
(11)

Acetonation of a mixture of 1-deoxy-5,6-O-isopropylidene-D-*arabino*- and -D-*xylo*-hexulose, obtained by hydroxylation of 1,3,4-trideoxy-5,6-O-isopropylidene-D-*glycero*-hex-3-enulose, gave 1-deoxy-2,3:4,5-di-O-isopropylidene-β-D-fructopyranose and 1-deoxy-2,3:4,6-di-O-isopropylidene-α-D-sorbofuranose. The same two products were formed when acetonation was carried out on a mixture of 1-deoxy-D-fructose and -D-sorbose obtained from hydroxylation of 1,3,4-tri-deoxy-D-*glycero*-hex-3-enulose.[7]

Isopropylidenation of D-mannitol using methyl isopropenyl ether in DMF at 20 °C with tosic acid catalysis gave the 1,2:5,6-diacetal as the major and 1,2:4,6- and 1,2:3,4-diacetals as minor products.[8] A detailed study of the same reaction using in this case propanone-sulphuric acid gave the 1,2:3,4:5,6-triacetal (63%), the 1,2:3,4-diacetal (23%), and the 1,2:5,6-diacetal (8%), together with traces of other acetals. The 1,2:3,4-diacetal could also be rapidly prepared from the major product in propanone by treatment with dilute sulphuric acid. Addition of glacial acetic acid to the propanone-sulphuric acid reagent gave, with D-mannitol, good yields of the triacetal (85%), with 12% 1,2:3,4-diacetal.[8] The synthesis and hydrogenolysis reactions of acetals, including isopropylidene examples, are described below.

## 2 Benzylidene and Related Acetals

Tetrafluoroboric acid has been recommended as an efficient catalyst in acetalations, e.g. methyl α-D-glucopyranoside with benzaldehyde dimethylacetal in DMF in the presence of tetrafluoroboric acid gave a 94% yield of the 4,6-O-benzylidene derivative. The same catalyst in aqueous acetonitrile solution is capable of rapidly and cleanly cleaving acetals, trityl ethers and TBDMS ethers.[9] Reaction of D-ribono-1,4-lactone with benzaldehyde-concentrated hydrochloric acid gave 3,4-O-(R)-benzylidene-D-ribono-1,5-lactone (12) and not the 3,5-acetal as previously suggested (see Vol. 2, p. 154). The regiochemistry of (12) was confirmed by X-ray crystal structure of its 2-O-acetyl derivative. Production of (12) in high yield as a single diastereoisomer is due to its insolubility. With zinc chloride as catalyst the products are the 2,3-O-(R)-benzylidene-D-ribono-1,4-lactone and its (S)-isomer.[10] Migration of a 1,2-O-benzylidene

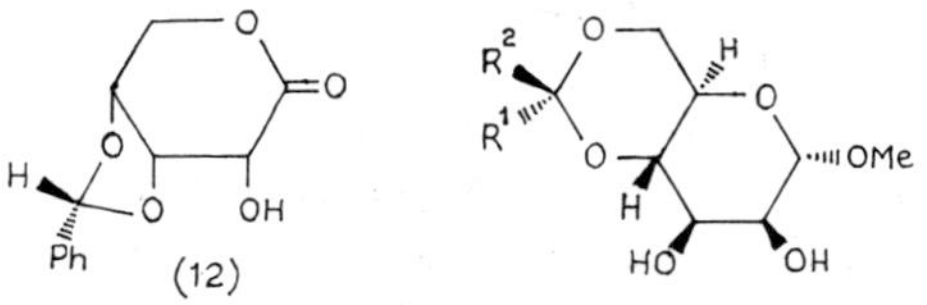

(12)

(13) $R^1 = Ph, R^2 = Me$
(14) $R^1 = Me, R^2 = Ph$

group under glycosidation conditions is mentioned in Chapter 3.

The photolysable O-nitrobenzylidene acetal has been used as an intermediate in the synthesis of a trisaccharide.[11] Reaction of methyl α-D-mannopyranoside with α-methoxystyrene and catalytic pyridinium tosylate in DMSO gave a 7:3 mixture of the 4,6-O-acetals (13) and (14).[12]

## 3 Other Acetals

An n.m.r. spectroscopic study has been carried out on ethylidene acetals to determine useful parameters for elucidating ring sizes (see also Chapter 21).[13] Allylglycosides rearrange to prop-1-enyl glycosides on treatment with trans-[$Pd(NH_3)_2Cl_2$], but in the presence of a neighbouring hydroxy group propylidene acetals are formed (Scheme 1).[14] Included in a study of relative stabilities of protecting groups towards iridium, rhodium, and palladium catalysts were some examples of prop-2-enylidene acetals.[15] The formation of 1,6-anhydro-3,4-O-[5-(hydroxymethyl)-2-furylidene]-β-D-galactopyranose (15) in the pyrolysis of lactose has been de-

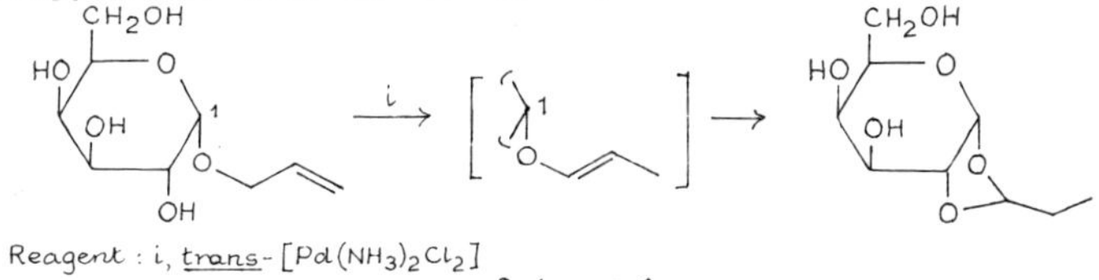


Scheme 1

scribed. The acetal, which has a strong bitter taste, was also formed by heating D-galactose, and synthesized from 1,6-anhydro-D-glucose and 5-hydroxymethyl-furfural.[16]

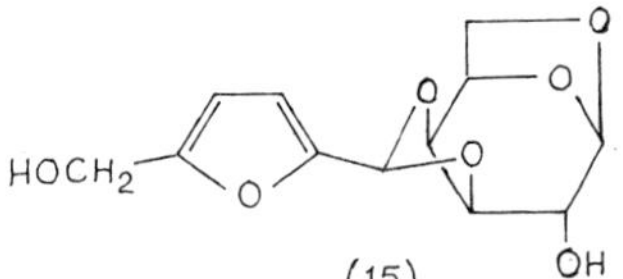


(15)

## 4 Acetal Cleavage Reactions

Reductive cleavage with lithium aluminium hydride-aluminium chloride, or with sodium cycloborohydride-hydrochloric acid, or with aluminium chloride-trimethylamine, of endo-3,4-O-(4-methoxybenzylidene) acetal rings of 1,6-anhydro-β-D-galactopyranose derivatives gives predominantly the axial 3-O-(4-O-methoxybenzyl)ether derivatives. Oxidative cleavage with DDQ affords the axial 3-O-(4-methoxybenzoyl) derivatives.[17] Catalytic transfer hydrogenation of 1,3-dioxolanes as a method for the selective removal of 5-membered ring benzylidene groups has been applied to methyl 2,3:4,6-di-O-benzylidene-α-D-mannopyranoside to yield the 4,6-benzylidene acetal in 86% yield. The same conditions applied to 1,3:2,4:5,6-tri-*O*-benzylidene-D-glucitol gave the 1,3:2,4-diacetal in 75% yield.[18] Hydrogenolysis of methylene, ethylidene and isopropylidene derivatives with lithium aluminium hydride-aluminium chloride gave axial ethers. *Endo*-phenyl isomers of the 1-phenylethylidene acetals derived from acetophenone gave axial 1-phenylethyl ethers as mixtures of the two diastereoisomers; the *exo*-phenyl isomers of arabinosides were stable to the reagent, while those of rhamnosides gave axial 2-(1-phenylethyl)-ethers in a slow reaction. Two examples are shown in Scheme 2.[19]

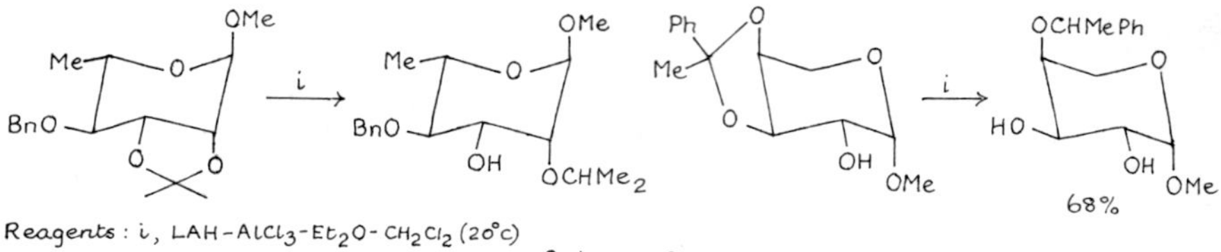


Scheme 2

As mentioned above aqueous tetrafluoroboric acid rapidly and cleanly cleaves acetals in acetonitrile solution.[9]

## References


1 K.N.Maloneyhuss, *Synth. Commun.*, 1985, 15, 273 (*Chem. Abstr.*, 1985, *103*, 88 133).
2 A.H.Al-Hakim, A.H.Haines, and C.Morley, *Synthesis*, 1985, 207.
3 M.Koos and H.S.Mosher, *Collect. Czech. Chem. Commun.*, 1985, *50*, 1994.
4 T.B.Grindley, C.J.P.Cote, and C.Wickramage, *Carbohydr. Res.*, 1985, *140*, 215.
5 G.Aslani-Shotorbani, J.G.Buchanan, A.R.Edgar, and P.K.Shahidi, *Carbohydr. Res.*, 1985, *136*, 37.
6 L.Kilaas and T.Anthonsen, *Acta Chem. Scand., Ser. B*, 1985, *39*, 685.
7 F.J.Lopez Aparicio, I.Izquierdo Cubero, and M.Gomez Guillen, *An. Quim., Ser. C*, 1983, *79*, 307 (*Chem. Abstr.*, 1985, *102*, 46 192).
8 J.Kuszmann and E.Tomori, *Carbohydr. Res.*, 1985, *137*, 276.
9 R.Albert, K.Dax, R.Pleschko, and A.E.Stutz, *Carbohydr. Res.*, 1985, *137*, 282.
10 N.Baggett, J.G.Buchanan, M.Y.Fatah, C.H.Lachut, K.J.McCullouth, and J.M. Webber, *J. Chem. Soc., Chem. Commun.*, 1985, 1826.
11 F.C.Baines, P.M.Collins, and F.Farnia, *Carbohydr. Res.*, 1985, *136*, 27.
12 M.Pietraszkiewicz and J.Jurczak, *J. Carbohydr. Chem.*, 1985, *4*, 429.
13 T.B.Grindley and C.Wickramage, *J. Carbohydr. Chem.*, 1985, *4*, 171.
14 T.Bieg and W.Szeja, *J. Carbohydr. Chem.*, 1985, *4*, 441.
15 M.Kloosterman, J.H.Van Boom, P.Chatelard, P.Boulanger,, and G.Descotes, *Tetrahedron Lett.*, 1985, *26*, 5045.
16 T.Vrashima, K.Suyama, and S.Adachi, *Carbohydr. Res.*, 1985, *135*, 324.
17 M.Kloosterman, T.Slaghek, J.P.G.Hermans, and J.H.Van Boom, *Recl. J. R. Neth. Chem. Soc.*, 1984, *103*, 335 (*Chem. Abstr.*, 1985, *102*, 149 649).
18 T.Bieg and W.Szela, *Carbohydr. Res.*, 1985, *140*, C7.
19 A.Liptak, Z.Szurmai, V.A.Olah, J.Harangi, L.Szabo, and P.Nanasi, *Carbohydr. Res.*, 1985, *138*, 1.

# 7
# Esters

A review on the regioselective manipulation of hydroxy groups using organotin derivatives includes a section on esterification.[1] Determination of the position of ester groups in glycosphingolypids by 2D $^{1}H$ n.m.r. spectroscopy has been described.[2]

## 1 Carboxylic Esters

Peracetates of aldoses, disaccharides, and methyl glycosides have been obtained in good yields by reaction of the trimethylsilylated sugars with iron(III) chloride-acetic anhydride without change in their stereochemistry. The same reaction on ketoses produced dehydration products only.[3]

Selective acetylation of benzyl α-D-xylopyranoside using four equivalents of acetic anhydride in the presence of sodium acetate-acetic acid gave 15.7% 2,4-diacetate, 4.2% 3,4-diacetate, and 22% 2,3,4-triacetate. The 2,3,4-triacetate of benzyl α-D-mannopyranoside was obtained by prior tritylation, acetylation, and detritylation. The $^{1}H$ and $^{13}C$ n.m.r. spectra for these compounds were assigned.[4] 2-O-Acetyl and -benzoyl derivatives of methyl α-D-galactopyranoside and methyl 2-amino-2-deoxy-α-D-galactoside have been synthesized from the corresponding 3,4-O-isopropylidene-6-O-trityl sugars and by direct acylation of the amino sugar, and the products tested for their binding affinities to isolectins.[5] The preparation of some 1,6-di-O-acetyl-D-galactopyranose derivatives carrying acetyl, benzyl, or allyl groups at O-2, O-3, and O-4 by acetolysis of the corresponding 1,6-anhydro-D-galactopyranoses has been reported.[6] An improved conversion of 6-O-trityl-D-galactose to 1,2,3,4-tri-O-acetyl-β-D-galactopyranose has been achieved by acetylation at RT using acetic anhydride-sodium acetate, followed by warming to 100 °C, and then detritylation using iodotrimethylsilane. Use of pyridine encourages formation of furanose forms. The product was treated with bromoacetyl bromide to yield 1,2,3,4-tetra-O-acetyl-6-O-bromoacetyl-β-D-galactopyranose. A previously reported synthesis of 1,2,3,4-tri-O-acetyl-β-D-galactopyranose (see Vol.10, p.15) was shown to yield the

1,2,3,6-isomer.[7]

Partial deacetylation of 3,3',4',6'-tetra-O-acetyl-1',2:4,6-di-O-isopropylidene sucrose with aluminium oxide impregnated with potassium carbonate followed by deacetalation gave 3'-O-acetylsucros and 3,6'-di-O-acetylsucrose, the structures of which were confirmed by $^{13}C$ n.m.r. spectroscopy and m.s. of their deuterioacetyl derivatives. Reference compounds were used to show that partial deacetylation of octa-O-acetylsucrose gave a similar mixture of partially acetylated sucroses.[8] The hexa-O-acetyl-sucroses formed b aluminium oxide-potassium carbonate deacetylation of sucrose octaacetate were shown to be 1',2,3,4,6,6'- and 2,3,4,4',6,6'-hexa-O-acetylsucrose. Tosylation of the former product yielded the 3',4'-ditosylate (1), which on treatment with TPP-DEAD gave the 3',4'-anhydro-*lyxo*-derivative (2). Sodium methoxide with (1), followed by acetylation gave the *ribo*-epoxide analogue (3). The 1',2:3',4'-dianhydro-derivative (4) was obtained from the tosylation mother liquors.[9]

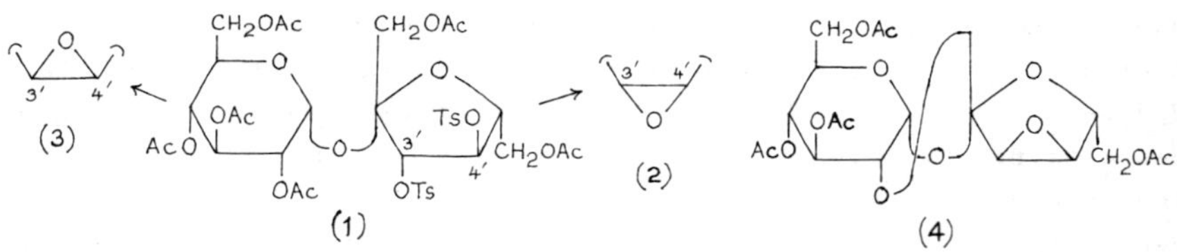


The acetylation of L-rhamnal and L-fucal has been studied. When acetyl chloride, N-acetylimidazole, benzoyl chloride, or N-benzoylimidazole were used, the esterification of L-rhamnal occurred primarily at the allylic hydroxy group. With acetic anhydride-pyridine it was mainly the homoallylic hydroxy group which esterified. Selective acylations of L-fucal could not be achieved since acyl migration from O-3 to O-4 occurred. The ratios of the products were studied under various conditions of solvent, temperature, and time.[10]

Potassium cyanide in 95% ethanol has been shown to be useful for removal of acetyl and benzoyl protecting groups from base-sensitive β-D-galacto-pyranoside derivatives, and from α-D-galactopyranosides bearing acid- and base-sensitive substituents such as isopropylidene or tosyl groups.[11]

An efficient synthesis of 1,3,5-tri-O-benzoyl-α-D-ribofuranose (5) has been achieved by the route shown in Scheme 1. The residue from the crystallization of (5) was the 2,3,5-tribenzoate which could be reacetylated at the anomeric position and recycled. The overall yield of (5) was 81% after one recycling.[12]

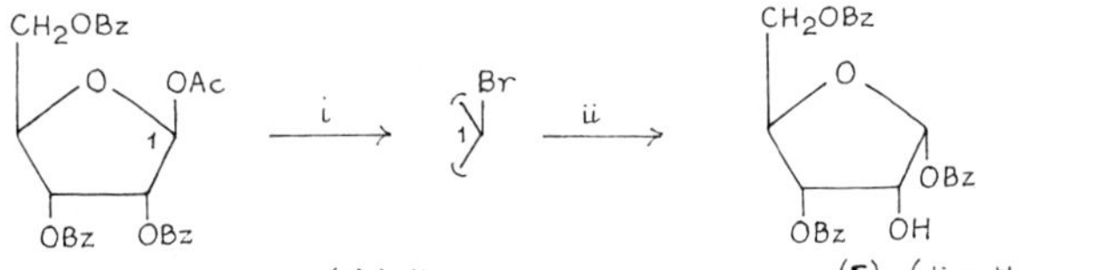

(5) (directly crystallized) 63%

Reagents: i, 1·1 mol. HBr - $CH_2Cl_2$ (0°C); ii, $H_2O$

Scheme 1

Methyl 2,4,6-tri-O-benzoyl-β-D-galactopyranoside and 1-O-acetyl-2,4,6-tri-O-benzoyl-β-D-galactopyranose have been synthesized via stannylene mediated 3-O-benzylation of methyl β-D-galactopyranoside, thus providing suitable intermediates for the synthesis of 3-linked oligosaccharides.[13] The 2,3-O-(4-methoxybenzylidene) derivative (6) of 1,6-anhydro-mannose is a versatile synthon for the preparation of 2-substituted glucose derivatives and the synthesis of 1,4-linked oligosaccharides. The 2-O-benzoyl and 2-O-laevulinoyl-glucose derivatives (7) and (8) were prepared as depicted in Scheme 2.[14] The oxidation step using DDQ is described in a separate publication.[15] A

(6)

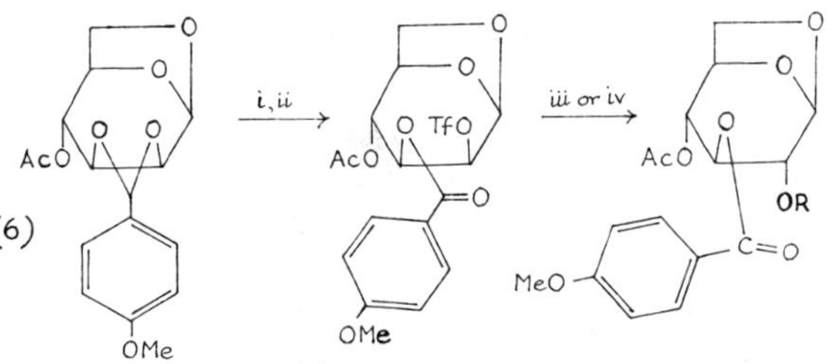

(7) R = Bz

(8) R = $CO(CH_2)_2COMe$

Reagents: i, DDQ; ii, $Tf_2O$; iii, BzOCs; iv, $MeCO(CH_2)_2CO_2Cs$

Scheme 2

series of papers describes a programme for synthesizing various partially benzoylated sucrose derivatives by use of limited benzoylating reagent on a variety of partially blocked sucrose species. Compounds described include 2,6,6'-, 6,3',6'-, 6,1',3'-, 2,6,1'-, 1',3',6'-, 2,1',6'-, 6,1',6'-, 3,3',6'-, and 3',4',6'-tri-O-benzoyl-sucrose.[16,17,18]

Three new dimeric, hydrolysable tannins have been isolated from Geum japonicum, one of which was the fully esterified D-glucose derivative (9). The other two were the compounds derived by hydrolytic loss of one or both 4,6-cyclic diesters.[19] The associations of polyphenols of the gallotannin and ellagitannin type (based upon

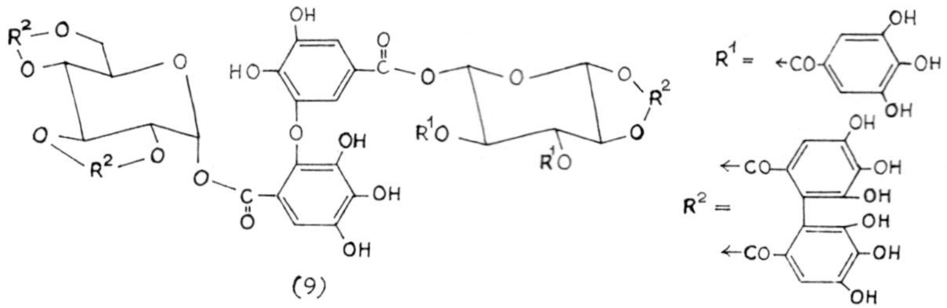

(9)

penta-O-galloyl-D-glucose) with the protein bovine serum albumin have been studied, and the results used to comment on the role of these tannins in nature.[20] The antiherpetic activity of hydrolysable tannins has been shown to be dependent upon the number of galloyl or related ester groups and on the degree of condensation, but not on the sugar moiety. The increased activity correlates with increased cytotoxicity.[21]

The preparation of O-aminoacyl derivatives as enzymatically removable protecting groups for sugars has been described. N-Benzyloxycarbonyl and N-(t-butoxycarbonyl)-glycine, alanine, valine, isoleucine, phenylalanine, lysine and asparagine were condensed with methyl 4,6-O-benzylidene-α-D-glucopyranoside and methyl 2,3-di-O-methyl-α-D-glucopyranoside. The aminoacyl groups were easily removed by enzymic hydrolysis using pronase E, trypsin and chymotrypsin, as well as by conventional alkaline treatment. The present work extends the previously described use of glycine as a protecting group (see Vol.18, p.66).[22] The methyl 2,3-di-O-aminoacyl-α-D-glycopyranosides prepared as described in the previous reference were evaluated for taste. Increase in the amino acid side-chain length correlated with the change from sweet to bitter and the presence of the sugar moiety intensified the taste of the amino acid. Methyl 2,3-O-alanyl-α-D-glucopyranoside was about fifteen times as sweet as sucrose.[23,24] Methyl 2,3-di-O-(L-α-aminobutyryl)-α-D-glucopyranoside is reported to be a new sweet substance about 50 times sweeter than sucrose.[25] The synthesis of L-arabinosyl esters of N-acyl amino acids has been reported. Thus 2,3,4-tri-O-benzyl-L-arabinopyranose condensed with N-acetylalanine in the presence of DCC-imidazole to yield the 1-O-ester (10) in which the trans α-anomer predominated. Analogously prepared were the 1-O-(N-acetyl) alanate ester of 2,3,5-tri-O-benzyl-N-α,β-L-arabinofuranose, and the 1-O-(N-tert-butoxycarbonyl-L-phenylalanyl) pyranose and furanose counterparts. Treatment of the 1,2-cis isomers with diazomethane in DMF or dry ethyl

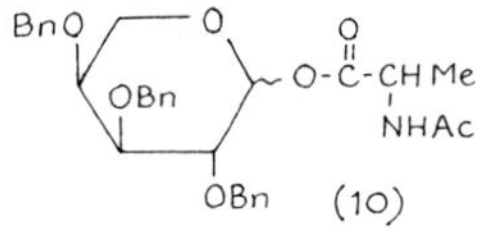
(10)

acetate caused migration of the acyl group to O-2 to yield the 1-O-acetyl-2-O-aminoacyl derivative.[26]

Selective acylation occurs at C-6 of methyl α-D-glucopyranoside on treatment with hexachloro- or pentachloro-propanone to yield trichloro- or dichloro-acetates respectively. The chlorinated ester group is easily removed by ammonia in ether and thus constitutes a base-labile complement to the acid-labile trityl function (Scheme 3).[27] Trifluoroacetylation of carbohydrates for g.c. analysis has

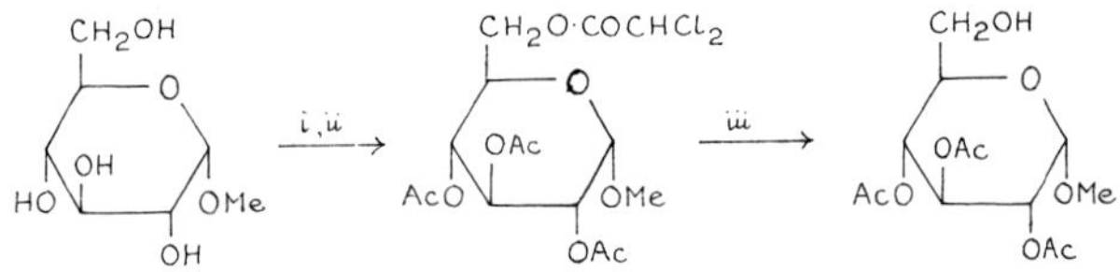

Reagents: i, $CHCl_2COCCl_3$-DMF; ii, $Ac_2O$-Py; iii, $NH_3$-$Et_2O$

Scheme 3

been achieved using N-methylbis(trifluoroacetamide) on hydrolysates under homogeneous conditions.[28]

6-O-Esters of D-galactose diethyldithioacetal have been prepared by reaction with $Me(CH_2)_nCOCl$ ($n$ = 10, 12, or 14)-pyridine and their structures studied by n.m.r. spectroscopy.[29] Various 6,6'-disubstituted α,α-trehalose esters of fatty acids such as linear $C_{30}$, α-branched $C_{30}$, α-branched β-hydroxy-$C_{32}$, α-branched β-hydroxy-$C_{48}$, and β-acyloxy-$C_{28}$ acids, and mycolic acid, as well as bis-N-acyl derivatives of 6,6'-diamino-6,6'-dideoxy-α,α-trehalose, have been synthesized as analogues of the mycobacterial glycolipid trehalose 6,6'-dimycolate. Their lethal and adjuvant activities have been examined.[30]

Taxifolin-3'-O-β-D-(6"-O-phenylacetyl)glucopyraoside (11), isolated from Pinus massoniana needles, is the first reported phenylacetylated sugar in nature.[31] β-O-[4-O-(p-Hydroxyphenylacetyl)-β-D-glucopyranosyl]-β-hydroxy-γ-butyrolactone (12) has been isolated from the roots of Taraxacum officinale.[32] 1,3-Di-O-sinapoyl-

(13), 1-O-sinapoyl-3-O-fernoyl- (14), and 1-O-acetyl-3-O-fernoyl- (15) sucrose have been isolated from the creeper Polygala chamaebuxus.[33] The glucuronide (16) is one of the metabolic products of

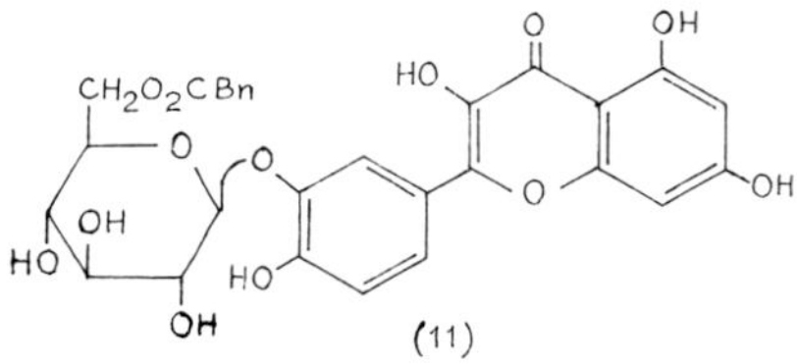

(11)

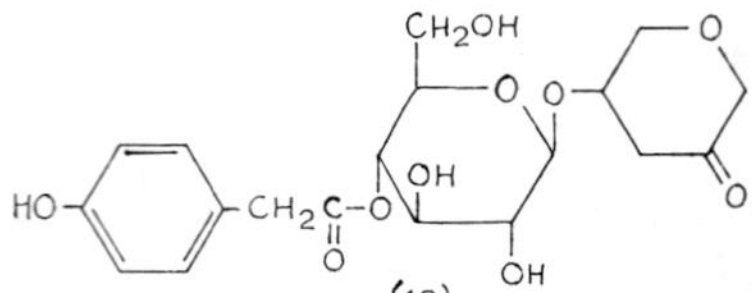

(12)

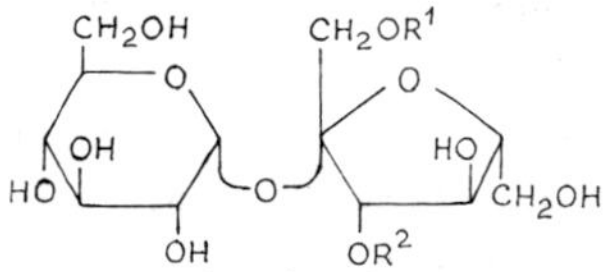

(13) $R^1 = R^2 = R^3$

(14) $R^1 = R^3$, $R^2 = R^4$

(15) $R^1 = Ac$, $R^2 = R^4$

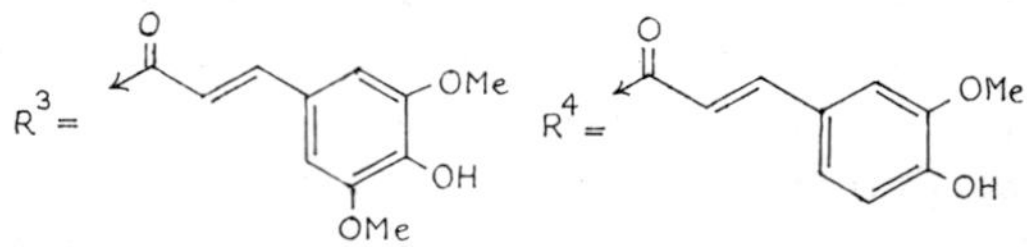

flunisal, an analgesic, found in the urine; it is suitable for a reversed-phase h.p.l.c. assay.[34] Reaction of O-(tetra-O-benzyl-α-D-glucopyranosyl)trichloroacetimidate with carboxylic acids gave

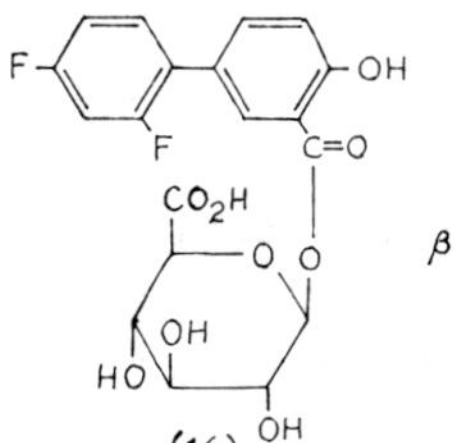

(16)

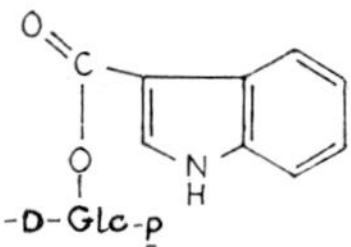

(17)

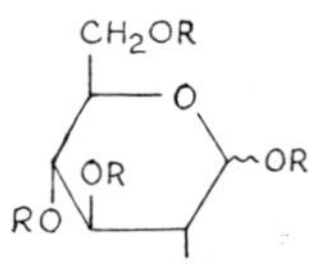

(18)

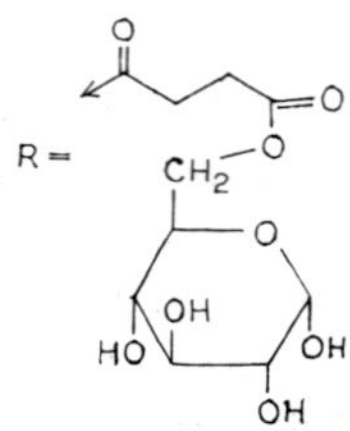

1-O-acyl-β-glucosides exclusively without other catalysts being required. The reaction was used in the resolution of racemic α-methoxyphenylacetic acid since its glycosylated derivatives were easily separable by chromatography. As an example, the benzpyrolic acid ester (17) was prepared in 55% overall yield after catalytic removal of the benzyl groups.[35] The glucosyl perester (18) has been synthesized as a high-calorific nutrient for intravenous administration.[36] Methanolysis of the potent new antibiotics PD 114, 759, and PD 115, 028, isolated from Actinomadura sp, produces 3'- and 4'-N-(2-methoxy-propenoyl)-4,5-dimethoxyanthranilate esters of methyl

2'-deoxy-α,β-L-fucopyranoside (19) and (20), which equilibrate with each other in aqueous ethanol.[37] Pregna-5,20-dien-3-β-yl 2-acetam-

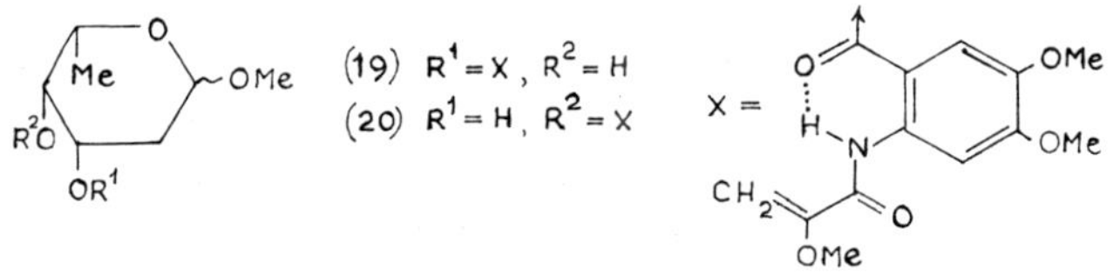


ido-2-deoxy-D-galactopyranosides with 4-, 6-, or 4,6-O-butyrate groups and the unesterified parent have been isolated from the gorgonian coral, Muricea futicosa.[38] Glycopyranosyl esters of 3-O-acetyloleanolic acid and octanoic acid, representatives of sterically hindered and less hindered acids, have been synthesized by standard methods from L-arabinose, D-xylose, D-glucose, and L-rhamnose, and some 1-2 linked disaccharides. The $^{13}$C n.m.r. spectra derivatives were studied to investigate the effect of the steric hindrance on the 2-O-glycosylation shifts.[39]

A series of analogues (21) of UDPG and UDPGlcNAc have been synthesized from 2,3,4,6-tetra-O-substituted-α-D-glucopyranose or 2-acetamido-3,4-tri-O-acetyl-2-deoxy-α-D-glucopyranose with 2',3'-O-isopropylideneuridine and chlorosulphonyl isocyanate, followed by deprotection. Some compounds show inhibitions of protein glycosylation and antiviral activity, particularly the benzoyl and benzyl derivatives which show good lipid/water partition.[40]

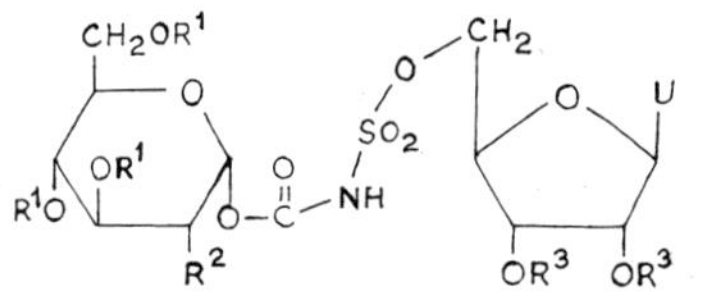


(21) $R^1$= Bn, Bz, Ac, $COC_{15}H_{31}$, $R^2$= $OR^1$ or NHAc ($R^1$=Ac) $R^3$= H or $R^3_2$ = >$CMe_2$

Silver salts have been investigated as catalysts for the synthesis of alkyl 1,2-orthoacetates of D-glucopyranose and D-galactopyranose from 1,2-cis and -trans-halides with methanol, propan-2-ol, and octan-1-ol in the presence of 2,4,6-trimethylpyridine. Although silver tosylate and the α-bromides gave 92% yields, best were the β-chlorides which gave near quantitative yields. Silver nitrate had some advantage in that the base is isolated as its inium nitrate which crystallizes out, leaving the orthoacetate in solution directly usable for further reaction.[41]

## 2 Phosphates and Related Esters

A review of phosphates and phosphonates of biochemical interest includes some carbohydrate examples.[42] A theoretical study has been carried out to determine the anomeric effect of phosphate groups (se also Chapter 21).[43] Enzymic syntheses of sugar phosphates are included in a review of enzymes in organic chemistry.[44]

Treatment of potato peels with phosphorylase gave D-glucose which was phosphorylated to yield D-glucose 1-phosphate of commercial grade.[45] A preparative enzymic synthesis of D-*threo*-2-pentulose 5-phosphate (D-xylulose 5-phosphate), following the route shown in Scheme 4, has been described.[46]

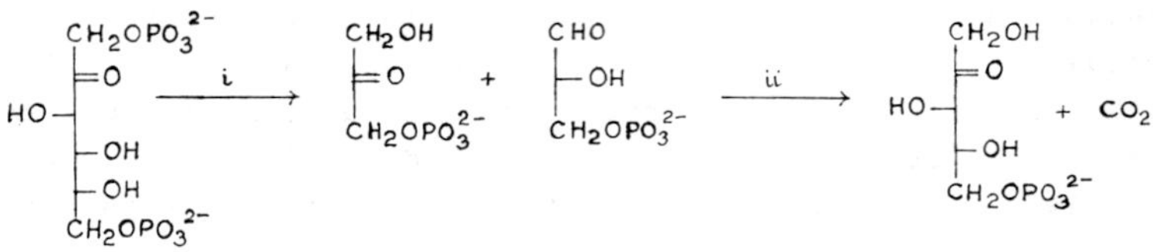

Reagents: i, aldolase; ii $CH_2(OH)COCO_2^-$, transketolase, $Mg^{2+}$, ThPP

Scheme 4

Fluorosugar phosphates (22) and (23), for sugar metabolism studies, have been synthesized by the method depicted in Scheme 5.[47] The glucose phosphate derivatives (24) have been synthesized

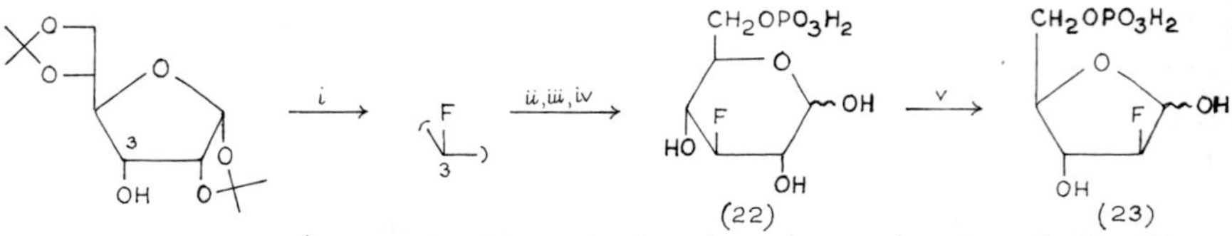

Reagents: i, DAST; ii, $H_3O^+$; iii, ATP-hexokinase; iv, Phosphoenolpyruvate-pyruvate kinase; v, $Pb(OAc)_4$-$H^+$

Scheme 5

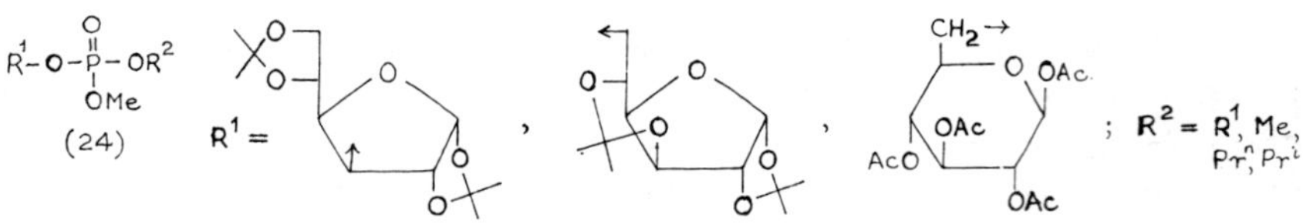

by the phosphate triester method.[48] A synthesis of the two anomers of D-arabinofuranose 1,5-diphosphate has been achieved (Scheme 6).[49] Dimethylphosphates (25) - (27) of deoxy sugars have been prepared by hydrogenation of enol phosphates.[50] An improved enzymic synthesis of octulose phosphates has been reported. D-*Glycero*-D-*altro*- and D-glycero-D-ido-octulose 1,8-diphosphates were prepared in high yield from D-fructose 1,6-diphosphate and either D-ribose 5-phosphate or D-

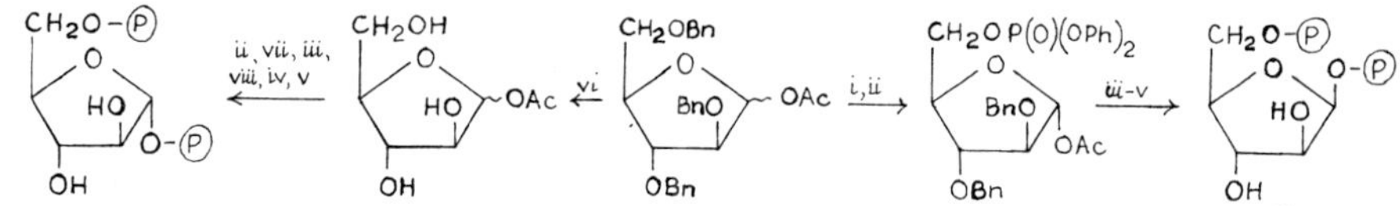

Reagents: i, $H_2$-Pd/C-AcOH-MeOH; ii, $(PhO)_2POCl$-Py; iii, HBr-$CH_2Cl_2$; iv, $(BnO)_2PO_2^-Et_3\overset{+}{N}H$; v, Li-$NH_3$; vi, $H_2$-Pd/C-HOAc; vii, $Ac_2O$-Py; viii, $AgBF_4$ Ⓟ = $P(=O)(OH)_2$

Scheme 6

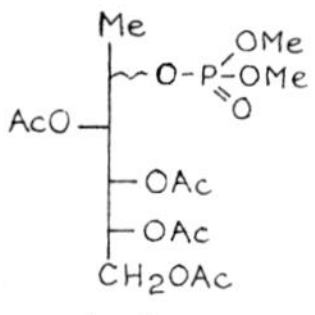

(25)

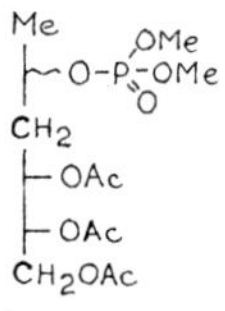

(26)

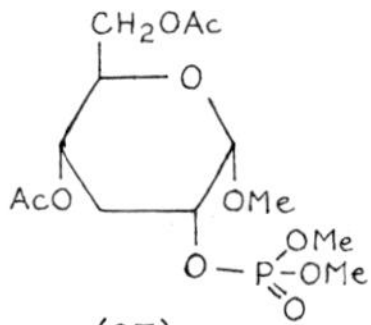

(27)

arabinose 5-phosphate in the presence of aldolase and triosephosphate isomerase. The 8-monophosphates were better prepared enzymically using D-glucose 6-phosphate or D-allose 6-phosphate and β-hydroxypyruvate than by selective hydrolysis of the diphosphates.[51]

Relatively fast phosphate migration has been observed in riboflavin phosphates at pH 2 and at temperatures above room temperature.[52] The artificial antigen 4-O-(2-acetamido-2-deoxy-D-glucopyranosyl)-D-ribitol phosphate derivative (28), bearing a spacer arm, has been synthesized to study the influence of the phosphodiester function in the immunogenicity of natural ribitol teichoic acid of Staphyllococcus aureus. For this study (28) was attached to bovine

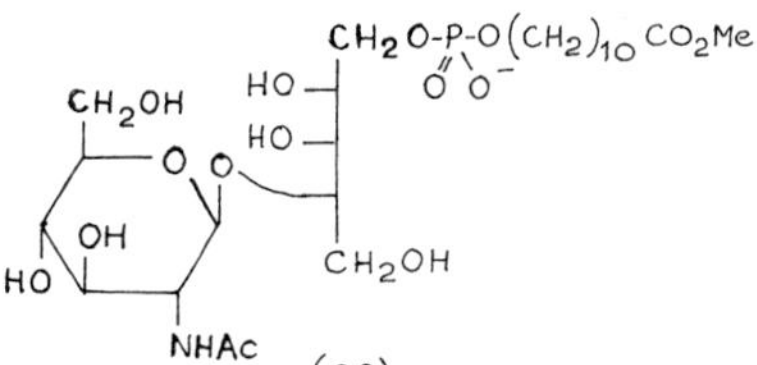

(28)

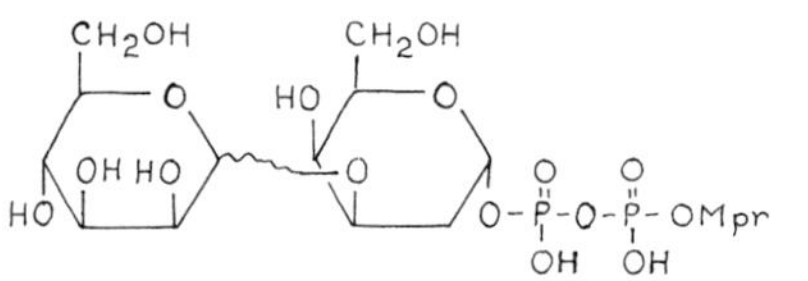

(29) α-anomer
(30) β-anomer

Mpr = moraprenyl

serum albumin.[53]

Morapreny1 β-D-glucopyranosyl phosphate has been prepared by condensing moraprenyl trichloroacetimidate, prepared in situ, with β-D-glucopyranosyl tri-n-octylammonium hydrogen phosphate.[54] Moraprenyl

3-_O_-α- and -β-D-mannopyranosyl-α-D-galactopyranosyl pyrophosphates (29) and (30), intermediates in the biosynthesis of _Salmonella_ O-antigenic polysaccharide serotypes $C_2$ and $C_3$, have been synthesized by phosphorylation of the disaccharide followed by reaction with moraprenylphosphoimidazole.[55]

Conventional methods have been used to synthesize the diglycosyl phosphate (31), of use in the assay of UDP-_N_-acetylglucosamine 1-phosphotransferase.[56] The syntheses of the _p_-nitrophenyl glycosidyl glycosyl phosphates (32) and (33) were achieved by condensation of the appropriate peracetyl glycosyl phosphate with _p_-nitrophenyl 2,3,4-tri-_O_-benzoyl-α-D-mannopyranoside followed by deprotection. Their conformations were investigated by $^{1}H$, $^{13}C$, and $^{31}P$ n.m.r. spectroscopy, and it was concluded that the phosphate group is stretched out rather than the two sugar units being folded towards each other. If this conformation were to persist in glycoconjugates then it would be a good recognition marker for the site of attachment of glycoproteins to the cell surface, since the phosphate diester terminal group is well exposed to the solvent.[57] Agrocinopine A and B, (34) and (35), have been isolated from tomato and sunflower tumours.[58]

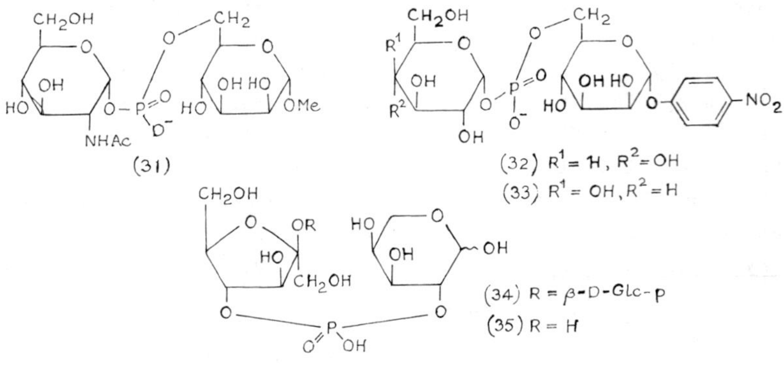


Two reviews on the synthesis and biological activity of lipid A analogues have appeared.[59,60] The synthesis of diglycosyl phosphates _via_ diglycosyl phosphites has been described in connection with an approach to the total synthesis of lipid A of _Salmonella_ lipopolysaccharide.[61] The total synthesis of _Escherichia_ _coli_ lipid A, the active principle of bacterial endotoxin, has been outlined.[62] The glucosamine derivative (36) has been used to prepare lipid X (37), the _Salmonella_ lipid A precursor.[63] A variety of mono- and disaccharide analogues of lipid A have been prepared; _O_-1 phosphorylation was effected by reaction of glycosyl halides with tributyltin-(dibenzyl or diphenyl)phosphate and O-4 phosphorylation with diphenyl

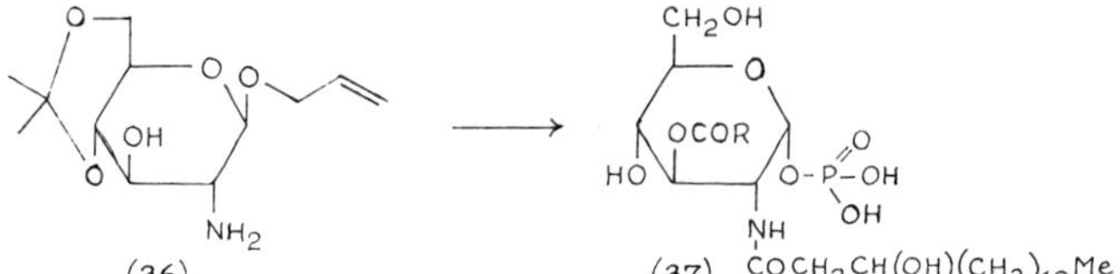


(36) (37)

phosphoryl chloride.[64] In an investigation of the molecular requirements for the biological activities of lipid A, a number of analogues (38) of the non-reducing sub-unit have been synthesized, among which that with R'=(R)-3-tetradecanoyloxytetradecanoyl, $R^2$= tetradecanoyl, $R^3$=phosphonyl, and $R^4$=H showed the most distinct and strong biological activity.[65] Another group have described the synthesis of the same or similar analogues (39), using a method in which the optically active fatty acid moiety was introduced in the last stage, and these were also shown to possess antitumour activity.[66,67]

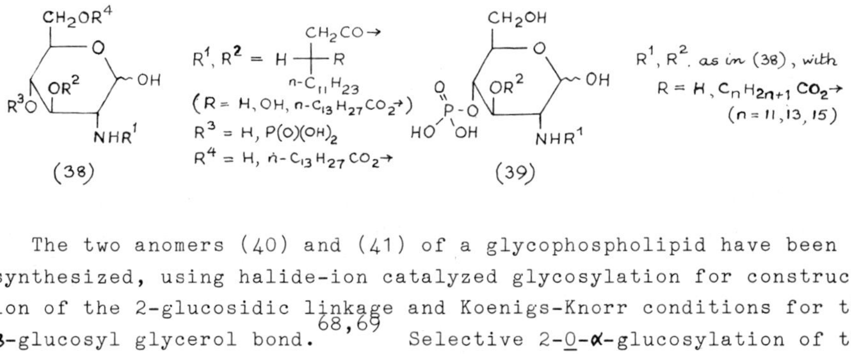


(38) (39)

The two anomers (40) and (41) of a glycophospholipid have been synthesized, using halide-ion catalyzed glycosylation for construction of the 2-glucosidic linkage and Koenigs-Knorr conditions for the β-glucosyl glycerol bond.[68,69] Selective 2-O-α-glucosylation of the

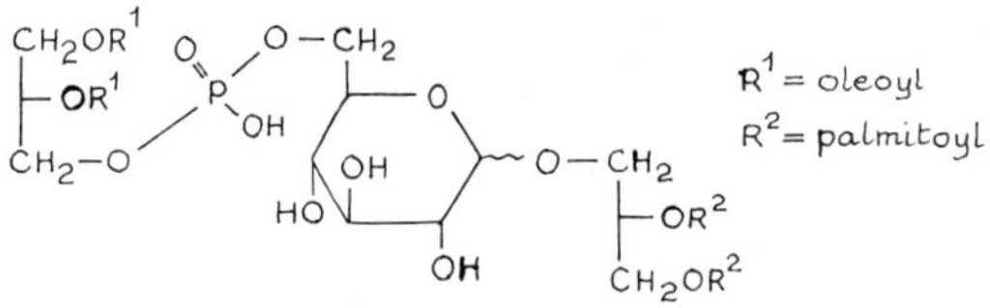


(40) α- anomer

(41) β- anomer

4,6-O-tetraisopropylsilyloxysilyl-protected glucoside (42) has been used in the preparation of the Streptococci glycophospholipids (43) -

(45).[70]

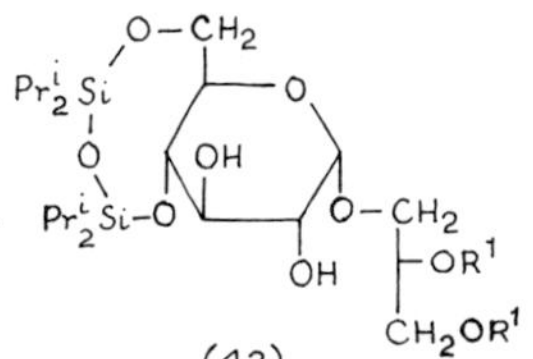

(42)

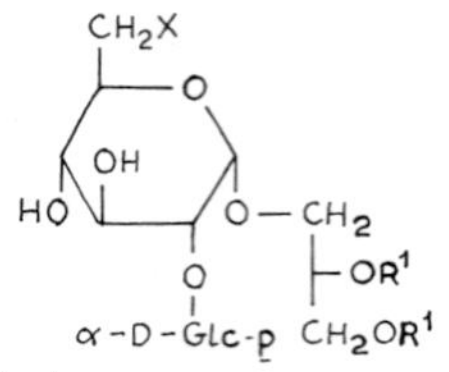

$R^1$ = palmitoyl

(43) X = H
(44) X = stearoyl
(45) X = $-O{\cdot}P(O)(OH)CH_2CH(OR^2)CH_2OR^2$ $R^2$ = oleoyl

2-Deoxy-α-D-ribopyranose 1,3,4-bicyclophosphite (46) is the product of the phosphorylation of the parent sugar with $P(NMe_2)_3$ or $PCl_3$ in dioxan.[71] New homolytic reactions of 1,2-O-alkylidene-α-D-glucofuranose 3,5,6-bicyclophosphites (47) have been reported. The 3,5-cyclic phosphates (48) and (49) are the products of reaction with hydrogen peroxide, and benzoyl peroxide respectively. The (2-methyl-2-cyano)ethyl ester (50) was also prepared.[72]

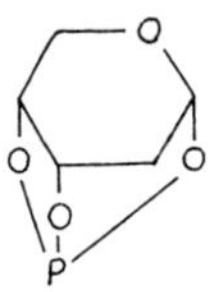

(46)

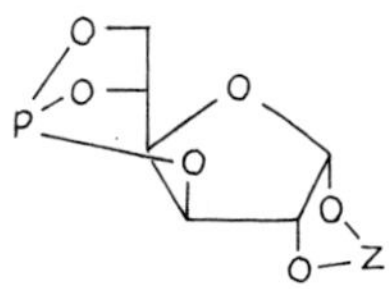

(47) Z = $Me_2C$, [cyclohexylidene], $CCl_3CH$.

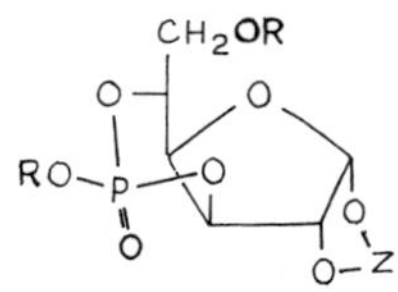

(48) R = H Z = as in (47)
(49) R = Bz
(50) R = $Me_2CCN$

Reaction of the bicyclophosphane (51) with polyols gave the corresponding peralkoxybicyclophosphoranes (52) which are in tautomeric equilibrium with the phosphites (53). Polyols used were D-arabinitol and galactitol.[73]

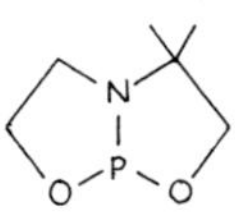

(51)

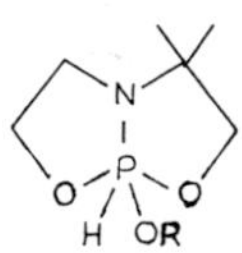

(52)

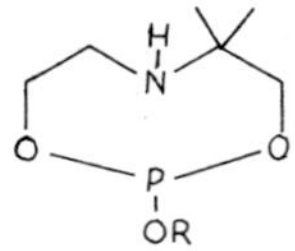

(53)

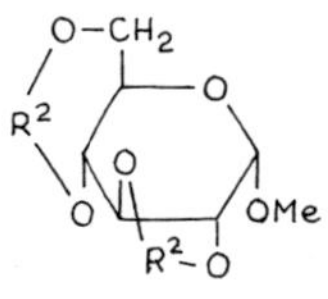

(54)

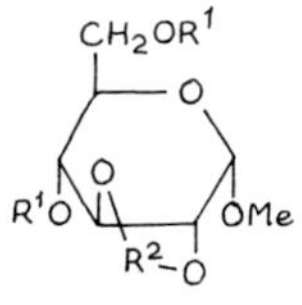

(55)

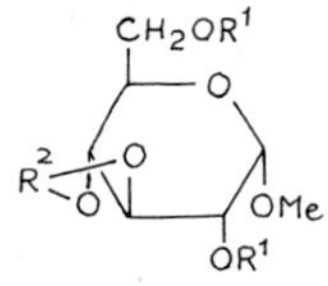

(56)

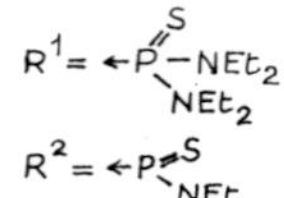

Condensation of methyl α-D-glucopyranoside with tris(diethyl)phosphamide and sulphur in benzene-ether gave two diastereoisomeric thiophosphates (54) and the thiophosphates (55) and (56).[74]

## 3 Other Esters

Ring contractions have been recorded in reductions[75] and displacements[76] of sulphonate esters. Treatment of the 6-deoxy-ditosylate (57) with lithium triethylborohydride gave the furanose (58), as well as the 3-reduced pyranoses (59) and (60), which arose via the mechanisms in Scheme 7.[75] Substantial amounts of the DL-diginofuranoside (61) and the DL-rhodinofuranoside (62) were obtained by reaction of sodium benzoate in DMF with the 4-O-mesylates of DL-oleandropyranoside (63) and DL-amicetopyranoside (64) (Scheme 8).[76]

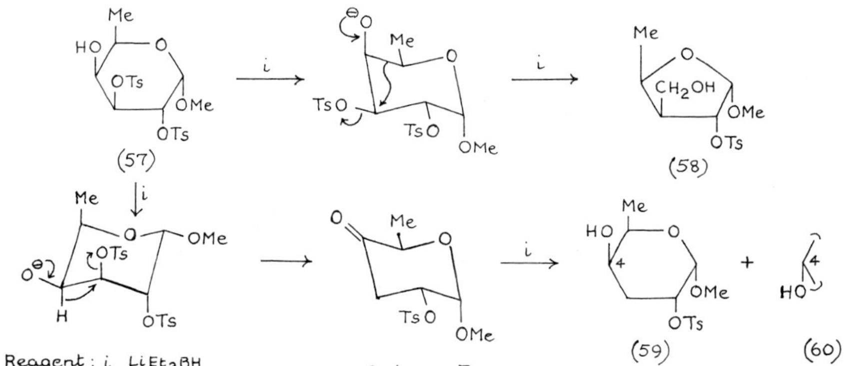


Scheme 7

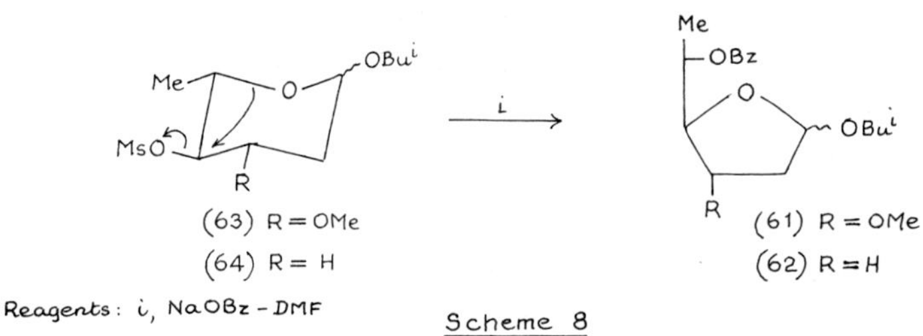


Scheme 8

Coupling of sugar moieties via the primary hydroxy group to proteins has been demonstrated in the attachment of several sugar derivatives to one molecule of bovine serum albumin using 6-O-(N-bromoacetylsulphanilyl) groups (65). These couple to proteins through amino acid imidazole and amino groups by bromine substitution in (65). Similar couplings were achieved with the corresponding 6-O-(p-bromoacetamidobenzoyl) derivatives.[77]

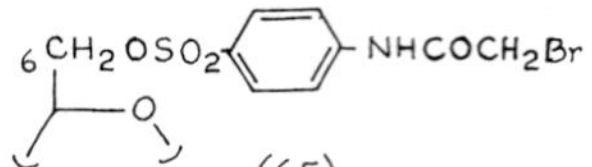

(65)

Cyclic sulphites of sugars may be regioselectively prepared *via* dibutylstannylene derivatives (Scheme 9). *Exo-endo* isomers of several sugars were obtained which were sometimes separable, *e.g.* methyl 2-O-benzoyl-β-L-arabinopyranoside (66) gave a 96% of the 3,4-O-cyclosulphite (67) in which the *exo-endo* ratio was 3:1.[78]

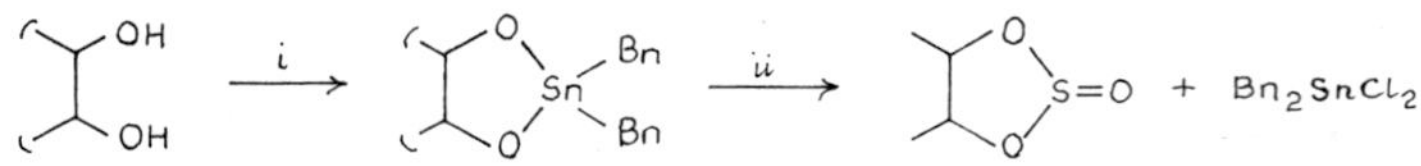

Reagents: i, $Bn_2SnO$; ii, $SOCl_2$

Scheme 9

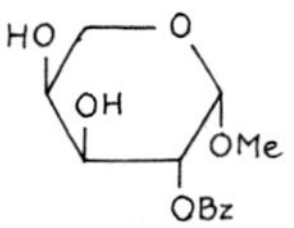

(66)

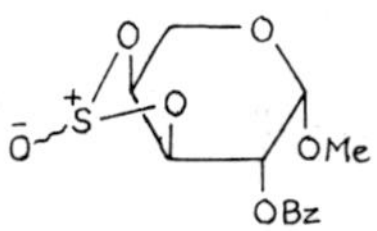

(67)

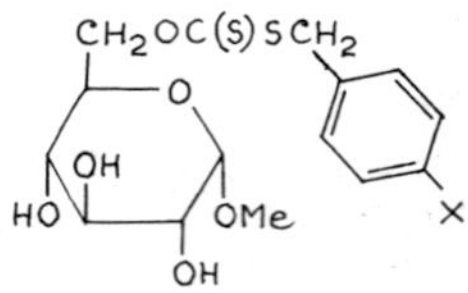

(68) X = H

(69) X = $NO_2$

A study of hydrolysis and aminolysis of xanthate esters included the two 6-substituted glucosides (68) and (69).[79]

Complexes of D-mannose, D-glucose, D-xylose, and L-arabinose with boric acid have been studied by $^{13}C$ n.m.r. spectroscopy and equilibrium data; they were considered to have the sugar in the furanose form.[80] The nature and corresponding association constants of the borate esters formed by a variety of polyhydroxy compounds (including alditols, aldonic acids, and aldaric acids) in aqueous solution at pH 11 has been probed by $^{13}C$ n.m.r. spectroscopy. The exchange rate between borate and its mono- and di-esters is slow on the $^{11}B$ n.m.r. timescale at room temperature, so that discrete signals for each of the components at equilibrium can be detected. Several general rules were devised to explain trends in the stability of the complexes.[81] The use of xylitol, sucrose, and glycerol concentration gradients in combination with borate buffers for creating stable artificial pH gradients for use in isoelectric focusing of proteins has been described.[82]

Sugar nitrates have been prepared by triflate displacements with nitrate ion: acetals and epoxides were unaffected, enabling nitrates (70) - (72) to be obtained.[83]

Reference to the stability of allyloxycarbonyl groups towards various heavy-metal reducing agents is made in Chapter 5, and to glycosylation *via* borate esters in Chapter 17.

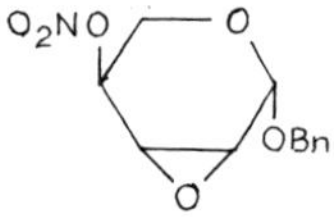


(70)

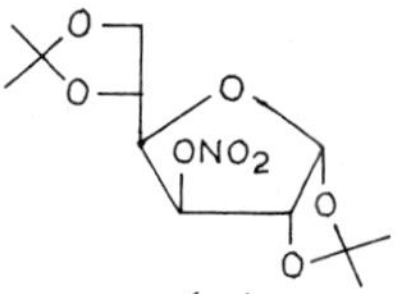


(71)

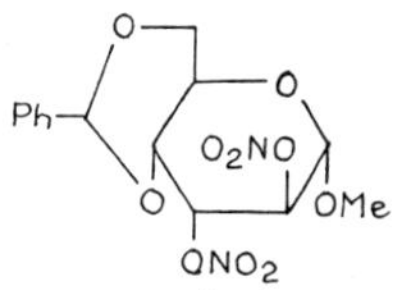


(72)

## References


1 S.David and S.Hanessian, Tetrahedron, 1985, 41, 643.

2 S.Gasa, M.Nakamura, A.Makita, M.Ikura, and K.Hikichi, Carbohydr. Res., 1985, 137, 244.

3 P.A.Garcia Ruiz, A.Soler Martinez, and M.A.Vidal Ros, An. Quim., Ser. C, 1983, 79, 240 (Chem. Abstr., 1985, 102, 46 191).

4 E.Lee, J.P.O'Reilly, J.O'Callaghan, and A.Bruzzi, Proc. R. Ir. Acad., Sect. B, 1984, 84B, 35 (Chem. Abstr., 1985, 103, 88 137).

5 R.Kaifu, L.S.Plantefaber, and I.J.Goldstein, Carbohydr. Res., 1985, 140, 37.

6 C.Subero, E.Correa, and M.Martin-Lomas, An. Quim., Ser. C, 1985, 81, 71 (Chem. Abstr., 1985, 103, 215 664).

7 P.Kovac and C.P.J.Glaudemans, Carbohydr. Res., 1985, 140, 313.

8 K.Capel, T.Vydra, J.Capkova, M.Rany, M.Blahova, and P.Sedmera, Collect. Czech. Chem. Commun., 1985, 50, 1039.

9 K.Kapek, T.Vydra, M.Rany, and P.Sedmera, Collect. Czech. Chem. Commun., 1985, 50, 2191.

10 D.Horton, W.Priebe, and O.Varela, Carbohydr. Res., 1985, 144, 317.

11 H.A.El-Shenawy and C.Schuerch, J. Carbohydr. Chem., 1985, 4, 215.

12 P.R.Brodfuehrer, C.Sapino, and H.G.Howell, J. Org. Chem., 1985, 50, 2597.

13 P.Kovac and C.P.J.Glaudemans, Carbohydr. Res., 1985, 138, C10.

14 M.Kloosterman, M.P.De Nijs, and J.H.Van Boom, Recl. Trav. Chim. Pays-Bas, 1985, 104, 126 (Chem. Abstr., 1985, 103, 196 332).

15 M.Kloosterman, T.Slaghek, J.P.G.Hermans, and J.H.Van Boom, Recl. J. R. Neth. Chem. Soc., 1984, 103, 335 (Chem. Abstr., 1985, 102, 149 649).

16 D.M.Clode, W.A.Laurie, D.McHale, and J.B.Sheridan, Carbohydr. Res., 1985, 139, 147.

17 D.M.Clode, W.A.Laurie, D.McHale, and J.B.Sheridan, Carbohydr. Res., 1985, 139, 161.

18 D.M.Clode, D.McHale, J.B.Sheridan, G.G.Birch, and E.B.Rathbone, Carbohydr. Res., 1985, 139, 141.

19 T.Yoshida, T.Okuda, M.Usman Memon, and T.Shingu, J. Chem. Soc., Perkin Trans. I, 1985, 315 .

20 J.E.Beart, T.H.Lilley, and E.Haslam, Phytochemistry, 1985, 24, 33.

21 M.Takechi, Y.Tanaka, M.Takehara, G.Nonaka, and I.Nishioka, Phytochemistry, 1985, 24, 2245.
22 M.Tamura, K.Kinomura, M.Tada, T.Nakatsuka, H.Okai, and S.Fukui, Agric. Biol. Chem., 1985, 49, 2011.
23 M.Tamura, T.Nakatsuka, K.Kinomura, and H.Okai, Pept. Chem., 1984, (Pub. 1985) 22nd, 329 (Chem. Abstr., 1985, 103, 193 891).
24 M.Tamura, T.Nakatsuka, K.Kinomura, H.Okai, and S.Fukui, Agric. Biol. Chem., 1985, 49, 891.
25 M.Tamura, M.Skoji, T.Nakatsuka, K.Kinomura, H.Okai, and S.Fukui, Agric. Biol. Chem., 1985, 49, 2579.
26 J.Horvat and D.Keglevic, Carbohydr. Res., 1985, 141, 255.
27 A.H.Haines and E.J.Sutcliffe, Carbohydr. Res., 1985, 138, 143.
28 P.Engelmeier, Carbohydr. Res., 1985, 144, 177.
29 J.Fernandez-Bolanos, R.Guzman, J.F.Guzman Bolanos, An. Quim., Ser. C, 1983, 79, 238 (Chem. Abstr., 1985, 102, 46 190).
30 F.Numata, K.Nishimura, H.Ishida, S.Ukei, Y.Tone, C.Ishihara, I.Saiki, I. Sekikawa, and I.Azuma, Chem. Pharm. Bull., 1985, 33, 4544.
31 Z.Shen and O.Theander, Phytochemistry, 1985, 24, 155.
32 H.-W.Rauwald and J.-T.Huang, Phytochemistry, 1985, 24, 1557.
33 M.Hamburger and K.Hostettmann, Phytochemistry, 1985, 24, 1793.
34 D.G.Musson, J.H.Lin, K.A.Lyon, D.J.Tocco, and K.C.Yeh, J. Chromatogr., 1985, 337, 363.
35 R.R.Schmidt and J.Michel, J. Carbohydr. Chem., 1985, 4, 141.
36 S.Hanessian, C.Hoomaert, A.Pernet, and A.M.Nadzan, Carbohydr. Res., 1985, 137, C14.
37 J.H.Witton, G.C.Hokanson, and J.C.French, J. Chem. Soc., Chem.Commun., 1985, 919.
38 M.M.Bandurraga and W.Fenical, Tetrahedron, 1985, 41, 1057.
39 K.Mizutani, K.Ohtani, R.Kasai, O.Tanaka, and H.Matsuura, Chem. Pharm. Bull., 1985, 33, 2266.
40 M.-J.Camarasa, P.Fernandez-Resa, M.-T.Garcia-Lopez, F.G.De Las Heras, P.P. Mendez-Castrillon, B.Alarcon, and L.Carrasco, J. Med. Chem., 1985, 28, 40
41 D.S.K.Tsui and P.A.J.Gorin, Carbohydr. Res., 1985, 144, 137.
42 D.W.Hutchinson, Organophosphorus Chem., 1985, 15, 144 (Chem. Abstr., 1985, 103, 208 968).
43 P.Kaliannan, S.Vishveshwara, and V.S.R.Rao, Int. J. Quantum Chem., 1985, 27, 181 (Chem. Abstr., 1985, 102, 185 355).
44 G.M.Whitesides, Angew. Chem., Int. Edn., 1985, 24, 617.
45 M.Szulczynski and F.Domka, Pr. Nauk. Akad. Ekon. im. Oskara Langego Wroclawiu, 1984, 267, 347 (Chem. Abstr., 1985, 102, 62 527).
46 A.Mocali, D.Aldinucci, and F.Paoletti, Carbohydr. Res., 1985, 143, 288.
47 D.G.Drueckhammer and C.-H.Wong, J. Org. Chem., 1985, 50, 5912.
48 C.Tu, S.Cai, and Z.Yang, Huaxue Xuebao, 1985, 43, 64 (Chem. Abstr., 1985, 102, 185 361).
49 B.E.Maryanoff, A.B.Reitz, G.F.Tutwiler, S.J.Benkovic, P.A.Benkovic, and S.J.Pilkis, J. Am. Chem. Soc., 1984, 106, 7851.
50 J.Thiem and D.Rasch, Liebigs Ann. Chem., 1985, 536.
51 M.Kapuscinski, F.P.Franke, I.Flanigan, J.K.MacLeod, and J.F.Williams, Carbohydr. Res., 1985, 140, 69.
52 P.Nielsen, P.Rauschenbach, and A.Bacher, Flavins Flavoproteins, Proc. Int. Symp., 8th, 1984, 71 (Chem. Abstr., 1985, 102, 204 218).
53 P.Boulanger, C.Andre, and G.Descotes, J. Carbohydr. Chem., 1985, 4, 279.
54 S.D.Mal'tsev, L.L.Danilov, V.N.Shibaev, and N.K.Kochetkov, Bioorg. Khim., 1985, 11, 138 (Chem. Abstr., 1985, 102, 185 359).
55 V.I.Torgov, K.A.Panosyan, A.T.Smelyanskii, and V.N.Shibaev, Bioorg. Khim., 1985, 11, 83 (Chem. Abstr., 1985, 102, 185 377).
56 R.Madiyalakan, S.-H.An, R.K.Jain, and K.L.Matta, Carbohydr. Res., 1985, 145, 89.
57 O.P.Srivastava and O.Hindsgaul, Carbohydr. Res., 1985, 143, 77.

58 M.H.Ryder, M.E.Tate, and G.P.Jones, J. Biol. Chem., 1984, 259, 9704 (Chem. Abstr., 1985, 102, 6979).
59 T.Shiba and S.Kusumoto, Handb. Endotoxin, 1984, 1, 284 (Chem. Abstr., 1985, 102, 149 643).
60 T.Shiba, S.Kusumoto, M.Inage, H.Chaki, M.Imoto, and T.Shimamoto, Bact. Endotoxin, 1984, 29 (Chem. Abstr., 1985, 102, 149 632).
61 T.Ogawa and A.Seta, Bact. Endotoxin, 1984, 51 (Chem. Abstr., 1985, 102, 167 063).
62 M.Imoto, H.Yoshimura, S.Kusumoto, and T.Shiba, Proc. Jpn. Acad., Ser. B, 1984, 60, 285 (Chem. Abstr., 1985, 102, 167 059).
63 T.Takahashi, C.Shimizu, S.Nakamoto, K.Ikeda, and K.Achiwa, Chem. Pharm. Bull., 1985, 33, 1760.
64 M.Kiso and A.Hasegawa, Bact. Endotoxin, 1984, 39 (Chem. Abstr., 1985, 102, 185 395).
65 M.Kiso, H.Ishida, T.Kito, S.Tanaka, and A.Hasegawa, Gifu Daigaku Nogakubu Kenkyu Hokoku, 1984, 171 (Chem. Abstr., 1985, 102, 204 207).
66 S.Nakamoto, T.Takahashi, K.Ikeda, and K.Achiwa, Chem. Pharm. Bull., 1985, 33, 4098.
67 T.Shimizu, S.Akiyama, T.Masuzawa, Y.Yanagihara, S.Nakamoto, T.Takahashi, K.Ikeda, and K.Achiwa, Chem. Pharm. Bull., 1985, 33, 4621.
68 C.A.A.Van Boeckel and J.H.Van Boom, Tetrahedron, 1985, 41, 4545.
69 C.A.A.Van Boeckel, G.M.Visser, and J.H.Van Boom, Tetrahedron, 1985, 41, 4557.
70 C.A.A.Van Boeckel and J.H.Van Boom, Tetrahedron, 1985, 41, 4567.
71 E.E.Nifant'ev, M.P.Koroteev, A.A.Zharkov, and A.I.Lutsenko, Zh. Obshch. Khim., 1984, 54, 1206 (Chem. Abstr., 1985, 102, 6963).
72 E.E.Nifant'ev, M.P.Koroteev, N.M.Pugashova, and N.A.Rytenkova, Zh. Obshch. Khim., 1985, 55, 315 (Chem. Abstr., 1985, 103, 88 131).
73 D.Houalla, A.Munoz, H.Ramdane, and R.Wolf, Tetrahedron Lett., 1985, 26, 2003.
74 L.I.Gurarii, A.N.Chernov, A.A.Musina, E.T.Mukmenev, and B.A.Arbuzov, Zh. Obshch. Khim., 1984, 54, 2150 (Chem. Abstr., 1985, 102, 95 931).
75 R.W.Binkley, J. Org. Chem., 1985, 50, 5646.
76 G.Berti, G.Catelani, F.Colonna, M.Ferretti, and L.Monti, Gazz. Chim. Ital., 1985, 115, 85 (Chem. Abstr., 1985, 102, 204 188).
77 M.Ohnishi, H.Sugimoto, H.Yamada, T.Imoto, K.Zaitsu, and Y.Ohkura, Chem. Pharm. Bull., 1985, 33, 674.
78 A.Guiller, C.H.Gagnieu, and H.Pacheco, Tetrahedron Lett., 1985, 26, 6067.
79 E.Humeras, M.Klug, and V.Soldi, Actas Simp. Iberoam. Catal., 9th, 1984, 1, 436 (Chem. Abstr., 1985, 102, 166 054).
80 J.Liu and L.Jiang, Youji Huaxue, 1984, 428 (Chem. Abstr., 1985, 102, 149 645).
81 M.Van Duin, J.A.Peters, A.P.G.Kieboom, and H.Van Bekkum, Tetrahedron, 1985, 41, 3411.
82 G.V.Troitsky and G.Yu. Azhitsky, J. Chromatogr., 1985, 324, 285.
83 N.Afza, A.Malik, F.Latif, and W.Voelter, Liebigs Ann. Chem., 1985, 1929.

# 8
# Halogeno-sugars

A review on stereo- and regio-controlled halogenation and their use in glycosidation reactions has appeared.[1]

(Diethylamino)sulphur trifluoride (DAST) has been used to prepar glycosyl fluorides from the corresponding anomeric hydroxy compounds.[2,3] The yields and α,β ratios are given in the formulae (1) to (4). Application of the DAST procedure to benzyl 3-azido-4,

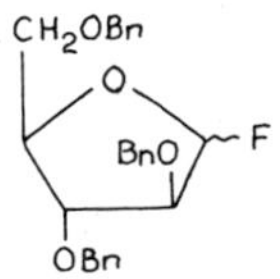

78% α:β, 3:1

(1)

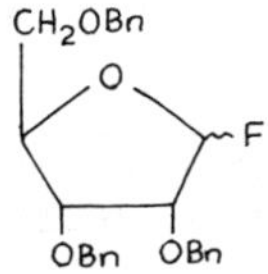

90% α:β, 1:10

(2)

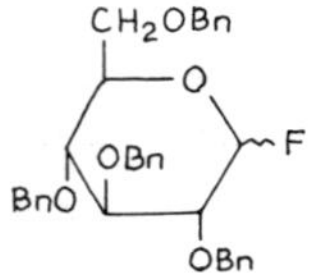

91% α:β, 1:4

(3)

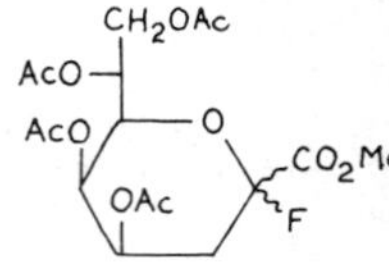

73% α;β, 2:1

(4)

O-benzylidene-3-deoxy-α-D-altropyranoside (5) gave the three fluoro-sugars (6) - (8), the first of which was converted to the 6-deoxy-

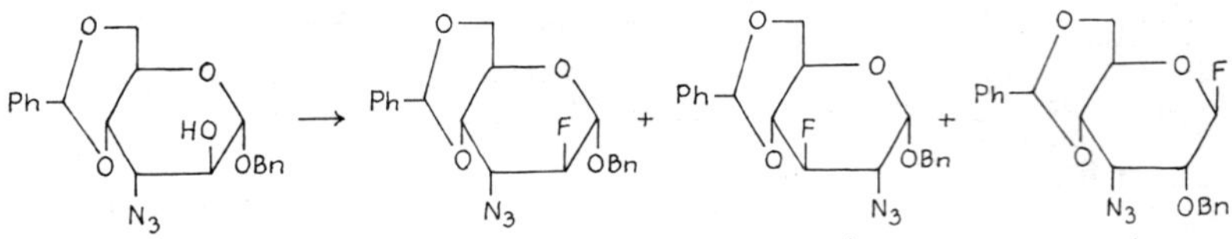

(5) (6) (7) (8)

fluoro-sugar (9), a 2-fluoro analogue of daunosamine, as shown in Scheme 1.[4] The glycosyl fluoride (10), required for the synthesis of the antibiotic efrotomycin, has been synthesized.[5,6]

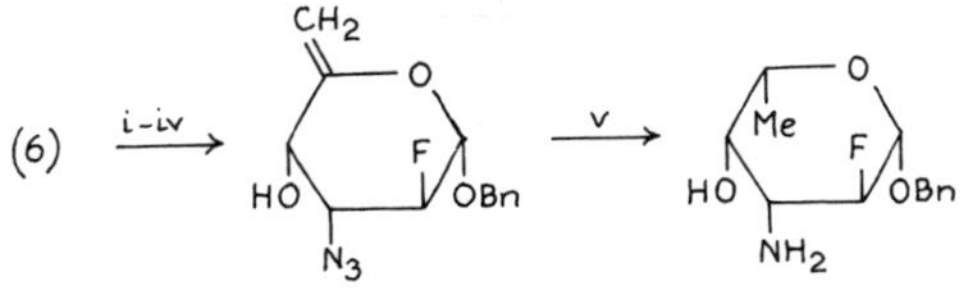

(9)

Reagents: i, $H^+$; ii, MsCl; iii, NaI; iv, AgF-Py; v, $H_2$-Pd/C

Scheme 1

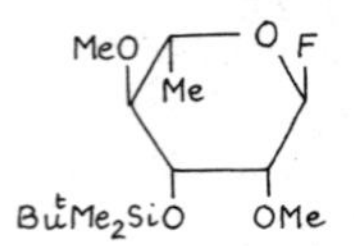

(10)

The 3-O-benzyl-α-D-galactosyl chloride (11) has been synthesized in 80% yield from the corresponding methyl β-D-galactoside derivative on controlled exposure to dichloromethyl methyl ether with catalytic zinc chloride. Prolonged exposure led to cleavage of the benzyl ether and formation of formate (12) and dichloromethyl ether (13).[7]

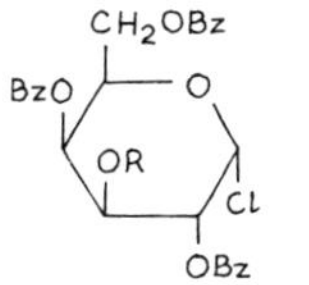

(11) R = Bn
(12) R = CHO
(13) R = $CHCl_2$

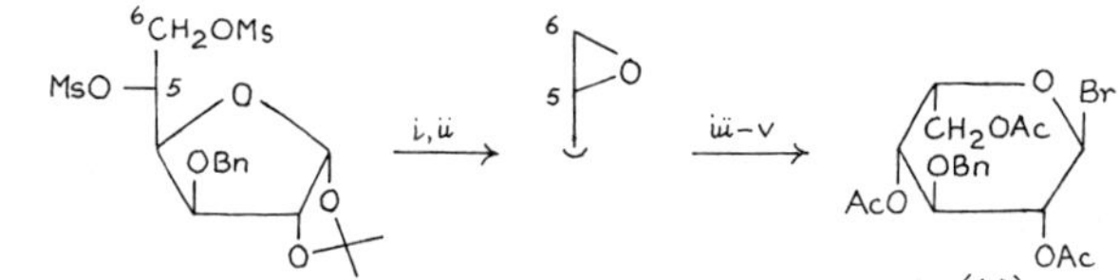

(14)

Reagents: i, KOAc - 18-Crown-6; ii, $KOBu^t$; iii, $H_3O^+$; iv, $Ac_2O$ - Py; v, $TiBr_4$

Scheme 2

The preparation of 2,3,4,6-tetra-O-methyl-α-D-glucopyranosyl bromide from 2,3,4,6-tetra-O-methyl-α-D-glucose has been improved using hydrogen bromide in dichloromethane.[8] The L-idosyl bromide (14), used in the synthesis of a pentasaccharide related to antithrombin III, has been synthesized as shown in Scheme 2.[9]

The 2-fluoro-arabinose derivative (15) has been prepared from the ribose N-pyridinyl sulphonate (16) using potassium hydrogen difluoride.[10] The fluoro-sugar phosphates (17) and (18) have been synthesized (Scheme 3) for sugar metabolism studies.[11] A synthesis

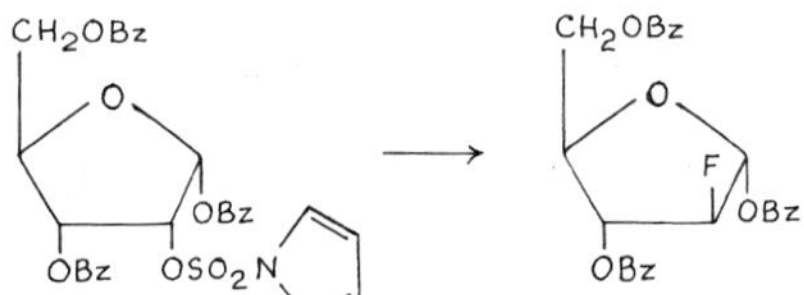

(16) (15)

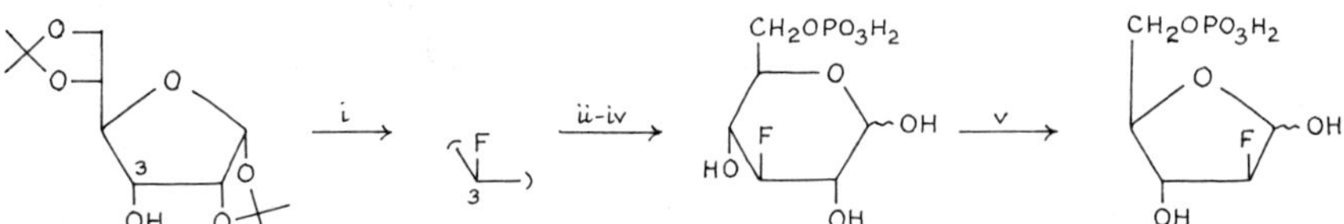

(17) (18)

Reagents: i, DAST; ii, $H^+$; iii, ATP-hexokinase; iv, Phosphoenolpyruvate-pyruvate kinase; v, $Pb(OAc)_4$-$H^+$

Scheme 3

of 1,3,4-tri-O-acetyl-2-deoxy-2-fluoro-β-D-ribopyranose (19)(obtained together with the corresponding arabinose derivative) using commercially available ethyl α-fluoroacetate with (R)-glyceraldehyde acetonide is outlined in Scheme 4.[12]

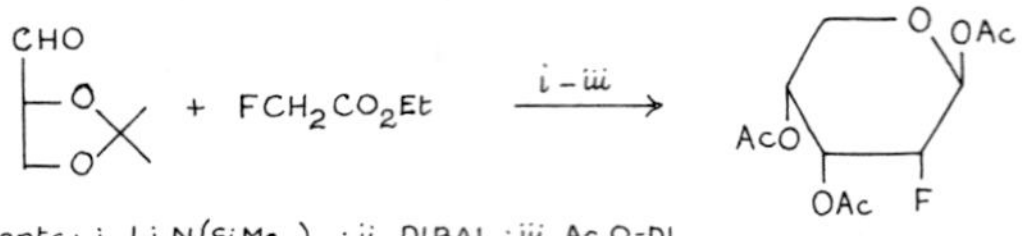

(19)

Reagents: i, $LiN(SiMe_3)_2$ ; ii, DIBAL ; iii, $Ac_2O$-DI
iii, $Ac_2O$-DMAP-$NEt_3$

Scheme 4

Reaction of D-glucal with fluorine in water, followed by acidic hydrolysis of any glycosyl fluoride, yields 2-deoxy-2-fluoro-D-glucose and -mannose in the ratio 4:1, which were separated by h.p.l.c. Similarly D-galactal gave 2-deoxy-2-fluoro-D-galactose an -talose in the ratio 5.3:1. The speed of these reactions (*ca*.45 mi makes them suitable for the preparation of [$^{18}$F]-labelled sugars, which was also realized.[13] A production system with remote control for the routine synthesis of 2-[$^{18}$F]fluoro-2-deoxy-D-glucose from fluorination of 3,4,6-tri-*O*-acetyl-D-glucal with labelled acetyl hypofluorite has been described.[14] The synthesis of this radio-labelled fluoro-sugar from D-glucal has been re-examined. A 2:1 mixture of the D-gluco and D-manno epimers was obtained with [$^{18}$F]-fluorine in water followed by hydrolysis. Using acetyl hypofluorit (from [$^{18}$F]fluorine with solid sodium acetate trihydrate) as re-agent, fluoroglucose could be obtained with 95% radiochemical purity.[15] The acetyl hypofluorite procedure has been developed for routine production, leading to a product of 98% purity and 28% radic chemical yield in a total synthesis time of 50 minutes from the end of the fluorine bombardment.[16] Addition of [$^{18}$F]acetylhypo-fluorite in acetic acid to D-glucal gave 2-deoxy-2-[$^{18}$F]fluoro-D-mannose as the main product, whereas with 3,4,6-tri-*O*-acetyl-D-gluca the *gluco* isomer predominated. A t.l.c. system using silica impreg-nated with sodium dihydrogen phosphate was developed for separating the C-2 epimers.[17] Sulphonate displacements have also been used to synthesize 2-deoxy-2-fluoro-D-glucose. Methods starting from methyl 3-*O*-benzyl-4,6-*O*-benzylidene-2-*O*-triflyl-β-D-mannopyranoside[18] and 1,6-anhydro-2-*O*-tosyl-D-mannopyranose[19] are shown in Schemes 5 and 6 respectively. The $^{19}$F n.m.r. spectrum of 2-deoxy-2-fluoro-D-glucose and its metabolites, the 6-phosphate, δ-phosphogluconolacton and 6-phosphogluconate have been reported; the results suggest that n.m.r. spectroscopy may be used to identify metabolic products that may be applicable to *in vivo* metabolism studies of the fluoro-sugar.[20] Deoxyfluoro-sugars have been synthesized by displacement of primary or secondary triflates using tris(dimethylamino)sulphonium difluorotrimethylsilicate; the reaction is fast (10 min at 0 °C) and

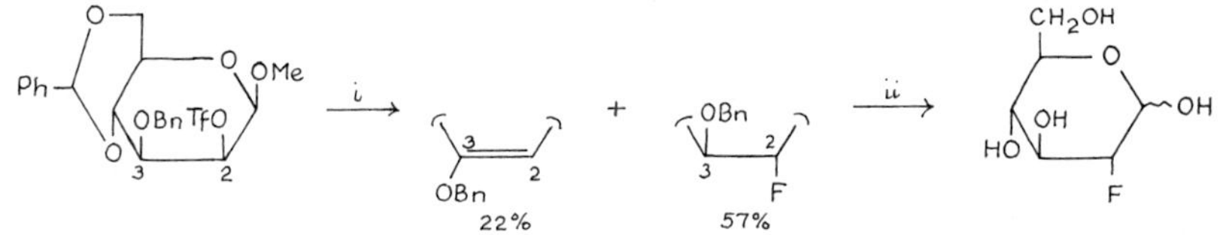

Reagents: i, $Et_4NF$-MeCN; ii $MeSO_3H$

Scheme 5

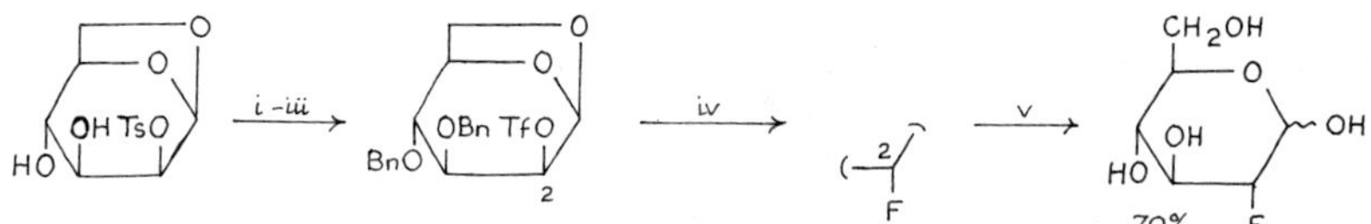

Reagents: i, BnBr-$Ag_2O$; ii, KOH; iii, $Tf_2O$-Py
iv, $Bu_4NF$; v, $MeSO_3H$-$H_2O$

Scheme 6

is thus suitable for the preparation of $^{18}$F-labelled radio-pharmaceuticals. Examples of fluoro-sugars prepared are the two glucose derivatives (21) and (22) and the 6-deoxy-6-fluoro-galactose derivative (23). Attempted displacement of the triflyl group of 1,2:5,6-di-O-isopropylidene-3-O-triflyl-α-D-glucopyranose gave only the 3,4-eno-sugar.[21] Methyl and benzyl 3-benzamido-2,3-dideoxy-

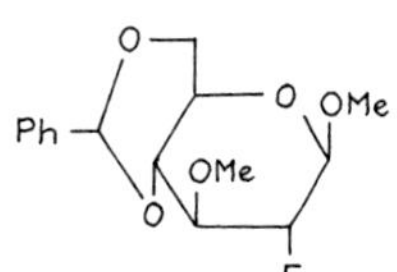

(21)

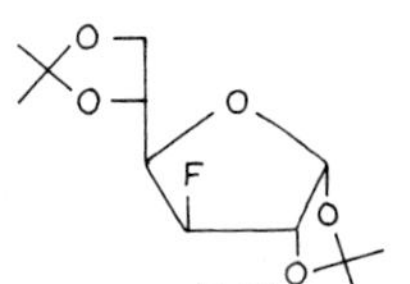

(22)

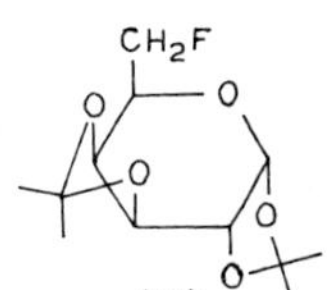

(23)

4,6-O-benzylidene-α-D-altropyranoside (24) has been synthesized from the 3-tosyl glucosamine derivatives (25) (Scheme 7) and converted by conventional methods into the corresponding glycosides of 3-benzamido-2,3,6-trideoxy-2-fluoro-β-L-galactopyranoside (26).[22] A

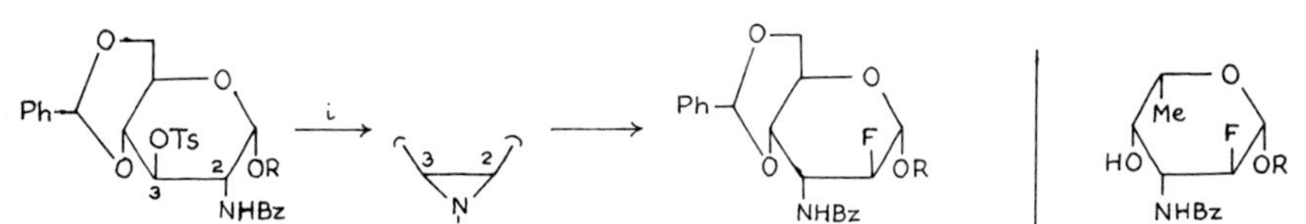

(25) R = Me, Bn (24) R = Me, Bn (26) R = Me, Bz

Reagents: i, $Bu_4NF$-HMPA

Scheme 7

similar route has been described by other workers, leading to the (S)-2-fluoro-L-daunosamine derivative (26), and its D-*altro* counterpart, (S)-2-fluoro-D-ristosamine.[23] The unprotected 2-fluoro-L-daunosamine hydrochloride (27) has been obtained as shown in Scheme 8, the key step of which was triethylamine - hydrogen fluoride

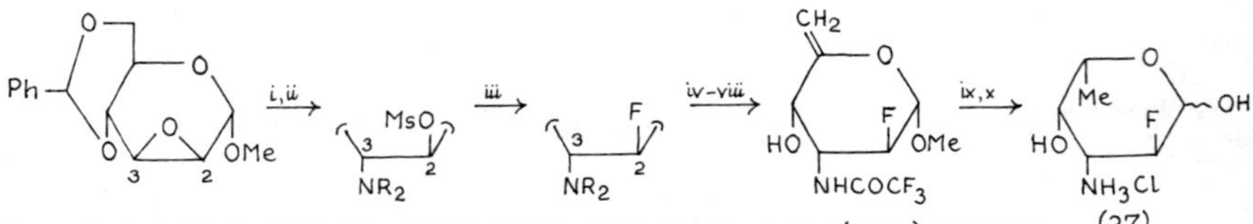

(27)

Reagents: i, $R_2NH$; ii, $MsCl-NEt_3$; iii, $NEt_3$-3HF; iv, Pd/C - $NaHCO_3$-$H_2O$; v, $(CF_3CO)_2O$; vi, NBS; vii, AgF-Py; viii, MeONa-MeOH; ix, $H_2$-Pd/C; x, $H_3O^+$

Scheme 8

complex displacement of a mesylate with retention of configuration *via* an aziridinium intermediate.[24,25] The reaction was also applied to methyl 2,3-anhydro-4-deoxy-α,β-D,L-*erythro*-pentopyranoside (28) to yield the 3-fluoro sugar (29). The intermediate aziridine was also treated with water to yield the trideoxy-*threo*-pentoside (30) and with lithium chloride to give the chloro derivative (31).[25]

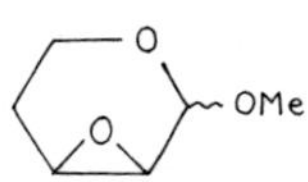

(28) (DL)

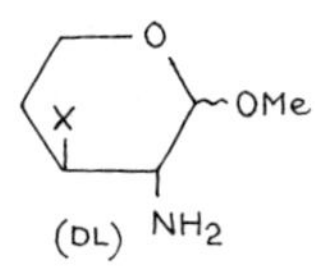

(DL)

(29) X = F
(30) X = OH
(31) X = Cl

The synthesis of radiolabelled halogeno-nucleosides is included in Chapter 20, while the use of halogeno-sugars in synthesis is referred to in Chapters 12 and 13.

A mechanistic study of the photolysis of deoxyiodo-sugars has been carried out. With secondary iodides epimerization was shown to precede reductive de-iodinations, while in the presence of oxygen a variety of different products are formed (Scheme 9).[26] The synthesis

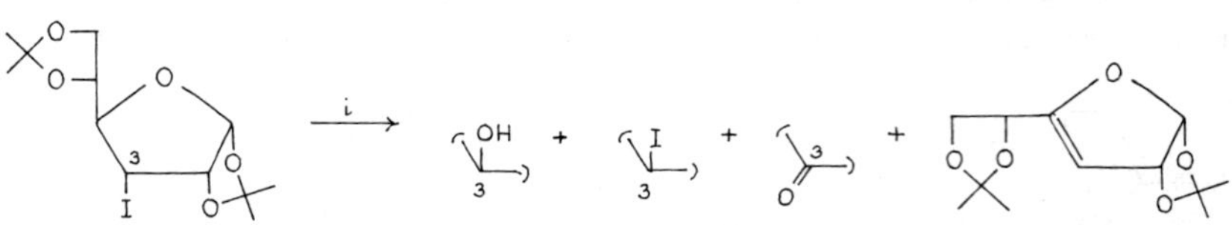

Reagent: i, $O_2$-hν

Scheme 9

and X-ray crystal structure of methyl [benzyl-2-(benzyloxycarbonyl)-

amino]-2,3,4-trideoxy-5-fluoro-α-D-erythro-hex-3-enopyranosid]uronate (32) has been reported. Two successive allylic substitutions were employed (Scheme 10). Treatment of (32) with methanolic ammonia rapidly gave the diacetal (33).[27]

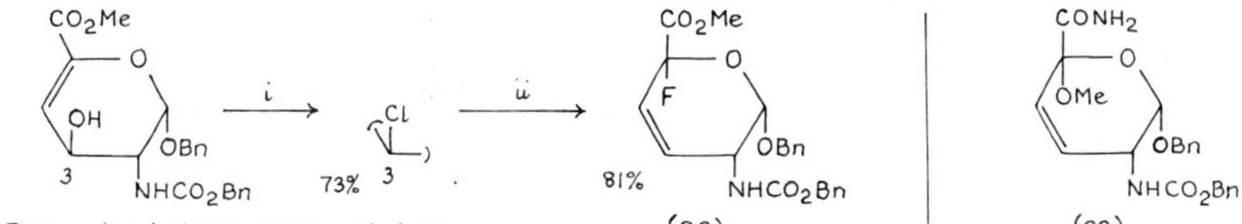

Reagents: i, $SOCl_2$-EtOAc; ii, AgF (32) (33)

Scheme 10

The epoxide-assisted displacement of triflyl groups by fluoride ion is reported to be an efficient approach to deoxyfluoro-sugars. Benzyl 2,3-anhydro-4-*O*-triflyl-pyranosides undergo inversion at C-4 with tetra-n-butylammonium fluoride (TBAF) at room temperature in good yield; thus treatment of the triflate (34) with TBAF in benzene for 24h gave the 4-fluoro-sugar (35) in 60% yield. The mild conditions avoid many of the difficulties encountered in fluoride-displacement reactions of carbohydrates.[28] The $^{13}C$ n.m.r. spectra

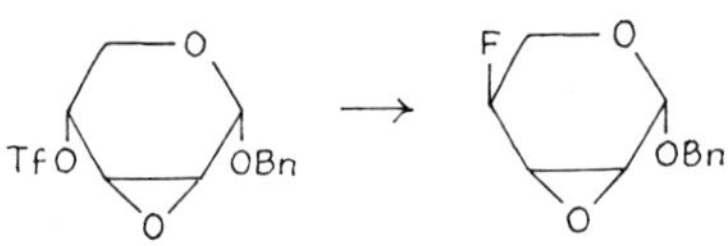

(34) (35)

of chlorinated sugars and their parent sugars have been compared and shifts caused by the presence of chlorine examined. 3-Chloro-deoxy-sugars show shifts different from the previously reported α-effects.[29] Triphenylphosphine-*N*-chlorosuccinimide has been shown to be effective for the direct chlorination of reducing sugars. In this way, maltose was selectively chlorinated at the primary position of the reducing moiety.[30] 6-Bromo-6-deoxy, 6,6'-dibromo-6,6'-dideoxy-, 6,1',6'-tribromo-6,1',6'-trideoxy-, and 6,6'-dichloro-6,6'-dideoxy-sucrose have been prepared by a modified procedure of Anisuzzaman and Whistler (see Vol. 12, p. 68). The products were examined as inactivators for dextransucrase but were found instead to be weak, reversible inhibitors.[31]

References


1 R.Johansson, Chem. Commun., Univ. Stockholm, 1984, (Monograph) 29 pp (Chem. Abstr., 1985, 102, 95 909).
2 W.Rosenbrook, D.A.Riley, and P.A.Lartey, Tetrahedron Lett., 1985, 26, 3.
3 G.H.Posner and S.R.Haines, Tetrahedron Lett., 1985, 26, 5.
4 S.Castillon, A.Dessinges, R.Faghih, G.Lukacs, A.Olesker, and T.T.Thang, J. Org. Chem., 1985, 50, 4913.
5 R.E.Dolle and K.C.Nicolaou, J. Am. Chem. Soc., 1985, 107, 1691.
6 R.E.Dolle and K.C.Nicolaou, J. Am. Chem. Soc., 1985, 107, 1695.
7 P.Kovac, N.F.Whittaker, and C.P.J.Glaudemans, J. Carbohydr. Chem., 1985, 4, 243.
8 H.A.Bates and C.Selick, J. Carbohydr. Chem., 1985, 4, 273.
9 C.A.A.Van Boeckel, T.Beetz, J.N.Vos, A.J.M.de Jong, S.F.Van Aelst, R.H. Van den Bosch, J.M.R.Mertens, and F.A.Van der Vlugt, J. Carbohydr. Chem., 1985, 4, 293.
10 C.H.Tann, P.R.Brodfuehrer, S.P.Brundidge, C.Sapino, and H.G.Howell, J. Org. Chem., 1985, 50, 3644.
11 D.G.Drueckhammer and C.-H.Wong, J. Org. Chem., 1985, 50, 5912.
12 J.T.Welch and S.Eswarakrishan, J. Chem. Soc., Chem. Commun., 1985, 186.
13 M.Diksic and D.Jolly, J. Carbohydr. Chem., 1985, 4, 265.
14 M.Haaparanta, J.Bergmann, O.Solin, and D.Roeda, Nuklearmedizin, Suppl. (Stuttgart), 1984, 21, 823 (Chem. Abstr., 1985, 102, 185 365).
15 G.T.Bida, N.Satyamurthy, and J.R.Barrio, J. Nucl. Med., 1984, 25, 1327 (Chem. Abstr., 1985, 102, 149 646).
16 M.J.Adam, T.J.Ruth, S.Jivan, and B.D.Pate, Int. J. Appl. Radiat. Isot., 1984, 35, 985 (Chem. Abstr., 1985, 102, 113 803).
17 C.J.S.Van Rijn, J.D.M.Herscheid, G.W.M.Visser, and A.Hoekstra, Int. J. Appl. Radiat. Isot., 1985, 36, 111 (Chem. Abstr., 1985, 102, 204 180).
18 T.Haradahira, M.Maeda, Y.Kai, H.Omae, and M.Kojima, Chem. Pharm. Bull., 1985, 33, 165.
19 T.Haradahira, M.Maeda, Y.Kai, and M.Kojima, J. Chem. Soc., Chem. Commun., 1985, 364.
20 T.Nakada, I.L.Kwee, G.A.Rao, and C.B.Conboy, Biochem. Arch., 1985, 1, 163 (Chem. Abstr., 1985, 103, 209 691).
21 W.A.Szarek, G.W.Hay, and B.Doboszewski, J. Chem. Soc., Chem. Commun., 1985, 663.
22 L.H.B.Baptistella, A.J.Marsaioli, J.D.de S.Filho, G.G.de Oliveira, A.B.de Oliveira, A.Dessinges, S.Castillon, A.Olesker, T.T.Thang, and G.Lukacs, Carbohydr. Res., 1985, 140, 51.
23 H.H.Baer and A.Jaworska-Sobiesiak, Carbohydr. Res., 1985, 140, 201.
24 D.Picq and D.Anker, J. Carbohydr. Chem., 1985, 4, 113.
25 D.Picq, I.Drivas, G.Carret, and D.Anker, Tetrahedron, 1985, 41, 2681.
26 R.C.Roth and R.W.Binkley, J. Org. Chem., 1985, 50, 690.
27 J.Kiss, P.C.Wyss, G.Flesch, W.Arnold, K.Noack, and P.Schonholzer, J. Carbohydr. Chem., 1985, 4, 347.
28 F.Latif, A.Malik, and W.Voelter, Z. Naturforsch., B: Anorg. Chem., Org. Chem., 1985, 40 B, 317 (Chem. Abstr., 1985, 103, 6627).
29 I.A.Toufeili and S.Z.Dziedzic, Aust. J. Chem., 1985, 38, 1425.
30 S.Z.Dziedzic, E.B.Rathbone, and G.G.Birch, Food Chem., 1984, 15, 51 (Chem. Abstr., 1985, 102, 6983).
31 M.K.Bhattacharjee and R.M.Mayer, Carbohydr. Res., 1985, 142, 277.

# 9
# Amino-sugars

## 1 Natural Products

Steroid glycosides with 2-acetamido-2-deoxy-β-D-glucopyranosyl moieties have been isolated from fruit pulp of the Nigerian plant *Tetrapleura tetraptera*,[1] and as the predatory-fish repellent constituents of the sole *Pardachirus marmoratus*,[2] while the gorgonian coral *Muricea fruticosa* has yielded a steroidal 2-acetamido-2-deoxy-D-galactopyranoside and analogues with unprecedented 4-, 6-, or 4,6-di-*O*-butyrate substituents on the sugar.[3] 6-Amino-6-deoxy-glucose has been reported to be a constituent sugar of a flavonoid diglycoside from *Myoporum tenuifolium* leaves.[4]

5-Acetamido-3,5,7,9-tetradeoxy-7-[(*R*)-3-hydroxybutyramido]-L-*glycero* L-*manno*-nonulosonic acid has been identified as a component of some O-specific polysaccharides of *Shigella boydii* and *Pseudomonas aeruginosa*.[5] 1,4- And 1,5-imino-hexitols (plant alkaloids and glycosidase inhibitors) are covered in Chapter 18.

## 2 Synthesis

The four stereoisomers of 3-amino-2,3,6-trideoxy-L-hexose, *i.e.* daunosamine (1), acosamine (2), ristosamine (3), and 3-epidaunosamine (4), have continued to be major synthetic targets, primarily because the first three are components of potent antibiotic substances. Syntheses of compounds (1)-(4), their racemic analogues,

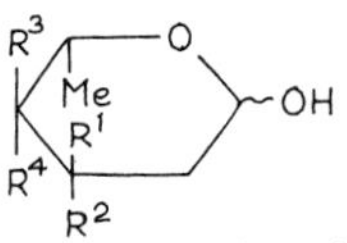


| | $R^1$ | $R^2$ | $R^3$ | $R^4$ |
|---|---|---|---|---|
| (1) | H | $NH_2$ | H | OH |
| (2) | H | $NH_2$ | OH | H |
| (3) | $NH_2$ | H | OH | H |
| (4) | $NH_2$ | H | H | OH |

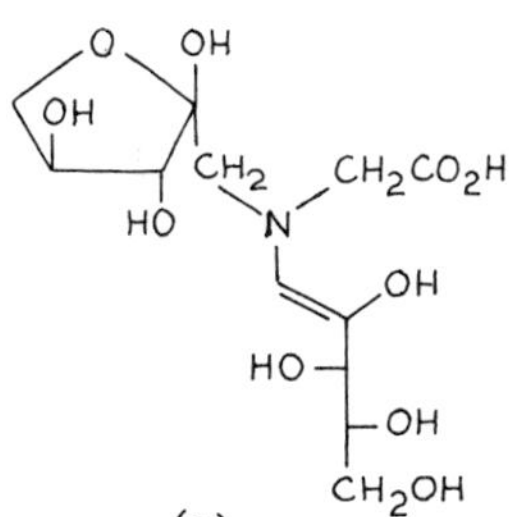


(5)

and derivatives are to be found throughout this section, which has been arranged according to the starting point for introduction of the amine functionality.

Weak catalysis by metal halides of the Amadori rearrangement of D-glucopyranosylamines has been reported.[6] The enaminol (5) was the major product from reaction of equimolar amounts of D-xylose and glycine (in $D_2O$) for 6 weeks.[7] 2-Alkylamino-2-deoxy-D-glucoses, of interest as surfactants, have been synthesized from D-fructose and fatty amines (with 5,7,9,11, and 13 carbon chains), via hydrolysis of 2-alkylamino-2-deoxy-D-glucosylamines.[8]

The 2-amino-2-halo-D-altrosides (6) and 2-fluoro-L-daunosamine (7) have been obtained by extension of the strategy (Vol.18, p.89) by which an epoxide is converted into the inverted N,N-diallylaziridinium ion intermediate (e.g. 8) which is opened with a nucleophile. Products in Scheme 1 resulted from retention of the nitrogen atom on C-3 on opening the aziridinium ion, while the

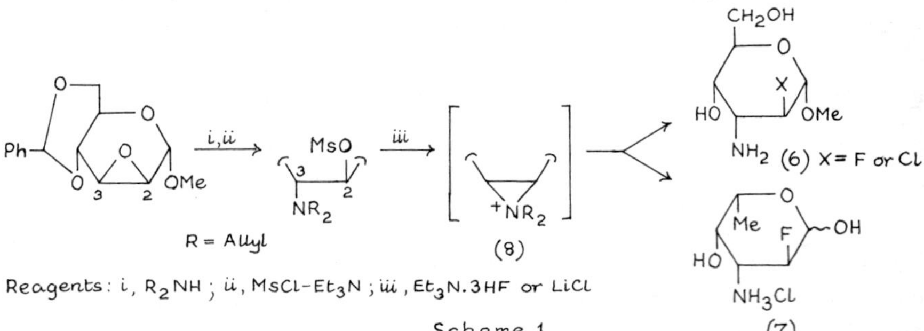

Scheme 1

racemic threo-pentopyranosides (9) in Scheme 2 resulted from a 1,2-shift of nitrogen on formation and opening of aziridinium ion

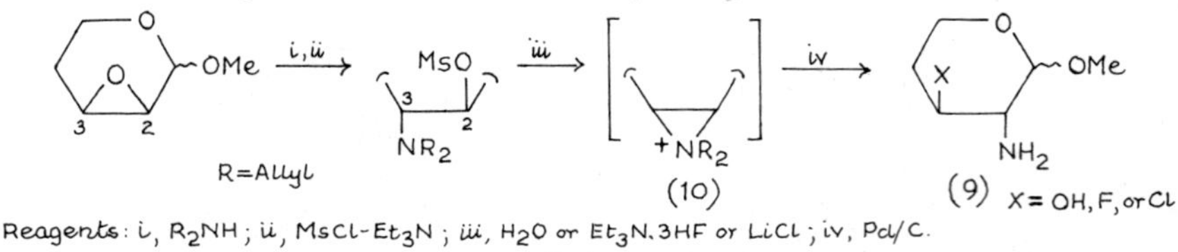

Scheme 2

(10).[9,10] The inversion of C-6 required in the synthesis of compound (7) was achieved by hydrogenation of a 5,6-alkene. Very similar syntheses of (S)-2-fluoro-L-daunosamine and -D-ristosamine are covered in Chapter 8. Derivative (11) of the disaccharide unit from an oligostatin α-glucosidase inhibitor has been synthesized, by coupling the corresponding aminocyclitol derivative with 2-O-benzyl-1,6:3,4-dianhydro-β-D galactopyranose, and converted into the internal glycoside (12) (Scheme 3).[11] In the synthesis of a pentasaccharide with three 2-amino-sugar moieties (see

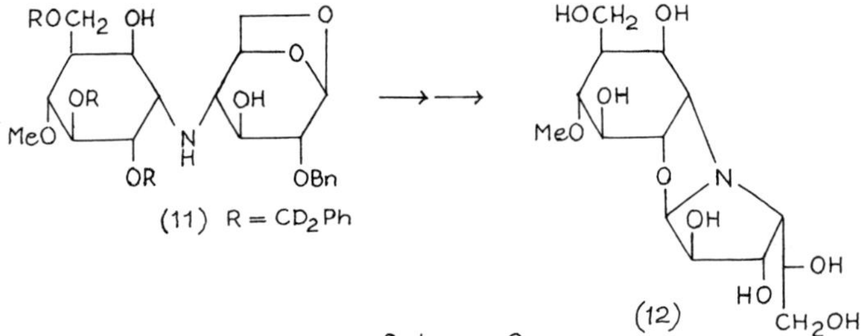

Scheme 3

Chapter 4), the amino function was introduced into each of the required subunits by azide-opening of the epoxide ring of a 1,6:2,3-dianhydro-β-D-mannopyranose derivative.[12] 3-Amino-3-deoxy-α-D-mannopyranosyl 3-amino-3-deoxy-α-D-mannopyranoside has been synthesized from α,α-trehalose *via* a bis-2,3-epoxide.[13]

Displacement of sulphonate by azide has been used to introduce the amino functionality in several syntheses. 3-Amino-3-deoxy-α-D-mannopyranosyl α-D-mannopyranoside (13), a further analogue of trehalose (*c.f.* ref.13), has been obtained from ditriflate (14) (Scheme 4).[14] Brimacombe and co-workers have reported convenient

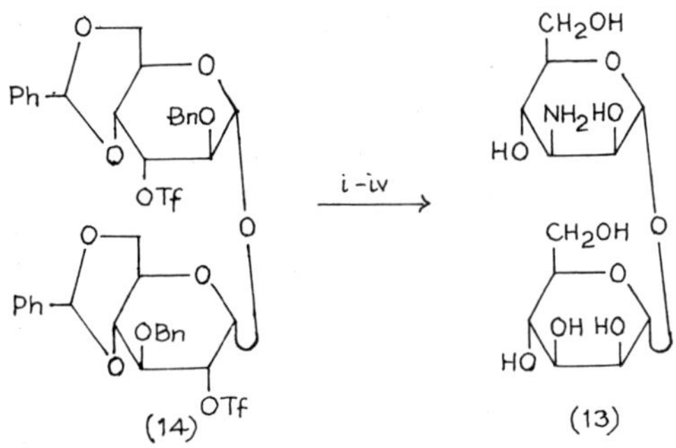

Reagents: i, $NaN_3$; ii, NaOBz; iii, NaOMe; iv, $H_2$-Pd/C-$HCO_2H$

Scheme 4

syntheses of *N*-acetyl-L-acosamine (15) and L-daunosamine (1) as its hydrochloride from L-rhamnose *via* the known glycosid-3-ulose (16) (Scheme 5). Conversion of azide (17) to L-daunosamine (1) required inversion at C-4, and this was effected by various methods.[15] Voelter and co-workers have converted isomeric 2,3-anhydro-4-*O*-trifluoromethanesulphonyl-pentopyranosides (*e.g.* 18) into the corresponding 4-amino-4-deoxy-sugars (*e.g.* 19) or their

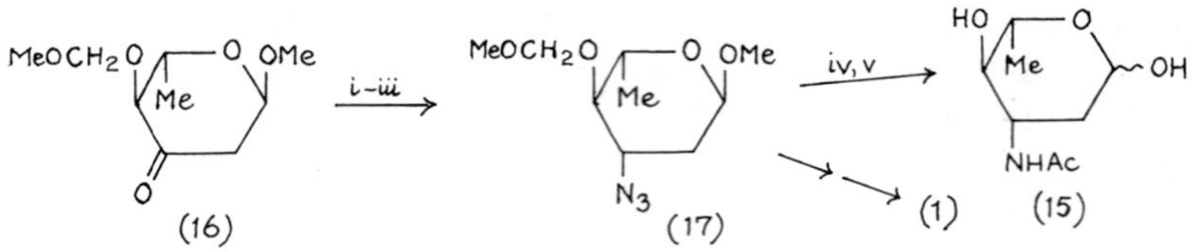


Reagents: i, $NaBH_4$; ii, MsCl–Py; iii, $NaN_3$; iv, $H_2$–Pd–HOAc; v, HOAc–$H_2O$

Scheme 5

4-aminoacid analogues (e.g. 20) with inversion at C-4 (Scheme 6) (c.f. Vol.18, p.90 ).[16,17]

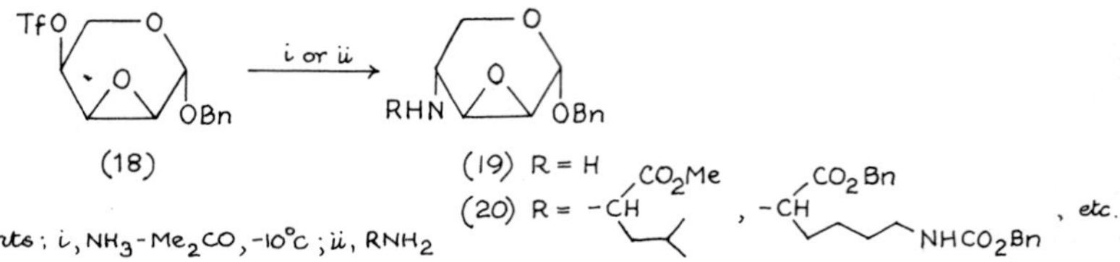


Reagents: i, $NH_3$–$Me_2CO$, –10°C; ii, $RNH_2$

Scheme 6

Displacement of halide by azide has been employed in the syntheses of the 6-amino-6-deoxy-α-D-glucosides (21) and (22).

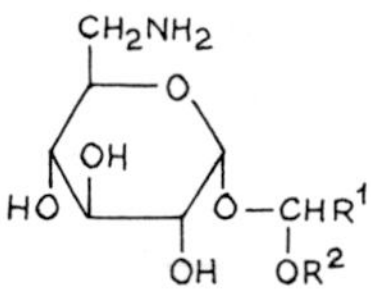


(21) $R^1 = CH_2NH_2$ ; $R^2 = Me$
(22) $R^1 = Me$ ; $R^2 = Et$

These glycosides, while being hydrolyzed under relatively mild acidic conditions, were resistant to yeast α-D-glucosidase.[18] A convenient synthesis of 6-acetamido-6-deoxy-D-glucose from D-glucose in 56% overall yield involved conversion to the known 1,2,3,4-tetra-O-(trimethylsilyl)ether derivative (4 eq. $MeCONHSiMe_3$-py), thence to the 6-bromide (NBS-$Ph_3P$) and 6-azide ($NaN_3$).[19] The C-6 epimeric 6-amino-6-deoxy-heptoses (23) and -hepturonates (24) have been synthesized by independent conversions of the separable major products (25) obtained from condensation of a sugar aldehyde with ethyl nitroacetate (Scheme 7).[20] A reinvestigation of the aminonitrile synthesis is covered in Chapter 10.

Derivatives of 2-azido-2-deoxy-D-mannopyranose and the corresponding uronic acid have been synthesized conventionally by azidonitration of D-glucal triacetate, for use as building blocks for bacterial polysaccharide sequences.[21] The α-L-ristosamine

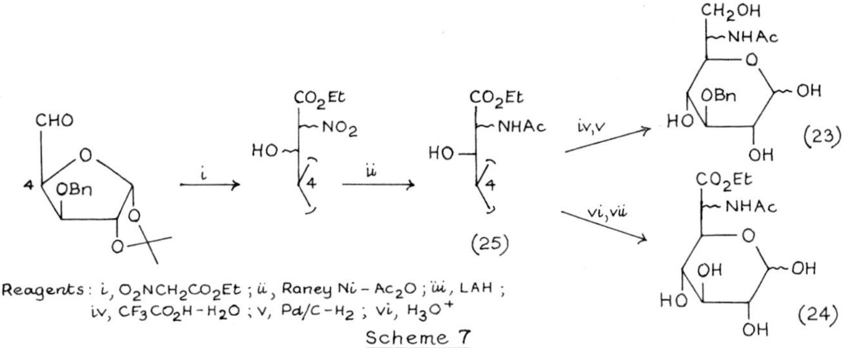

Reagents: i, $O_2NCH_2CO_2Et$; ii, Raney Ni - $Ac_2O$; iii, LAH; iv, $CF_3CO_2H-H_2O$; v, Pd/C - $H_2$; vi, $H_3O^+$

Scheme 7

derivative (26) has been synthesized from L-rhamnal diacetate (27) as shown in Scheme 8, the intermediate (28) being obtained by stereoselective reduction of an enone precursor.[1] The analogous derivative of 3-epidaunosamine (4) was obtained by inversion at C-4 of the allylic azide (28) through sulphonate displacement.[22]

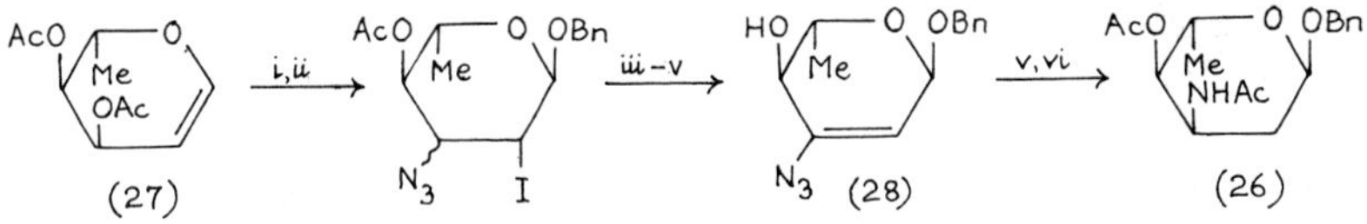

Reagents: i, $NaN_3-BF_3.OEt_2$; ii, NIS - BnOH; iii, NaOMe - MeOH; iv, PDC; v, $NaBH_4$; vi, $Ac_2O$-Py

Scheme 8

A relatively straightforward synthesis (Scheme 9) of the methyl 2-amino-2-deoxy-β-D-mannoside (29) involved selective oxidation of the β-D-glucoside (30) at C-2 with bromine, conversion to the

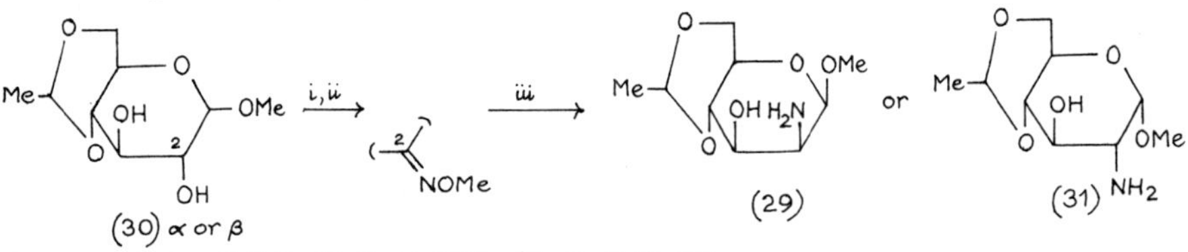

Reagents: i, $Br_2-H_2O$ (pH 7); ii, $MeONH_2$; iii, $H_2$-Pd/C-HCl

Scheme 9

stable O-methyl oxime, chromatographic separation of the minor product resulting from oxidation at C-3, and reduction. The same sequence on the α-D-glucoside led to the methyl 2-amino-2-deoxy-α-D-glucoside (31) as the major product.[23] A related oxidation-oximation-reduction sequence for the synthesis of β-glycosidically linked oligosaccharides containing 2-acetamido-2-deoxy-D-mannose residues is covered in Chapter 4. Methyl sibirosaminide (32), identical with a sample prepared from the antibiotic sibiromycin, has been synthesized from the known 3-C-methylglycosid-4-ulose (33) via oxime reduction, and N-methylation by reduction of an N-carbomethoxy derivative (Scheme 10) (see also Scheme 14).[24]

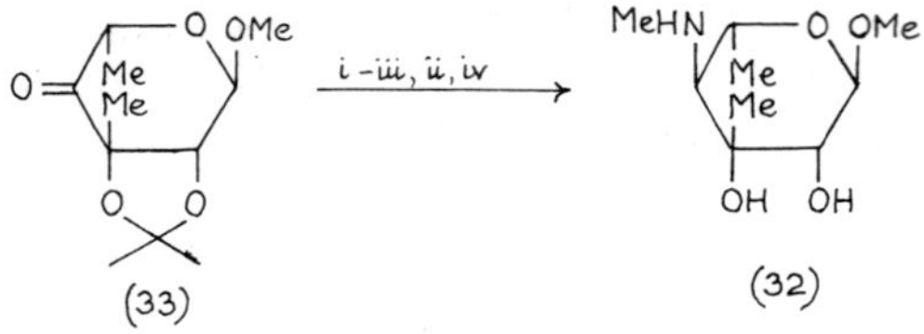

Reagents: i, $NH_2OH$-Py ; ii, LAH ; iii, $ClCO_2Me$ ; iv, $HOAc$-$H_2O$

Scheme 10

Several amino-sugar syntheses have employed 3- and 4-carbon chiral starting materials. 2,3-O-Isopropylidene-D-glyceraldehyde (34) has been elaborated into the 3-acetamido-2,3-dideoxy-pyranoside (35) and the 3-nitro-furanosiduronic acid (36) by stereoselective allylation of the nitronate dianion derived from nitro derivative (37) (Scheme 11)[25] and into 2-amino-2-deoxy-D-arabinose

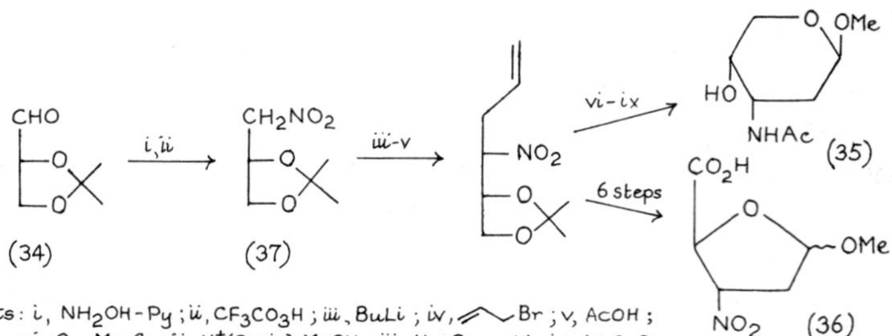

Reagents: i, $NH_2OH$-Py ; ii, $CF_3CO_3H$ ; iii, BuLi ; iv, allyl bromide ; v, AcOH ; vi, $O_3$-$Me_2S$ ; vii, $H^+$(Resin)-MeOH ; viii, $H_2$-Raney Ni ; ix, $Ac_2O$-Py

Scheme 11

(38) by nucleophilic addition of methyl isocyanoacetate, which occurred with high erythro selectivity (~100%), followed by intramolecular cyclization to give an 8:2 mixture of oxazolines (39),

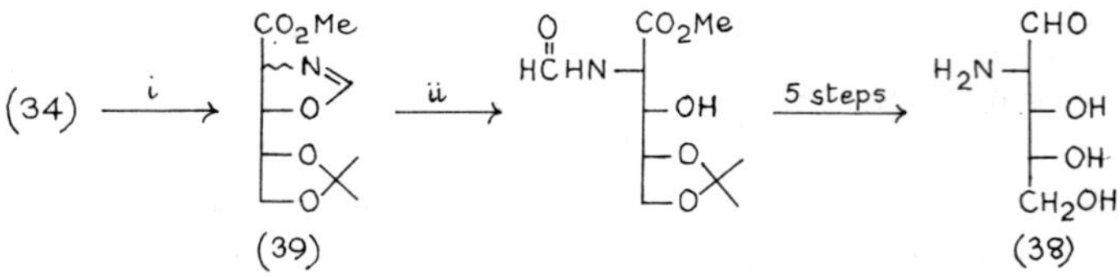

Reagents: i, $CNCH_2CO_2Me$-$Et_3N$-$PdCl_2$ ; ii, $HOAc$-$H_2O$

Scheme 12

the synthesis proceeding via the major isomer (Scheme 12).[26] A related synthesis of 2-amino-2-deoxy-D-arabinitol is covered in Chapter 18.

Two groups have used condensations of methyl 3-nitropropionate with chiral O-protected lactaldehydes for the synthesis of 3-amino-2,3,6-trideoxy-hexoses. Using the L-lactaldehyde derivative (40), Brandänge and Lindqvist obtained only two 3-nitrohexono-1,4-lactone isomers, (41) and (42), in ratios between 2:1 and 5:1 depending upon conditions; probably the other diastereoisomers isomerized at the nitro group to give the most stable products after lactonization (Scheme 13). Lactone (41) was converted into the L-

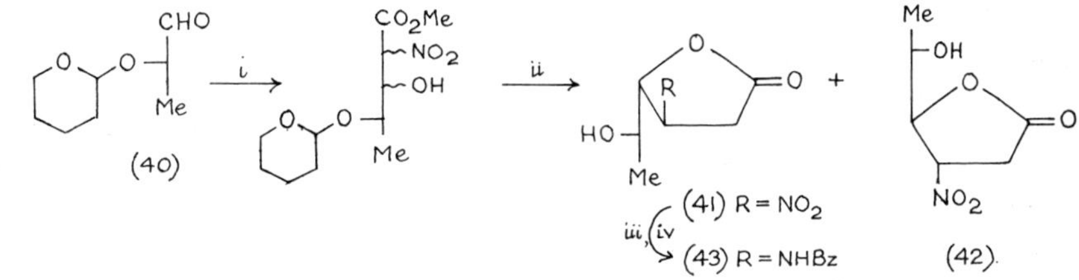

Reagents: i, $O_2N$~~$CO_2Me$ - $Bu^tOK$ or KF-$Bu_4NCl$ ; ii, Py.HOTs ; iii, $H_2$-Pd ; iv, BzCl-Py

Scheme 13

ristosamine precursor (43), while lactone (42) similarly gave a derivative of L-daunosamine (1).[27] Hanessian and Kloss investigated this condensation with O-benzyl-D- and L-lactaldehydes, which are both available from cheap ethyl L-lactate. By varying the catalyst (i.e. neutral alumina, $KOBu^t$-$ZnBr_2$, or $KOBu^t$-$MgBr_2$) the condensation could be influenced to yield the arabino-, ribo-, or xylo-isomer [i.e. precursors of sugars (2), (3) or (4)], respectively, as the major product. The N-benzoyl derivative of L-daunosamine (1), with the L-lyxo configuration, was synthesized by inversion at C-5 in a product with the D-ribo configuration.[28]

The N-phenylsulphonyl derivative (44) of sibirosamine (c.f. Scheme 10) has been synthesized from the L-allothreonine derivative (45) as shown in Scheme 14, involving stereospecific addition of

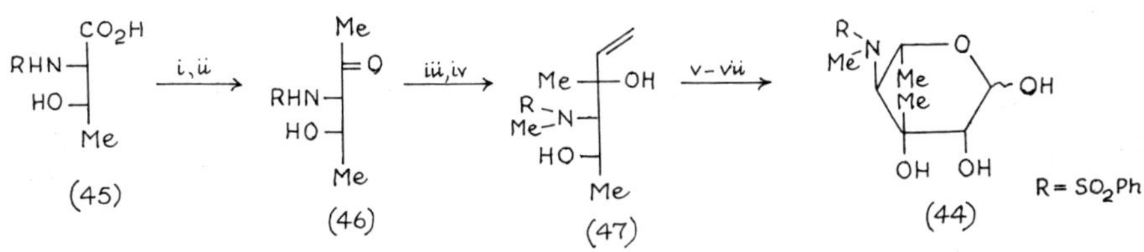

Reagents: i, MeLi; ii, MeMgBr ; iii, ⁄⁄~MgBr ; iv, MeI-$K_2CO_3$-$Pr^iOH$ ; v, $OsO_4$-(N-methylmorpholine N-oxide) ; vi, Pt-$O_2$ ; vii, $Bu^i_2AlH$

Scheme 14

vinyl Grignard to ketone (46) and stereoselective osmylation of alkene (47) (which gave a separable mixture of products in a 4:1 ratio). A related 13-step synthesis of the analogous 3-episibirosamine derivative was also reported.[29] 4-Amino-2,4,6-trideoxy-L-<u>lyxo</u>-, L-<u>arabino</u>-, and L-<u>ribo</u>-hexoses, <u>i.e.</u> analogues of sugars (1)-(3) respectively, in which the 3-amino and 4-hydroxy functions are interchanged, have been prepared as their <u>N</u>-trifluoroacetates from the chiral synthon (48), derived previously from cinnamaldehyde and Baker's yeast. The strategy used is illustrated in Scheme 15 for the synthesis of the L-<u>ribo</u> isomer (49), in which an epoxide is intramolecularly opened by an adjacent carbamate moiety.[30]

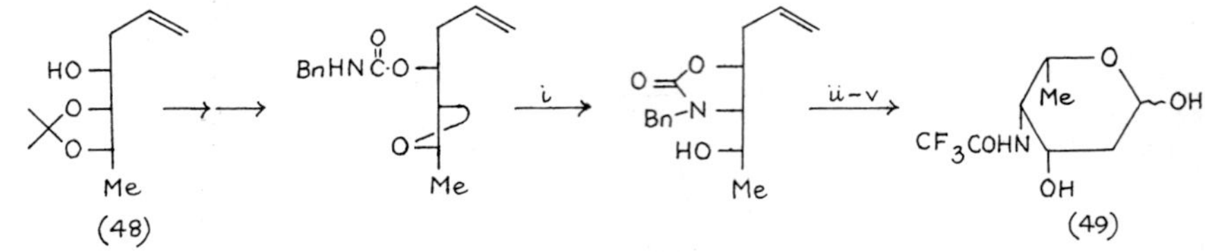

Reagents: i, $KOBu^t$; ii, $Li-NH_3$; iii, LiOH-EtOH; iv, $(CF_3CO)_2O$; v, $O_3-Me_2S$

<u>Scheme 15</u>

A variety of achiral, non-carbohydrate materials have been converted into amino-sugars, the products, with one exception, being racemic mixtures. Racemic daunosamine (1) has been synthesized from the furan adduct (50) by a process involving two Baeyer-Villiger oxidations (Scheme 16).[31] Danishefsky and co-workers

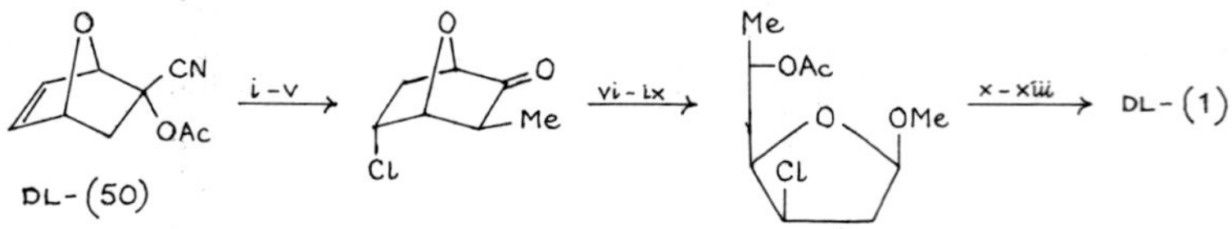

Reagents: i, PhSCl; ii, NaOMe; iii, $CH_2O$; iv, Raney Ni; v, $KOBu^t-MeI$; vi, $MCPBA-NaHCO_3$; vii, $MeOH-H^+$; viii, MeLi; ix, $CF_3CO_3H$; x, $NaN_3$; xi, $H_2-Pd/C$; xii, $NH_3-MeOH$; xiii, $H_3O^+$

<u>Scheme 16</u>

have employed their Lewis acid-catalyzed hetero-Diels-Alder strategy to synthesize dihydropyrones suitable for elaboration into the derivative (51) of DL-daunosamine (1) (Scheme 17)[32] and methyl

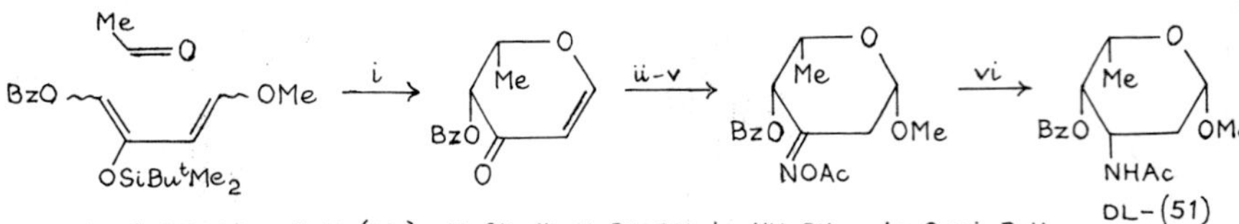

Reagents: i, $BF_3.OEt_2$; ii, $Hg(OAc)_2-MeOH$; iii, $NaBH_3CN$; iv, $NH_2OH$; v, $Ac_2O$; vi, $B_2H_6$

<u>Scheme 17</u>

β-DL-lincosaminide (52) (Scheme 18).[33] In the former synthesis the amine group was introduced by reduction of an oxime acetate. In the latter synthesis, a highly stereoselective "off-template"

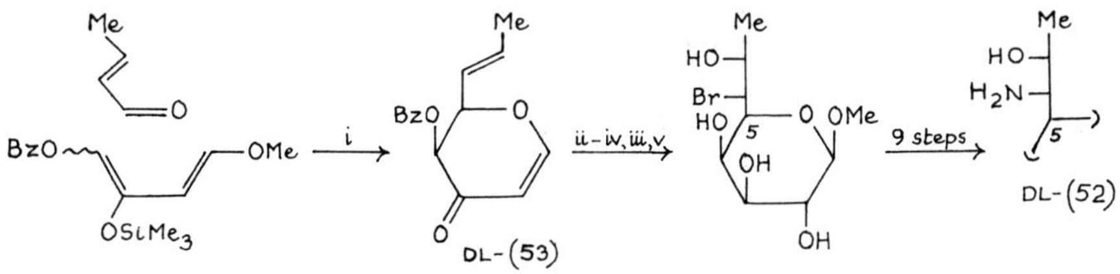

Reagents: i, $BF_3 \cdot OEt_2$ ; ii, $NaBH_4$-$CeCl_3$ ; iii, BzCl-DMAP-$Et_3N$ ; iv, MCPBA-MeOH ; v, NBS-$H_2O$

Scheme 18

functionalization of the 6,7-double bond in compound (53) was achieved (step v), although the subsequent introduction of the amine group required a multistep procedure via 6,7-epoxide and -N-phosphorylaziridine intermediates. The hetero-Diels-Alder reaction illustrated in Scheme 19 has provided an entry into branched-chain amino-sugars of the garosamine type.[34]

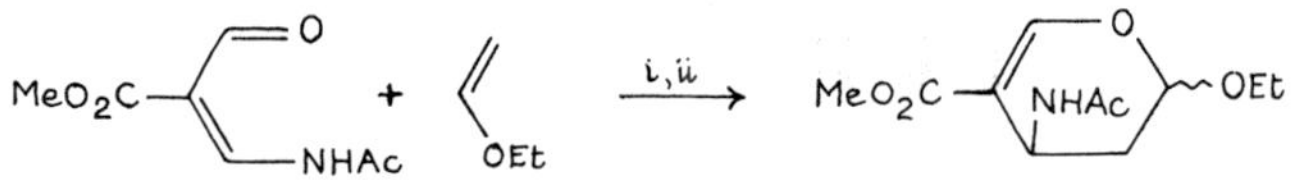

Reagents : i, Δ ; ii, $BF_3 \cdot OEt_2$

Scheme 19

Intramolecular dipolar cycloaddition of the nitrone (54), which has a chiral substituent on nitrogen, gave an 82:18 mixture of isoxazolidines (55) and (56), respectively. The major isomer (55) was converted into the methyl glycosides (57) of L-acosamine as shown in Scheme 20, or into the methyl glycosides of L-daunosamine

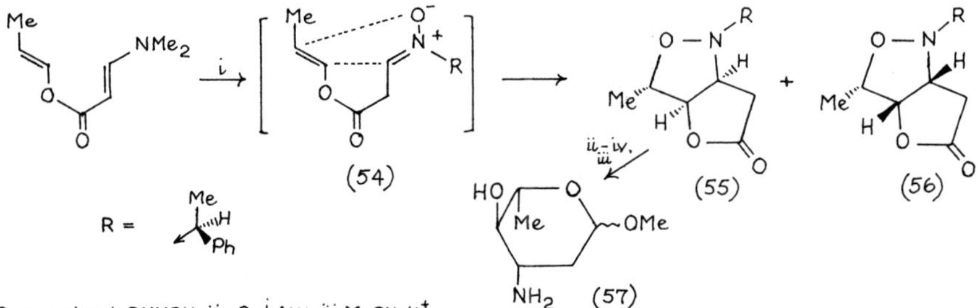

Reagents: i, $RNHOH$ ; ii, $Bu^i_2AlH$ ; iii, MeOH-$H^+$; iv, $H_2$-Pd/C

Scheme 20

by a related sequence including an inversion at C-4 through mesylate displacement.[35] Intermolecular dipolar cycloadditions of furan with nitrile oxides have been employed in syntheses of racemic aminodeoxy-aldose (58) and (59), and -dialdose (60) derivatives (Scheme 21).[36,37]

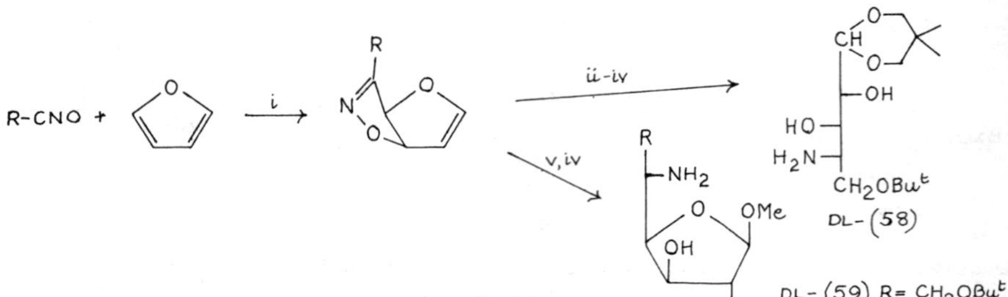

Reagents: i, $BF_3.OEt_2$ ; ii, $O_3$ ; iii, $Me_2C(CH_2OH)_2$-$H^+$; iv, LAH ; v, MCPBA-MeOH

Scheme 21

Racemic daunosamine (1) has been synthesized in eight steps, in >30% overall yield, from 1-(2-furyl)ethanol (61) as shown in Scheme 22; allylic imidate (62) was synthesized by a modified

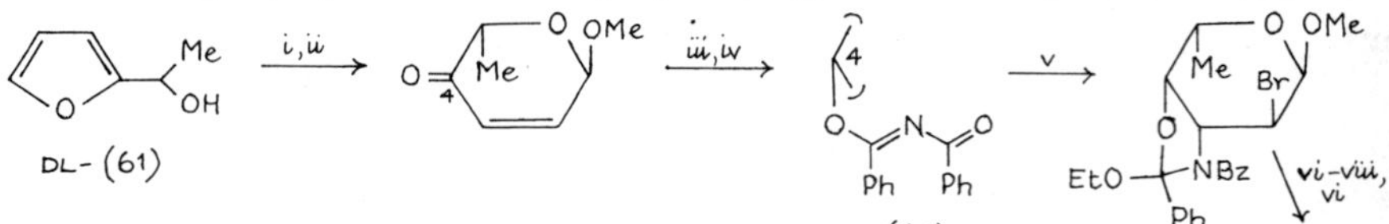

Reagents: i, $Br_2$-MeOH ; ii, $HCO_2H$-MeOH ; iii, $NaBH_4$ ; iv, $Ph_3P$-DEAD-$HNBz_2$ ; v, NBS-EtOH ; vi, $H_3O^+$ ; vii, $Bu_3SnH$ ; viii, $OH^-$

Scheme 22

Mitsunobu reaction and used to introduce the amine group by intramolecular cyclization. Since both enantiomers of (61) can be prepared, this procedure could be used for the preparation of optically active material.[38]

Three groups have elaborated derivatives of sorbic acid (i.e. hexa-2,4-dienoic acid) into amino-sugars. Hirama and co-workers synthesized all four racemic 2-acylamino-2,3,6-trideoxyhexoses (1)-(4) from ethyl sorbate (63). The double bonds were functionalized in sequence by standard stereoselective hydroxylation procedures followed by intramolecular Michael addition of a 3- or 4-O-carbamoyl group as outlined in Scheme 23.[39,40] The racemic

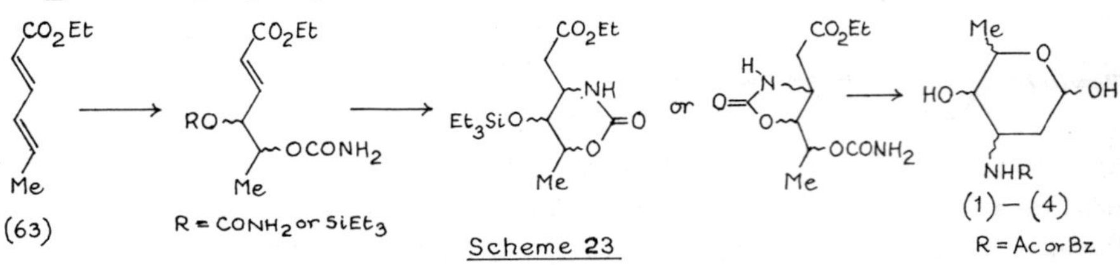

Scheme 23

isopropylidenated erythro-diol (64), synthesized from methyl sorbate by a similar strategy, has been converted into a separable mixture of the N-acetyl derivatives of DL-acosamine (2) and DLdesmethylholantosamine (65). The formation of this latter 4-

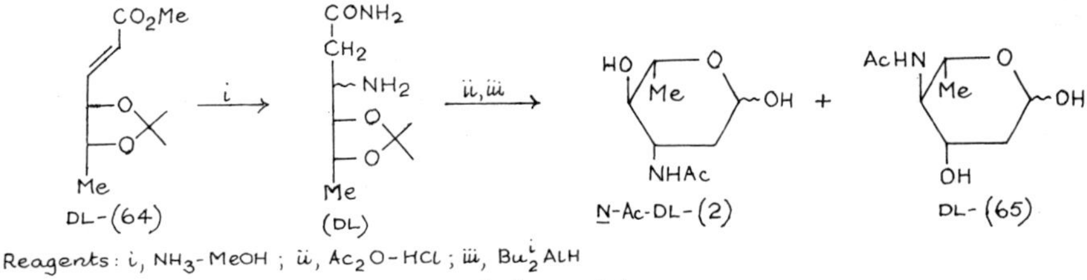

Scheme 24

amino-sugar involves an as yet unexplained intramolecular rearrange-ment during the lactone formation (step ii, Scheme 24), but its identity was confirmed through an X-ray structure determination of a derivative.[41] Methyl 3-deoxy-3-methylamino-DL-arabinopyranoside (66) has been synthesized from sorbic aldehyde via an intramolecular N-sulphinylcarbamate Diels-Alder adduct (67) (Scheme 25), a

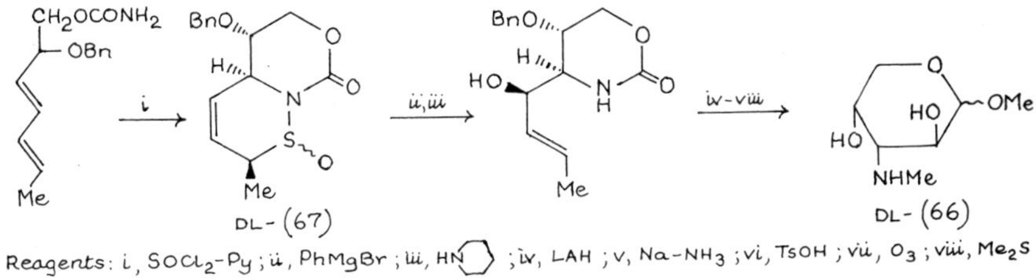

Scheme 25

strategy used previously for the synthesis of desosamine (Vol.18, p. 96).[42]

The synthesis of an optically active isoxazole derivative, which would be suitable for conversion into amino sugars, is detailed in Chapter 10.

## 3 Reactions

Several 4-alkylamino-4,6-dideoxyhexose derivatives of D-glucose, and of maltose, 2'-deoxymaltose, and 4-O-(α-D-galactopyranosyl)-D-glucose substituted in the non-reducing moiety, have been synthesized from the corresponding amino-sugars by reductive amination ($NaBH_3CN$) of ketones (e.g. acetone, cyclopentanone, and cyclohexanone) and aldehydes (e.g. benzaldehyde and 2,3:4,5-di-O-isopropylidene-D-arabinose), as part of a study on acarbose related α-glucosidase inhibitors.[43] $^{13}C$-N.m.r. data have been recorded for the Amadori rearrangement products 1-N-(p-methyl-, -ethyl-, and -methoxy-phenyl)amino-1-deoxy-D-fructoses, which are predominantly in the β-pyranose form.[44] Mass spectral investigations of certain dimethylamino-sugar derivatives are covered in Chapter 22.

Japanese workers have reported extensively on the synthesis of lipid A analogues, which contain 4-O-phosphoryl-2-amino-2-deoxy-D-glucose units N-acylated with fatty acids, 3-hydroxy- or 3-(fatty acyloxy)-fatty acid moieties. This work is covered in detail in Chapter 7, and to lesser extent in Chapter 3. N-Acyl derivatives of methyl 2-amino-2-deoxy-α-D-galactopyranoside have been synthesized conventionally from the free amine, in connection with a study of their binding affinities with isolectins.[45] Bis-N-(complex fatty acyl) derivatives of 6,6'-diamino-6,6'-dideoxy-α,α-trehalose have been synthesized as analogues of the mycobacterial glycolipid trehalose 6,6'-dimycolate and their lethal and adjuvant activities examined.[46] The oxazoline derivative (68) suffered elimination on glycosidation as shown in Scheme 26, whereas the unacetylated analogue (69) did not undergo elimination.[47]

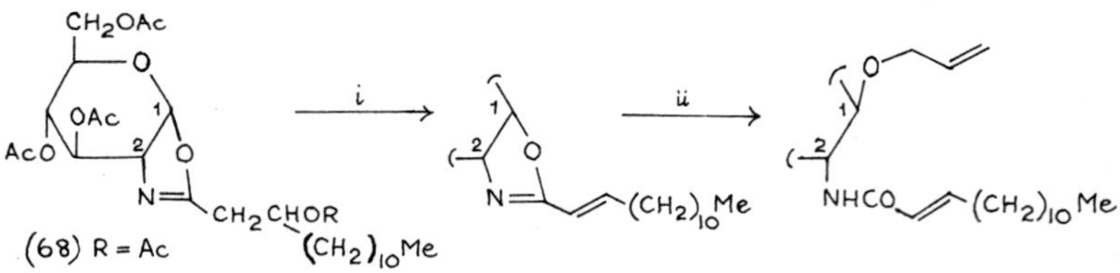


(68) R = Ac

(69) R = H

Reagents: i, TsOH; ii, allyl alcohol (CH2=CHCH2OH)

Scheme 26

Binding of gadolinium(III) ion to two core O-linked glycopeptides [i.e. β-D-Gal-(1 → 3)-α-D-GalNAc-Ser and -Thr] has been shown by $^{13}$C-n.m.r. and spin-lattice relaxation time data to be close to C-2 of the acetamido-sugar moieties.[48] As part of a study on the dissolution of chitin, the effect of lithium chloride on the $^{1}$H-n.m.r. spectra of 2-acetamido-2-deoxy-D-glucose and methyl β-D-chitobioside in N,N-dimethylacetamide has been examined.[49] The influence of trifluoroacetylation on the $^{13}$C-n.m.r. chemical shifts of 2-acetamidosugars is covered in Chapter 21.

The thiourea derivatives (70) have been synthesized by reaction of arylamines with the corresponding sugar isothiocyanate.[50]

Pyrolysis of 2-acetamido-2-deoxy-D-glucose *in vacuo* gives a tar from which 3-acetamidofuran, 3-acetamido-5-acetylfuran, and acetamidoacetaldehyde have been isolated; 3-acetamido-5-methylfuran, acetamido-substituted 2- and 4-pyrones, and hydroxydihydro-pyran-4-one have been tentatively identified, in addition to acetamide, by pyrolysis capillary g.c.-m.s.[51]

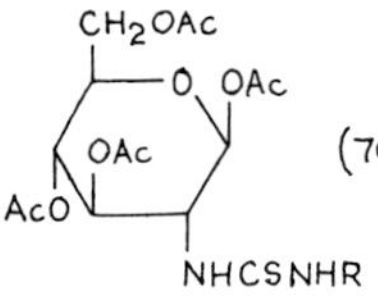

(70) R = $\cdot C_6H_4X$, or 2-naphthyl ; X = EtO, MeO, or $NO_2$

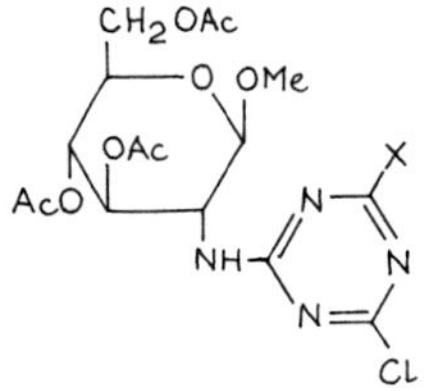

(71) X = Cl or

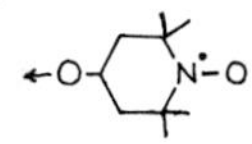

Using two model oligosaccharide alditols, factors affecting the reactivity of 2-acetamido-2-deoxy-hexopyranosides in the hydrazinolysis-nitrous acid deamination sequence have been investigated. Hydrazinolysis in the presence of hydrazinium sulphate led to virtually quantitative N-deacetylation.[52] Another reference to this procedure can be found in Chapter 10.

The 1,3,5-triazine derivatives (71), useful as spectroscopic probes, have been synthesized by displacement of chlorine from 2,4,6-trichloro-1,3,5-triazine.[53] A variety of 3-aryltriazen-1-yl sugar derivatives, such as the xylose derivative (72) or its N-methyl analogue (73), have been synthesized from the corresponding amino- and methylamino-sugars, and their tautomeric or rotameric

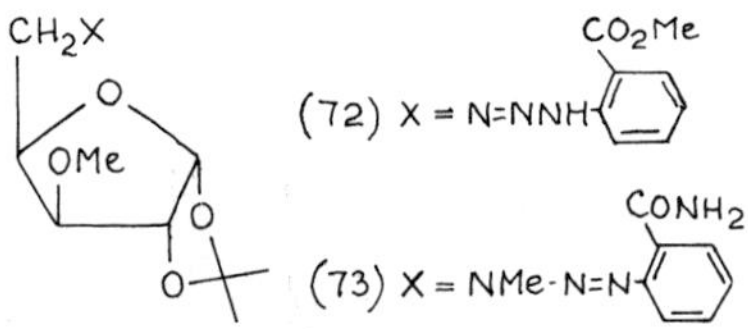

(72) X = N=NNH-

(73) X = NMe-N=N-

properties have been investigated by n.m.r. methods; none of these derivatives had anticancer properties (c.f. earlier work by Tronchet and Rachidzadeh, Helv. Chim. Acta, 1979, 62, 971).[54]

2-Amino-2-deoxy-D-gulose (74) has been synthesized from 2-amino-2-deoxy-D-glucose via the glycoside derivative (75) (Scheme 27).[55]

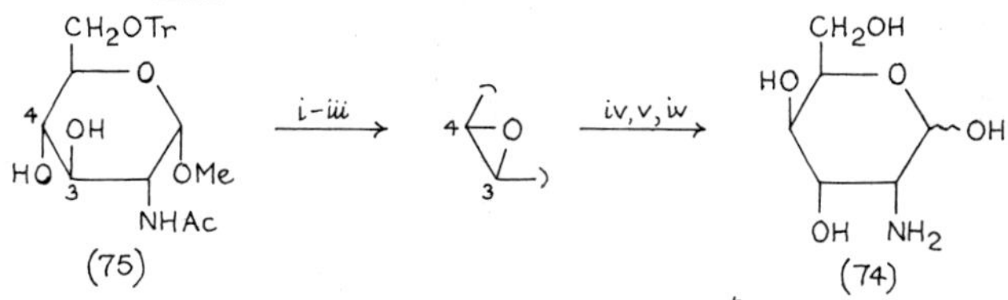

Reagents: i, BzCl-Py ; ii, MsCl-Py ; iii, NaH-DMF ; iv, $H_3O^+$ ; v, $Ac_2O$-Py

Scheme 27

2-Acetamido-1,6-anhydro-2,3-dideoxy-β-D-ribo-hexopyranose (76) has been synthesized from the corresponding 2-amino-2-deoxy-D-glucose derivative (77), via the thiazoline derivative (78) (Scheme 28).[56]

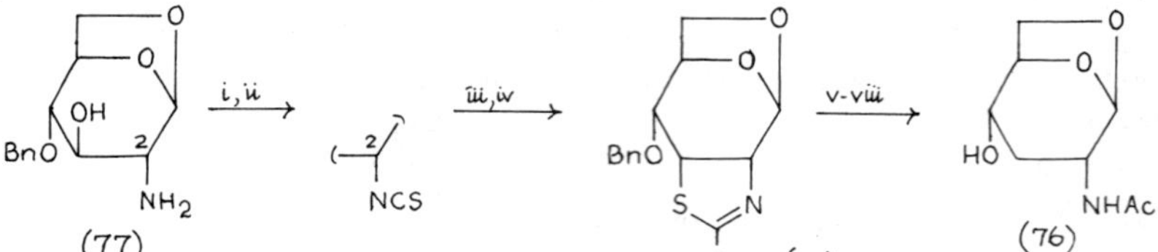

(77) (78) (76)

Reagents: i, $CS_2$; ii, $I_2$; iii, TsCl-Py; iv, $Et_3N$-MeOH; v, HCl-EtOH; vi, NaOH; vii, $Ac_2O$; viii, Raney Ni

Scheme 28

References to branched-chain amino-sugars and to amino-sugar nucleosides can be found in Chapters 19 and 20 respectively.

## 4 Diamino-Sugars

Azide displacement of the 3-O-mesylate group in the 2-amino-sugar (79) has yielded the azido-sugar derivatives (80)-(82) in 31, 7, and 18% yields respectively (Scheme 29), indicating that the C-2

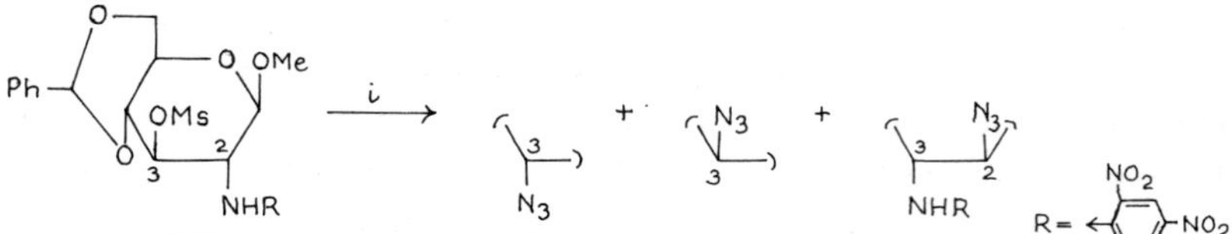

(79) (80) (81) (82)

Reagents: i, $NaN_3$-DMF

Scheme 29

substituent competes with azide as a nucleophile to form an intermediate 2,3-aziridine. Several related reactions and products were detailed.[57] Further details on the synthesis of racemic diaminosugars, e.g. (83) and (84), from 1,2-dihydropyridines and nitrosobenzene (c.f. Vol.16, p.94) have been published.[58]

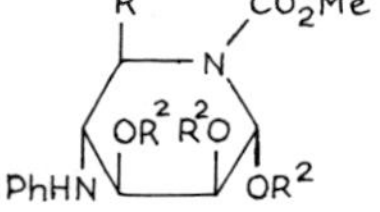

(83) $R^1 = R^2 = H$
(84) $R^1 = Me, R^2 = Ac$

## References


1 S.K.Adesina and J.Reisch, Phytochemistry, 1985, 24, 3003.
2 K.Tachibana, M.Sakaitani, and K.Nakanishi, Tetrahedron, 1985, 41, 1027.
3 M.M.Bandurraga and W.Fenical, Tetrahedron, 1985, 41, 1057.
4 M.T.Estan, Rev. Agroquim. Tecnol. Aliment., 1984, 24, 529 (Chem. Abstr., 1985, 102, 201 169).

5 Y.A.Knirel, E.V.Vinogradov, A.S.Shashkov, N.K.Kochetkov, V.L'Vov, and B.A.Dmitriev, Carbohydr. Res., 1985, 141, Cl.
6 S.Kolka, Pol. J. Chem., 1984, 58, 689.
7 L.Benzing-Purdie and J.H.Nikiforuk, J. Carbohydr. Chem., 1985, 4, 15.
8 J.Fernandez-Bolanos, J.Galbis Perez, and F.Zamora Mata, An. Quim., Ser. C, 1983, 79, 206 (Chem. Abstr., 1985, 102, 62 541).
9 D.Picq, I.Drivas, G.Carret, and D.Anker, Tetrahedron, 1985, 41, 2681.
10 D.Picq and D.Anker, J. Carbohydr. Chem., 1985, 4, 113.
10a D.Picq, G.Carret, D.Anker, and M.Abou-Assali, Tetrahedron Lett., 1985, 26, 1863.
11 H.Paulsen and W.Röben, Liebigs Ann. Chem., 1985, 974.
12 C.A.A.van Boeckel, T.Beetz, J.N.Vos, A.J.M.de Jong, S.F.van Aelst, R.H.van den Bosch, J.M.R.Mertens, and F.A.van der Vlugt, J. Carbohydr. Chem., 1985, 4, 293.
13 H.H.Baer, B.Radatus, and J.Defaye, Can. J. Chem., 1985, 63, 440.
14 H.H.Baer and B.Radatus, Carbohydr. Res., 1985, 144, 77.
15 J.S.Brimacombe, R.Hanna, and L.C.N.Tucker, Carbohydr. Res., 1985, 136, 419.
16 A.Malik, M.Roosz, and W.Voelter, Z. Naturforsch., B: Anorg. Chem., Org. Chem., 1985, 40B, 559 (Chem. Abstr., 1985, 103, 196 348).
17 W.Kowollik, A.Malik, N.Afza, and W.Voelter, J. Org. Chem., 1985, 50, 3325.
18 S.Jegge and J.Lehmann, Carbohydr. Res., 1985, 142, 47.
19 R.Csuk, Carbohydr. Res., 1985, 140, 167.
20 H.Paulsen, M.Brieden, and V.Sinnwell, Liebigs Ann. Chem., 1985, 113.
21 H.Paulsen, J.P.Lorentzen, and W.Kutschker, Carbohydr. Res., 1985, 136, 153.
22 J.Theim and D.Springer, Carbohydr. Res., 1985, 136, 325.
23 R.Andersson, I.Gouda, O.Larm, M.E.Riquelme, and E.Scholander, Carbohydr. Res., 1985, 142, 141.
24 J.Yoshimura, K.Sato, and R.B.Singh, Chem. Lett., 1985, 69.
25 T.M.Williams and H.S.Mosher, Tetrahedron Lett., 1985, 26, 6269.
26 Y.Yamamoto, M.Kirihata, I.Ichimoto, and H.Ueda, Agric. Biol. Chem., 1985, 49, 1435.
27 S.Brandänge and B.Lindgvist, Acta Chem. Scand. Ser. B, 1985, 39, 589.
28 S.Hanessian and J.Kloss, Tetrahedron Lett., 1985, 26, 1261.
29 P.J.Maurer, C.G.Knudsen, A.D.Palkowitz, and H.Rapoport, J. Org. Chem., 1985, 50, 325.
30 G.Fronza, C.Fuganti, P.Grasselli, and G.Pedrocchi-Fantoni, Carbohydr. Res., 1985, 136, 115.
31 A.Warm and P.Vogel, Tetrahedron Lett., 1985, 26, 5127.
32 S.J.Danishefsky and C.J.Maring, J. Am. Chem. Soc., 1985, 107, 1269.
33 S.J.Danishefsky, E.Larson, and J.P.Springer, J. Am. Chem. Soc., 1985, 107, 1274.
34 L.F.Tietze, E.Voss, K.Harms, and G.M.Sheldrick, Tetrahedron Lett., 1985, 26, 5273.
35 P.M.Wovkulich and M.R.Usković, Tetrahedron, 1985, 41, 3455.
36 V.Jäger and I.Müller, Tetrahedron, 1985, 41, 3519.
37 V.Jäger, I.Müller and E.F.Paulus, Tetrahedron Lett., 1985, 26, 2997.
38 P.G.Sammes and D.Thetford, J. Chem. Soc., Chem. Commun., 1985, 352.
39 M.Hirama, T.Shigemoto, Y.Yamazaki, and S.Itô, Tetrahedron Lett., 1985, 26, 4133.
40 M.Hirama, T.Shigemoto, and S.Itô, Tetrahedron Lett., 1985, 26, 4137.
41 J.T.Welch and B.-M.Svahn, J. Carbohydr. Chem., 1985, 4, 421.
42 S.W.Remiszewski, T.R.Stouch, and S.M.Weinreb, Tetrahedron, 1985, 41, 1173.
43 A.Köhn and R.R.Schmidt, Liebigs Ann. Chem., 1985, 775.
44 D.J.Nelson and M.J.Lavin, J. Carbohydr. Chem., 1985, 4, 91.
45 R.Kaifu, L.S.Plantefaber, and I.J.Goldstein, Carbohydr. Res., 1985, 140, 37.
46 F.Numata, K.Nishimura, H.Ishida, S.Ukei, Y.Tone, C.Ishihara, I.Saiki, I.Sekikawa, and I.Azuma, Chem. Pharm. Bull., 1985, 33, 4544.
47 C.Diolez, S.R.Sarfati, and L.Szabó, J. Chem. Soc., Perkin Trans. 1, 1985, 57.
48 K.Dill, H.K.Lannom, M.Denarie, J.M.Lacombe, and A.A.Pavia, Carbohydr. Res., 1985, 142, 11.

49 M.Vincendon, Makromol. Chem., 1985, 186, 1787 (Chem. Abstr., 1985, 103, 215 687).
50 J.Fuentes Mota, M.C.Ortiz Mellet, F.Segura Ramos, M.A.Pradera Adrian, and A.Cert Ventula, An. Quim., Ser. C, 1983, 79, 221 (Chem. Abstr., 1985, 102, 46 208).
51 R.A.Franich, S.J.Goodin, and A.L.Wilkins, J. Anal. Appl. Pyrolysis, 1984, 7, 91 (Chem. Abstr., 1985, 102, 149 678).
52 Y.Inoue and K.Kitajima, Carbohydr. Res., 1985, 135, 342.
53 M.J.Adam and L.D.Hall, Carbohydr. Res., 1985, 144, 23.
54 J.M.J.Tronchet, F.Barbalat-Rey, J.F.Tronchet, and F.Rachidzadeh, J. Carbohydr. Chem., 1985, 4, 193.
55 T.Suami, K.Tadano, Y.Iimura, and H.Tanabe, Carbohydr. Res., 1985, 135, 319.
56 T.Elbert and M.Cerny, Collect. Czech. Chem. Commun., 1985, 50, 2000.
57 W.Meyer zu Reckendorf and H.-J.Lenzen, Liebigs Ann. Chem., 1985, 477.
58 G.Augelmann, J.Streith, and H.Fritz, Helv. Chim. Acta, 1985, 68, 95.

# 10
# Miscellaneous Nitrogen Derivatives

## 1 Glycosylamines

2,3,4,6-Tetra-O-acetyl-β-D-gluco- and -galacto-pyranosylamines (1) have been obtained from the corresponding glycosyl azides (2) by Staudinger reaction followed by hydrolysis of the intermediate phosphinimines (3); alternative reactions of intermediates (3) with carbon dioxide yielded the 1,2-*trans*-carbamates (4) (Scheme 1).[1] Attempts to react the D-glucosyl derivative (1) with

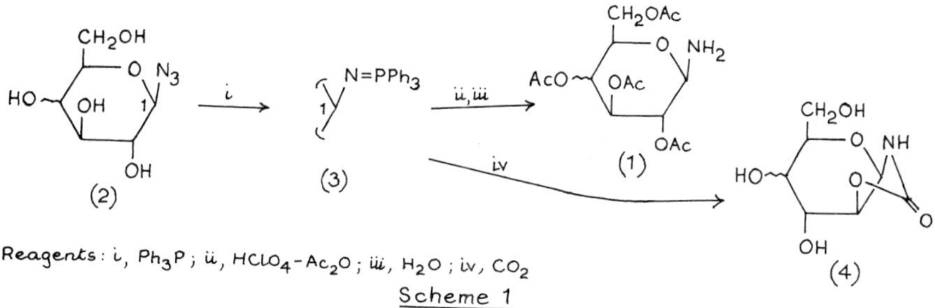

Reagents: i, $Ph_3P$; ii, $HClO_4$-$Ac_2O$; iii, $H_2O$; iv, $CO_2$

Scheme 1

α-halocarboxylic acid esters (with $NaHCO_3$ in $Me_2CO$) led instead to bis(2,3,4,6-tetra-O-acetyl-β-D-glucopyranosyl)amine. The same report described the Schiff bases formed by reaction of compound (1) with substituted benzaldehydes.[2] The synthesis of 1-N-alkyl-2-alkylamino-2-deoxy-D-glucopyranosylamines from D-fructose and fatty amines and their hydrolysis to the free 2-alkylaminosugars have been described.[3] Anomer effects in 2-substituted tetrahydropyrans have been determined by variable-temperature n.m.r. techniques; 2-methylamino substituent prefers the equatorial position, whereas other substituents (2-Cl, -OMe, or -OH) prefer the axial position.[4] $^{13}$C-N.m.r. data on N-(*m*- and *p*-substituted phenyl)-β-D-glucopyranosylamines and their peracetates have been reported.[5] A kinetic study on the hydrolysis of piperidine N-glycosides at pH 5.33 has shown the following order of decreasing stability: β-D-glc>α-D-man>α-D-gal>β-D-rib>arabinosides.[6] N-Aryl-β-D-glucosylamine peracetates and their corresponding methyl uronate analogues have been synthesized by trimethylsilyl triflate-catalyzed reaction of peracetylated β-glycopyranoses with N-acetyl-arylamines.[7] Previous work on the synthesis and conformation of N-phenyl-2,3,5-tri-O-(p-nitrobenzoyl)-β-D-ribofuranosylamine (Vol. 12, p.82) has been extended to include p-chloro- and m-nitrophenyl analogues and their N-acetyl derivatives.[8] Attempted displacements of the 2-mesyloxy groups in N-acetyl-N-aryl-β-D-xylopyranosyl-amines (5) with azide yielded products (6)-(8) due to participation of the N-acetyl group or by O-S bond cleavage (Scheme 2);

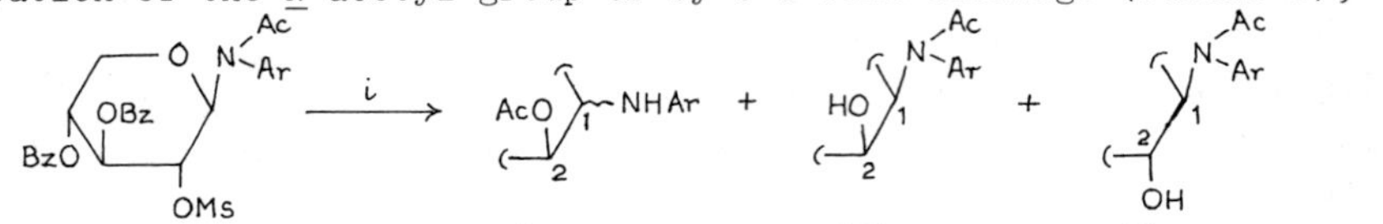


(5) (6) (7) (8)

Reagents: *i*, $NaN_3$-DMF

Ar = p-C₆H₄-X, X = Cl or OMe

Scheme 2

in contrast, conventional displacement with inversion occurred in a related 4-mesylate and in the C-2 epimer of (5).[9]

α-L-Arabinofuranosyl pyridinium salt, its tri-O-benzoate, and a variety of analogues with a 4-bromoisoquinoline or substituted pyridine moieties have been synthesized conventionally and their conformations studied. Hydrolysis of the 4-bromoisoquinolinium salt has been examined in detail, particularly under alkaline conditions.[10]

Nephritogenoside (9), the first example of a glycopeptide with

$$\alpha\text{-D-Glc-}(1\rightarrow6)\text{-}\beta\text{-D-Glc-}(1\rightarrow6)\text{-}\alpha\text{-D-Glc-}\underset{\mathbf{H}}{\text{N}}\overset{\overset{\text{O}}{\|}}{\text{C}}\text{-(peptide)} \qquad (9)$$

an α-glycosylamine linkage, has been isolated from the glomerular basement membrane of rats.[11] Model glycopeptides with N-(peracetylated 2-acetamido-2-deoxy-β-D-glucopyranosyl)ated aspartic acid moieties have been studied by $^1$H-n.m.r., but no specific intramolecular interactions were detected.[12] The hydrazinolysis-N-reacetylation of 2-acetamido-1-N-(L-aspart-4-oyl)-2-deoxy-β-D-glucopyranosylamine has been studied as a model for the analogous treatment of glycopeptides. The glycosylhydrazine (10) has been identified as the intermediate formed during the N-reacetylation step, which gives rise to 2-acetamido-2-deoxy-D-glucose on final mild hydrolysis.[13]

Reaction of O-(tetra-O-benzyl-α-D-glucopyranosyl)trichloroacetimidate with o-chlorobenzoic acid in acetonitrile gave, in

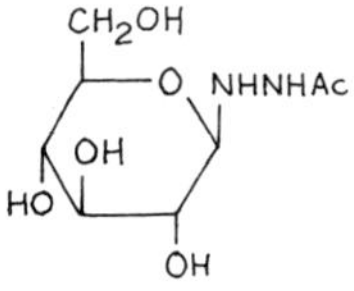

(10)

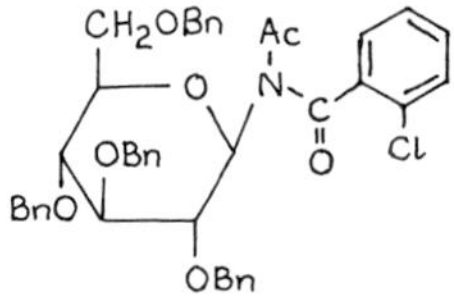

(11)

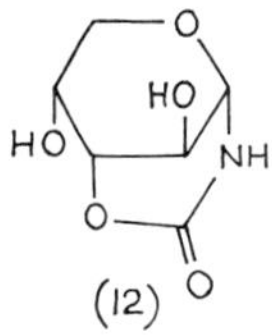

(12)

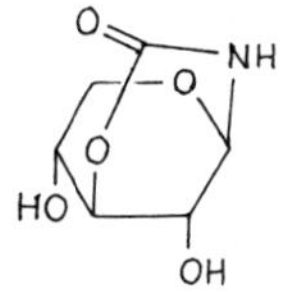

(13)

addition to the expected 1-O-acyl derivative (24% yield), the di-N-acylated glucosylamine derivative (11) (56% yield) incorporating a molecule of solvent.[14] The cyclic carbamates (12), (13), and D-gluco-(4) have been obtained on treatment of 3-glycosyl-1-methyl-1-nitrosoureas containing α-D-arabino-, β-D-xylo-, and β-D-gluco-pyranosyl moieties, respectively, with aqueous alkali.[15] Hydrolysis of the anhydronucleoside (14; X = O) led to the β-D-arabinofurano-oxazolidine derivative (15), whereas its thio-analogue (14; X = S) gave the imine (16) (Scheme 3).[16]

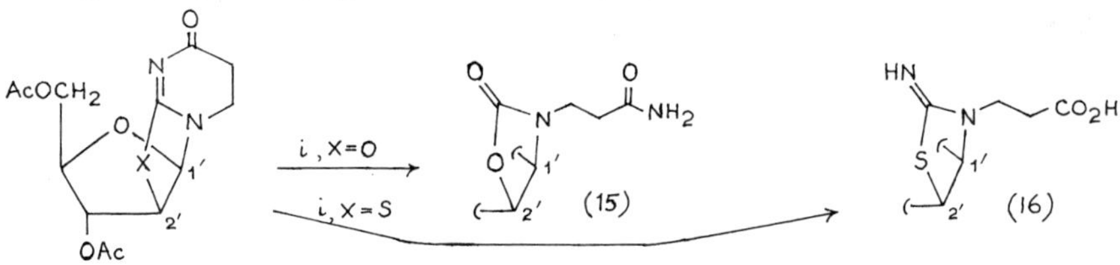

(14) X = O or S

Reagents: i, HOAc-$H_2O$

Scheme 3

The N-glycosides (17) have been obtained from the reaction of xylose or glucose with 4-aminopyrimidines,[17] and the thermal behaviour of their 5-nitroso derivatives (18) has been examined.[18] Exposure of DNA to the carcinogen (19) gives the 2'-deoxy analogue of compound (20) which is a ring-opened guanosine derivative. Compound (20) has therefore been synthesized by coupling a ribosylamine derivative to a suitable chloropyrimidine derivative, to pursue studies of the DNA-carcinogen reaction.[19] 6-(D-Glucopyranosylimino)-1,3,5-thiadiazine derivatives have been synthesized by reaction of the corresponding dithiobiurets [RNHC(SBn)=NC(S)NH-

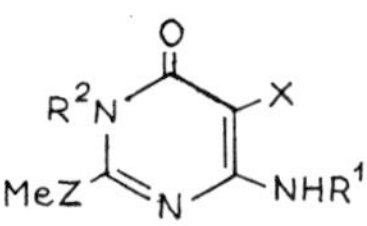

$R^1$ = β-D-Glc-p or β-D-Xyl-p.
$R^2$ = H or Me ;

(17) X = H
(18) X = NO
Z = O or S

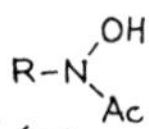

(19)

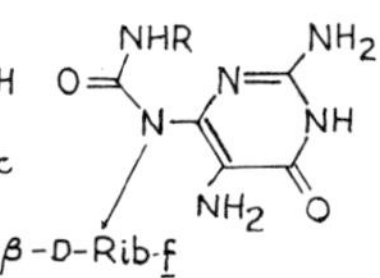

(20)

R = 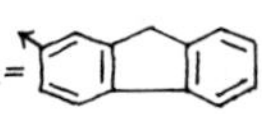

Glc] with phenyl isocyanide dichloride ($PhN{=}CCl_2$),[20] and an N-glucosylated xanthine derivative has been obtained by cyclization of a 6-acetamido-5-(β-D-glucopyranosylacetamido)-uracil derivative.[21] Other N-glycosylated heterocycles are covered in sections 5 and 6, and Chapter 20.

A triterpene glycoside isocyanate derivative derived from glycyrrhizic acid has been converted into a variety of corresponding urea and carbamate derivatives.[22]

N-Glycosyl-thiourea derivatives of D-glucose, D-galactose, and D-glucosamine have been synthesized by coupling peracetylated β-D-glycopyranosyl isothiocyanates with amines[23] and α-aminoketones;[24] with α-aminoacetone, cyclization occurred to give the imidazole nucleoside analogue (21), which could be desulphurized with Raney nickel. Attempted cyclization ($Me_2CO-H_3O^+$) of the isothiocyanate-derived thiourea (22), which was expected to yield a 5,6-dihydro-

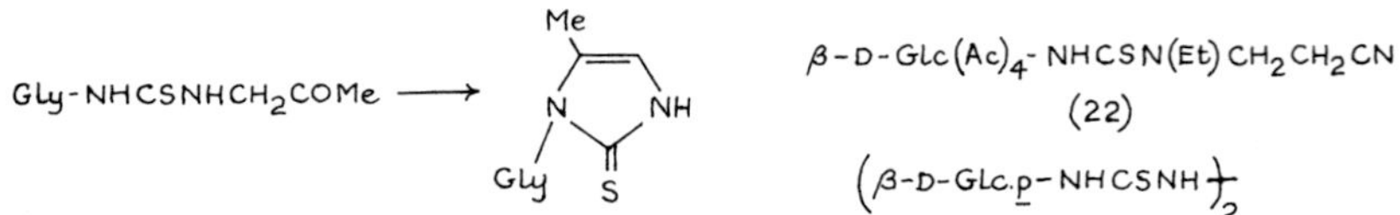

(21) (22) (23)

2-thiouracil derivative, was not successful, leading only to hydrolysis of the nitrile function to an amide.[25] Di-N,N'-β-D-gluco- and galacto-pyranosyl-dithiocarbamyl hydrazines, e.g. (23), have been synthesized by coupling peracetylated glycosyl isothiocyanates and 3-(glycosyl)thiosemicarbazides.[26] The method by which enantiomeric amines can be converted into diastereoisomeric thiourea derivatives for h.p.l.c. analysis by reaction with 2,3,4,6-tetra-O-acetyl-β-D-glucopyranosyl isothiocyanate (c.f. Vol.18, p.107) has been extended to the analysis of sulphate conjugates of adrenergic drugs such as 4'-hydroxypropranolol sulphate,[27] the antiarrhythmic drug mexiletine [1-(2,6-dimethylphenoxy)-2-aminopropane],[28] the enzymatic hydroxylation products of N-acetyldopamine,[29] and chiral oxirans following reaction with simple volatile alkylamines.[30]

Glycosyl isoselenocyanates have been synthesized from the isocyanides, as shown in Scheme 4 for the D-glucosyl derivative (24). They are air sensitive, unstable at room temperature, and could not be deacetylated. They could be converted into the somewhat more stable selenoureas (e.g. 25) or reduced to 1,5-anhydrohexitols (e.g. 26). 2,3,4,5,6-Penta-O-acetyl-D-gluconyl isoselenocyanate was obtained from the corresponding acid chloride and potassium

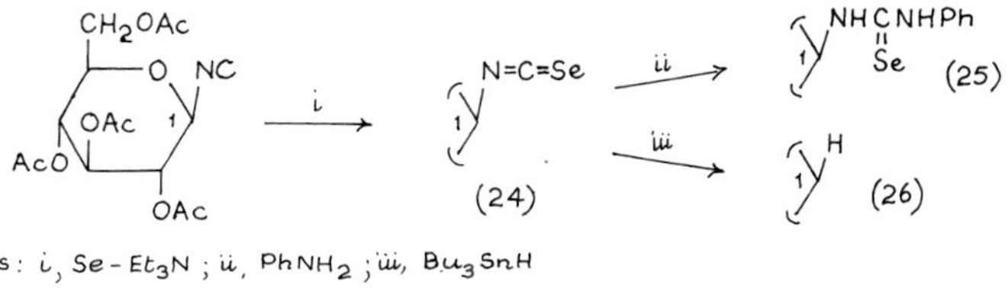


Scheme 4

selenocyanate.[31]

The formation and cleavage of four-membered ring lactams from [2+2]cycloaddition of p-toluenesulphonyl isocyanate to glycals under high pressure is covered in Chapters 13 and 14.

Amadori rearrangements of glycosylamines are covered in Chapter 9. Several studies have examined the Maillard reaction. Volatile products from reaction of D-glucose with butylamine[32] and with glycine[33] have been examined by g.c.-m.s. Melanoidins have been isolated from the reaction of glucose with glycine using strong anion exchange resins.[34] Those non-dialysable melanoidins formed from mono- and di-saccharides with L-tryptophan have been shown to be mutagenic only after nitrite treatment.[35] Maillard polymers have been synthesized from D-[1- and 6-$^{14}$C]glucose and [1- and 2-$^{14}$C]glycine, as well as [1- and methyl-$^{14}$C]methionine. Radioactivities of the isolated polymers were monitored as a function of time. D-Glucose carbons showed the largest incorporation, and C-1 of the aminoacids the lowest, with the evolved $CO_2$ arising primarily from this latter carbon atom.[36]

## 2 Azido- and azi-sugars

Azido-sugars are often intermediates in the synthesis of amino-sugars (see Chapter 9, and ref.9 above).

Reaction of O-(tetra-O-benzyl-α-D-glucopyranosyl)trichloroacetimidate with hydrazoic acid gave the α-glycosyl azide (61% yield).[14] 6-Deoxyhexopyranosyl azides have been made from the corresponding 6-deoxy-glycosyl acetates (with $Me_3SiN_3$-$SnCl_4$) and their conformations probed by $^1$H-n.m.r. spectroscopy.[37] The $^{13}$C-n.m.r. spectra of a number of glycosyl azides (some synthesized for the first time), including some 6-deoxy- and 2-acylamido-2-deoxy- examples, and of some 2- and 3-azido-monosaccharide derivatives have been unambiguously assigned using 2D-n.m.r. techniques. Values for $^1J_{C,H}$ were determined, coupling involving an equatorial hydrogen being larger by 4-9 Hz than that involving an axial hydrogen.[38]

1,6-Anhydro-2-azido-2-deoxy-β-D-glucopyranose derivatives have been synthesized from 1,6-anhydro-β-D-mannopyranose,[39] and 3-O-acylated 4-azido-2-O-benzyl-4,6-dideoxy-α-D-galactopyranosyl bromides (building blocks for oligosaccharide synthesis) from methyl 2-O-benzyl-4,6-O-benzylidene-D-glucopyranoside,[40] by conventional procedures involving sulphonate displacements with azide ion. The 3,4-diazide (27) was the major product (77%) from reaction of β-L-*ribo*-epoxy-tosylate (28) with azide ion, the minor 2,4-diazide (29) resulting from the alternative opening of the epoxide at C-2 (Scheme 5). Analogous reaction of the α-D-*ribo*-

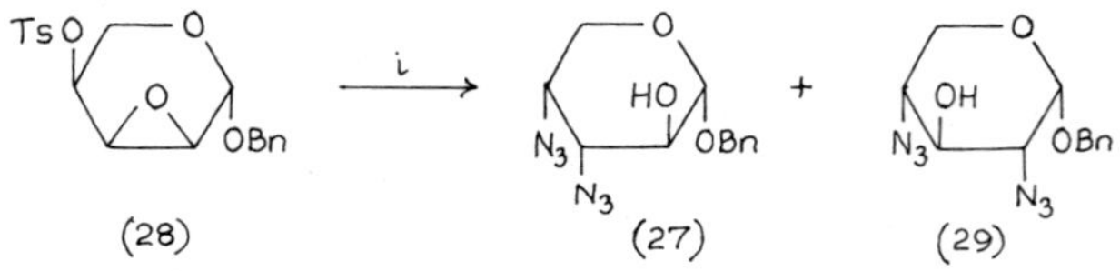


Reagents: i, $NaN_3$-DMF

Scheme 5

epoxy-tosylate [*i.e.* the α-anomer of (28), but in the D-series] gave similar amounts of 3,4- and 2,4-diazides.[41] α-Azidoalcohols have been obtained with high stereoselectivity by the substitution of cyclic sulphites (the synthesis of which is covered in Chapter 7) by azide ion ($NaN_3$-DMF, 110°C). Thus L-arabinoside (30) gave the 4-azide (31) in 92% yield, and the D-ribofuranoside (32) gave the 3-azide (33) in 63% yield.[42]

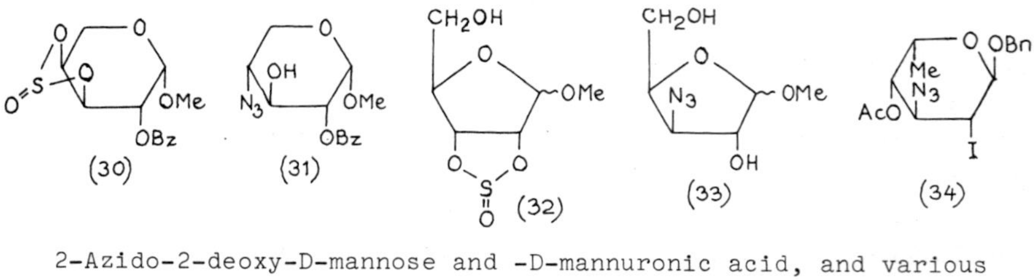


2-Azido-2-deoxy-D-mannose and -D-mannuronic acid, and various derivatives including glycosyl bromides, have been synthesized as building blocks for bacterial polysaccharide sequences *via* azidonitration of glycal derivatives.[43] Benzyl 3-azido-2,3-dideoxy-2-iodo-hexopyranoside derivatives [*e.g.* (34) from L-fucal diacetate] have been synthesized from 3-azidoglycals (available from addition-allylic rearrangement of acetylated glycals with azide; Vol.15, p.94) by reaction with benzyl alcohol-N-iodo-succinimide; conversion of these 2-iodides into 2,3-unsaturated derivatives is detailed in Scheme 8 of Chapter 9.[44] The synthesis of novel azido analogues of anthracycline antibiotics is covered in Chapter 19.

Reactions of 6-azido-6-deoxy-α-D-galactose and methyl 2-azido-2-deoxy-α-D-altropyranoside with triphenylphosphine (the Staudinger reaction) has been shown to yield unstable sugar phosphinimines analogous to those formed from the glycosyl azides in Scheme 1.[1]

4-Azido-4-deoxy-D-galactose, 4-azi-4-deoxy-D-xylo-hexopyranose (35), the related diazirino systems (36) and (37), and 4-O-(2-diazo-3,3,3-trifluoropropionyl)-D-glucose (38) have been synthesized

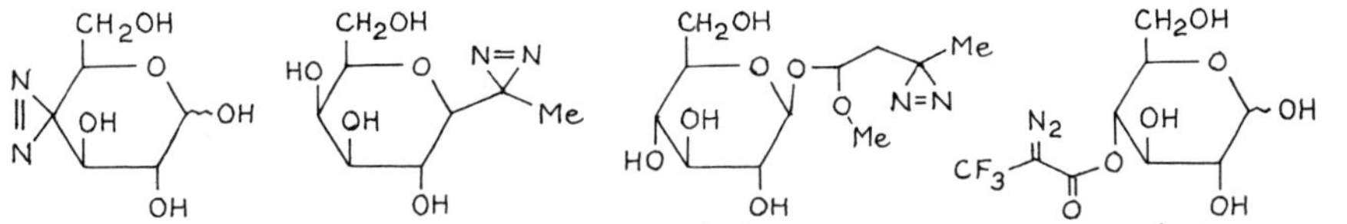

(35) (36) (37) (38)

and tested as photolabile sugar derivatives for affinity labelling of sugar-binding proteins; the best results were obtained with the diazirino systems (35)-(37), which were obtained from the carbonyl precursors (using $NH_3$-$NH_2OSO_3H$, then $I_2$-$Et_3N$).[45]

## 3 Nitro- and Nitroso-sugars

Nitro-sugars have been synthesized by addition of nitromethane to four aldehydo-sugar derivatives, and their stereochemistry investigated. The 6-nitro-sugar (39), obtained as a single isomer under mild conditions, was converted to the range of novel derivatives shown in Scheme 6, via the nitroenose (40).[46]

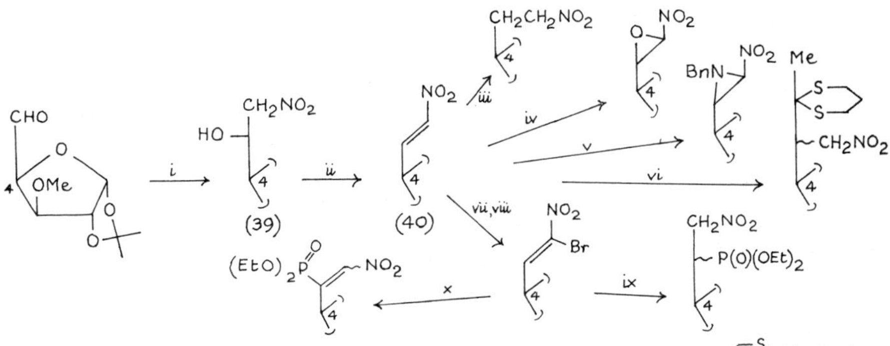

Reagents: i, $MeNO_2$-MeONa; ii, $Ac_2O$-NaOAc; iii, $NaBH_4$; iv, $H_2O_2$-NaOH; v, $BnNH_2$; vi, 2-methyl-1,3-dithiane, BuLi; vii, $Br_2$; viii, $Ag_2CO_3$; ix, $(EtO)_3P$ - PhMe; x, $(EtO)_3P$ - EtOH

Scheme 6

The branched-chain nitro-sugars L-rubranitrose (2,3,6-trideoxy-3-C-methyl-4-O-methyl-3-nitro-L-*xylo*-hexose, 41), its D-enantiomer, and a derivative of D-kijanose (2,3,4,6-tetradeoxy-4-methoxycarbonylamino-3-methyl-3-nitro-α-D-*xylo*-hexose, 42) have been synthesized as outlined in Schemes 7 and 8.[47,48] Other branched-chain

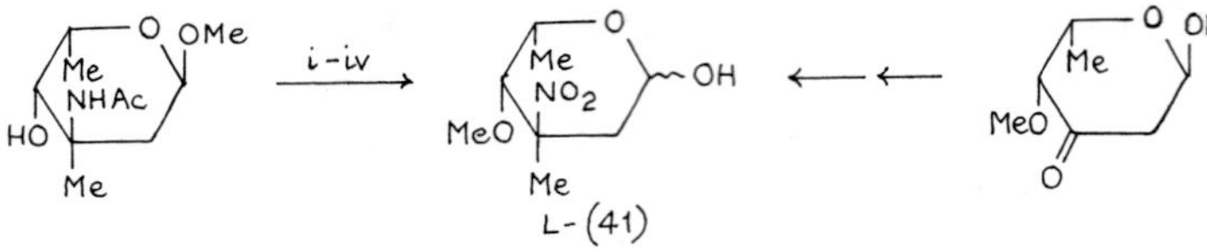

Reagents: i, MeI-NaH; ii, Ca-$NH_3$(l); iii, MCPBA; iv, $H^+$

Scheme 7

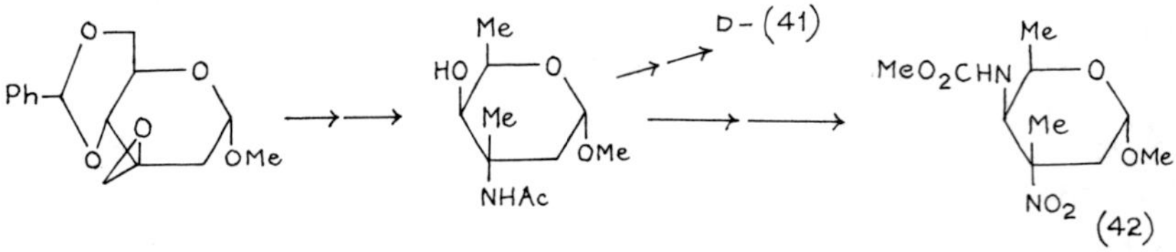

Scheme 8

nitro-sugars, including L-decilonitrose, are covered in Chapter 14.

Displacement of nitro-groups (with $Bu_3SnH$) in ketose derivatives such as compound (43) provided sugar radicals which have been condensed *in situ* with acrylonitrile to yield di-C-glycosides such as compound (44).[49]

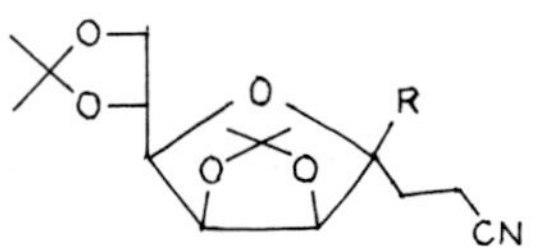

(43) R = $NO_2$

(44) R = ⌄⌄CN

Acyclic sugar nitro-olefins have been condensed with cyclohexane-1,3-diones to give bicyclic oxime products as shown in Scheme 9.[50] Various 1-deoxy-1-C-nitro-pentitols and -heptitols

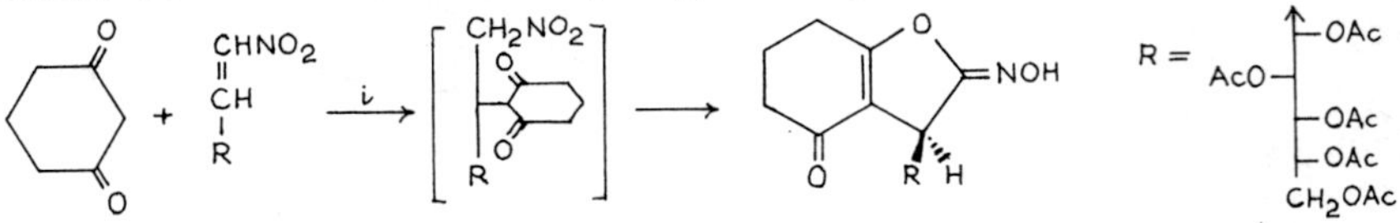

Reagents: i, $Et_3N$-MeOH

Scheme 9

and their benzylidene derivatives have been synthesized from D-glucose.[51]

Reactions of 1-C-nitroglycosyl halides and 1-C-nitroglycosyl sulphones (*e.g.* 45) with dialkylphosphite anions proceed by competing nucleophilic attack of the anion on the nitro group leading ultimately to the nitrile phosphates (e.g. 46) and single electron transfer to give 1-C-nitroglycosyl phosphonates (e.g. 47),

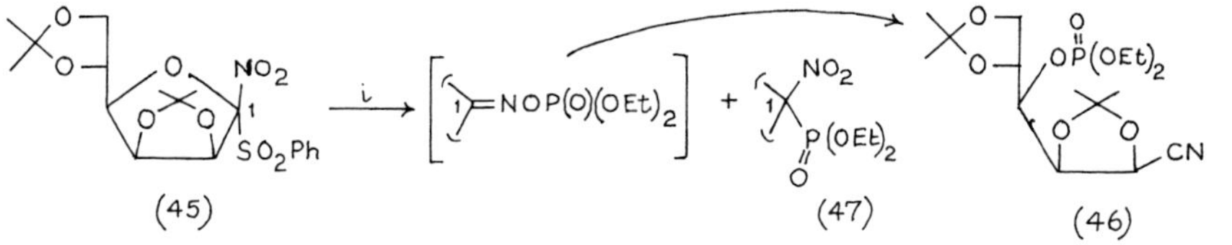

Reagents: i, $KP(O)(OEt)_2$-DMSO-hν

Scheme 10

as exemplified in Scheme 10; while sulphone (45) gave mainly the phosphonate (47), its anomer gave mainly the nitrile (46).[52]

## 4 Nitriles, Oximes, and Hydroxylamines

All four 2,3,5-tri-O-benzoyl-D-pentofuranosyl cyanides with the 1,2-trans-stereochemistry have been synthesized from the correspond-ing 1-O-acetates (with $Me_3SiCN$-$SnCl_4$); their c.d. spectra have been interpreted by the exciton chirality method.[53] The synthesis of α-aminonitriles by addition of hydrogen cyanide and alkylamines to D-glucose, D-galactose, and D-mannose has been reexamined, and new products have been obtained depending upon the conditions as well as the sugar. As shown in Scheme 11, the α-aminonitriles (48) can be transformed into the lactone-imines (49) or (50), or

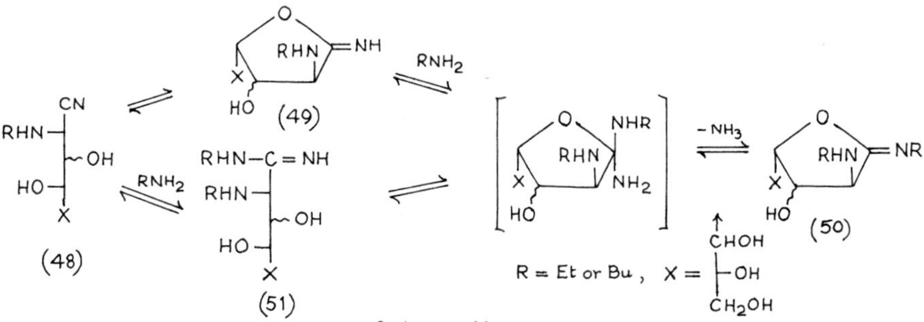

Scheme 11

into the heptonamidines (51).[54] The synthesis of an aldononitrile derivative is covered in the preceeding section on nitro-sugars.

A variety of oxidants (i.e. $MnO_2$, $Hg(OAc)_2$-$Na_2CO_3$, or $O_2$-$Cu_2Cl_2$-py) have been used to convert aldose oximes into aldono-lactone oximes. Free sugar oximes gave products such as the D-gluco example (52) with a Z-configuration, whereas 2,3:5,6-di-O-isopropylidene-D-mannose yields an E/Z mixture (c.f. Vol.16, p.111 and Vol.18, p.111). Beckmann rearrangements were demonstrated

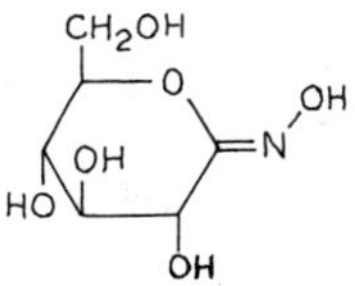


(52)

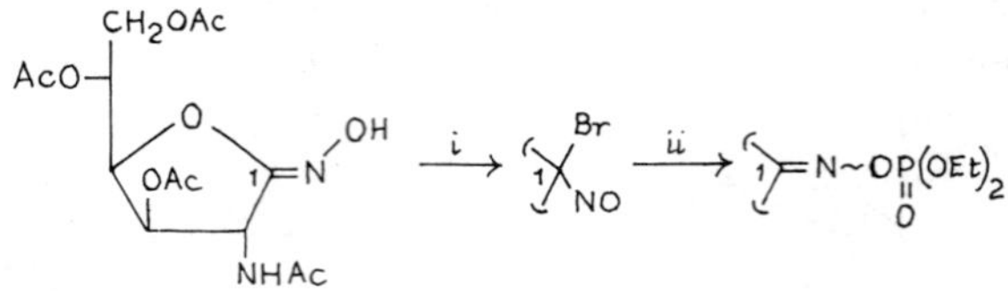


Reagents: i, NBS-NaOAc-$H_2O$; ii, NaP(O)$(OEt)_2$

Scheme 12

with two Z-isomers to yield ring-expanded products. Aldonolactone oxime phosphates were prepared via bromonitroso compounds as exemplified in Scheme 12.[55]

Vasella and co-workers have examined nucleophilic additions of dialkylphosphite anions [e.g. LiP(O)$(OMe)_2$] to N-glycosyl-nitrones, e.g. (53) and (54), leading to N-glycosyl-hydroxylamine deriva-

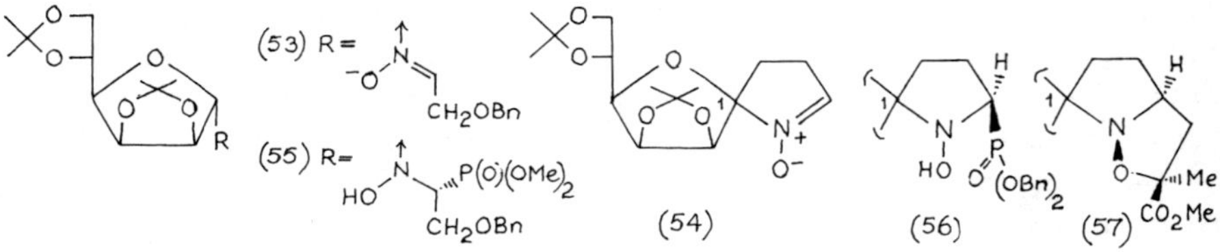


tives, e.g. (55)[56] and (56)[57] respectively, with high stereo-selectivity. Chiral α-aminophosphonic acids were synthesized from such adducts, e.g. (S)-phosphoserine from compound (55). 1,3-Dipolar cycloaddition of methyl methacrylate to nitrone (54) gave the tricyclic hydroxylamine (57).[57] These results indicated that reactants approach the nitrone so as to generate a lone pair on nitrogen syn-coplanar with the C(1)-O bond. Tronchet and co-workers have similarly stereoselectively synthesized α-hydroxy-amino-sugar phosphonates [e.g. (58)] from the sugar nitrone (59) (Scheme 13) and its C-3 epimer, which can be obtained from the

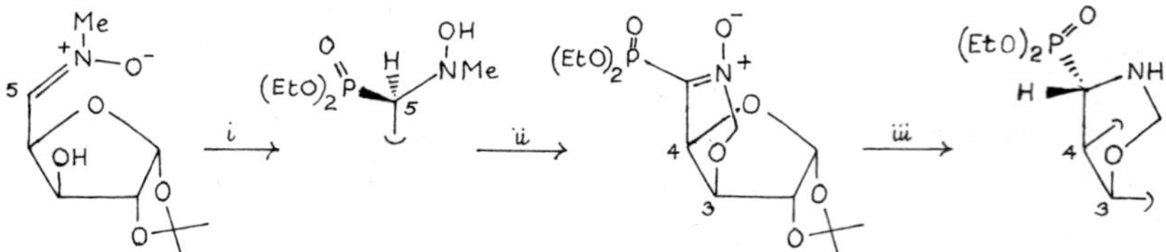


(59) (58) (60) (61)

Reagents: i, $(EtO)_2$P(O)H-NaOEt; ii, DDQ; iii, $NaBH_4$

Scheme 13

corresponding 5-aldehydo-sugar derivatives and N-methylhydroxylamine. Oxidation of hydroxylamine (58) yielded the cyclized nitrone (60), which gave the α-aminophosphonic acid derivative (61) on reduction. Air oxidation of hydroxylamine (58) and similar compounds gave nitroxide free radicals which were studied by e.s.r.[58]

Methyl 6-deoxy-6-hydroxyamino-α-D-glucopyranoside has been synthesized by oxidation of the 6-amino-glycoside to a 6-oxime ($H_2O_2$-Na tungstate), and reduction ($NaBH_3CN$); 6'-N-hydroxykanamycin A and 6'-N-hydroxydibekacin were similarly obtained.[59]

In connection with carcinogenicity studies, Yoshioka and coworkers have synthesized N-glycosylated N-arylhydroxylamines by direct condensation of the hydroxylamine with unprotected D-glucuronic acid salt or lactone.[60] The O-glycosylated analogues were obtained by the orthoester glycosidation method;[61,62] acid hydrolysis of the N-acetyl derivatives of the latter yielded D-glucono-1,5-lactone and the corresponding amine by redox fission.[63]

## 5 Hydrazines, Hydrazones, Osazones, and Derived Heterocycles

L-Gulose derivatives suitable for use in the synthesis of bleomycin have been obtained via 6-hydrazone derivatives of 6-aldehydo-D-glucose as shown in Scheme 14. The equilibrium between the D-

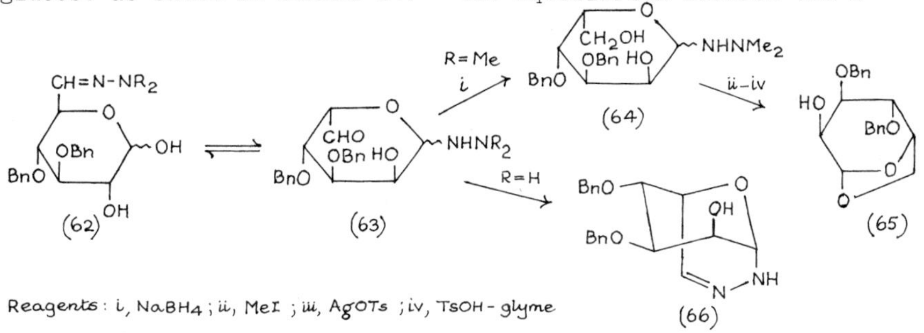


Scheme 14

gluco- and the L-gulo-derivatives (62) and (63) was shifted by reduction to the N-(L-gulosyl)-N',N'-dimethylhydrazine (64), which exists as a 7:3 mixture of the cyclic and acyclic forms, and was converted into the 1,6-anhydride (65). The unsubstituted hydrazone (62; R = H) formed the 1,6-(1-hydrazinyl-2-ylidene)gulose derivative (66).[64]

The glycosyl hydrazine encountered as an intermediate in the hydrazinolysis-re-N-acetylation of a model glycopeptide is covered in Section 1. N,N'-Di-(β-D-glucopyranosyl)-N-phenylhydrazine has been isolated in 70% yield from D-glucose phenylhydrazone in acetic acid.[65,66] The Amadori rearrangement product of glucose and valine, i.e. fructosyl valine, has been synthesized as a model compound for that formed from the haemoglobin β-chain (which is terminated by valine) and glucose; its phenylhydrazone derivative was shown to exist in the open-chain form in base and the cyclic form (67) in acid.[67]

A variety of 2-mono- and 2,3-di-hydrazone derivatives of dehydro -L-ascorbic acid have been synthesized, and some converted to bicyclic derivatives, e.g. (68), on oxidation with copper(II) chloride.[68] Copper(II) complexes of D-glucosone bisthiosemicarbazones have been synthesized and shown to have antitumour activity.[69]

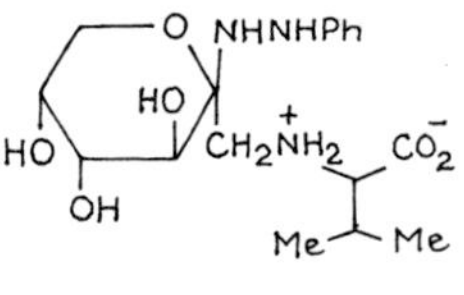


(67)

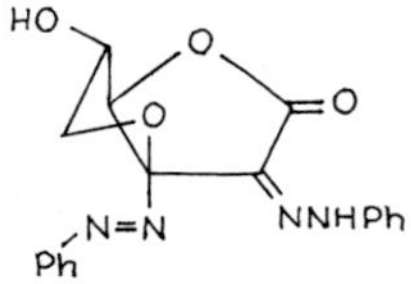


(68)

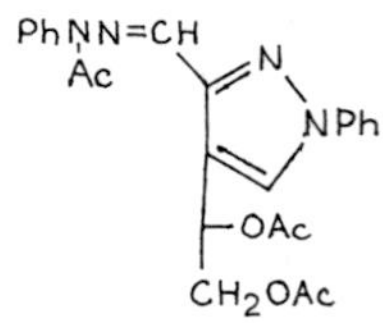


(69)

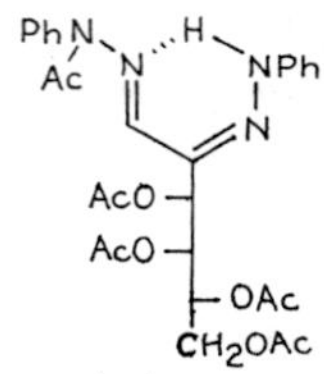


(70)

Conditions have been reported for mono- and di-N-acetylation of sugar osazones and related model compounds, the monoacetylation occurring on the non-chelated nitrogen.[70,71] Pyrazole derivative (69) has been obtained by reaction ($Ac_2O$-NaOAc) of the mono-N-acetylated osazone (70).[72] Other pyrazole derivatives, e.g. (71), have been synthesized by condensation of D-galactose phenylhydrazones with dimethyl acetylenedicarboxylate.[73]

## 6 Other Heterocyclic Derivatives

Alditol-1-yl-pyrazole derivatives have been synthesized by condensing aminosugars with β-diketones and related compounds. Further 2-(alditol-1-yl)-1-alkyl-4,5,6,7-tetrahydroindol-4-ones have been obtained from 2-alkylamino-2-deoxy-hexoses and -heptoses with 5,5-dimethyl-1,3-cyclohexadione (c.f. Vol.17, p.113).[74] While a 2-(D-arabino-tetritol-1-yl)-1-methylpyrazole derivative was obtained from the condensation of 2-deoxy-2-methylamino-D-glucose with ethyl acetoacetate (c.f. Vol.14, p.89), the 3-substituted pyrrole analogues (72) were obtained from 1-deoxy-1-methylamino-

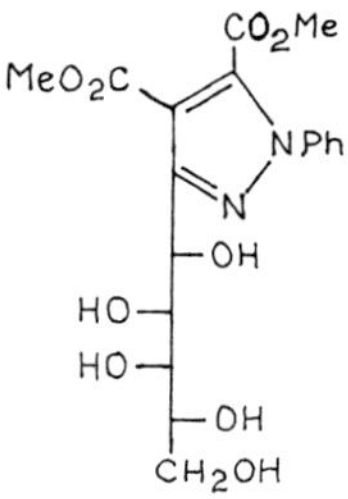

(71)

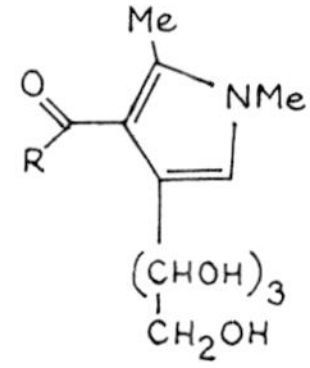

(72)

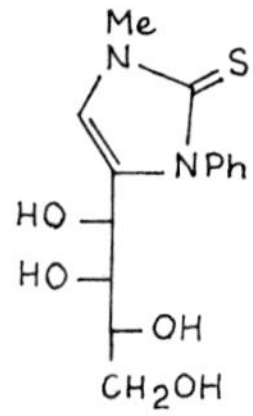

(73)

D-arabino- and -lyxo-hex-2-uloses with ethyl acetoacetate or pentane-2,4-dione.[75] Condensation of 1-amino-1-deoxy-D-lyxo-hex-2-ulose with phenyl isothiocyanate yielded the acyclic imidazole-2-thione (73) and its anomeric furanosyl isomers (74).[76] The imidazoles (75) have been obtained by desulphurization of previously reported imidazole-2-thiones [Vol.18, p.113].[77]

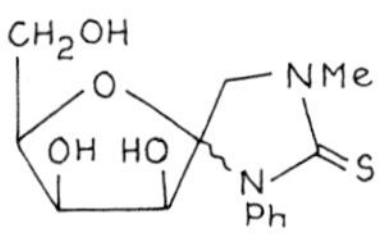

(74)

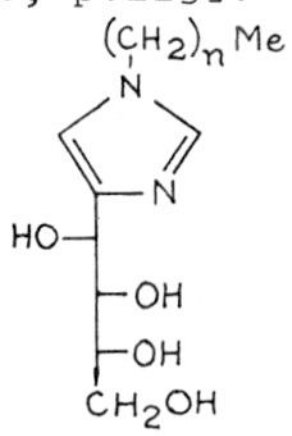

(75) n = 5,7,⋯15

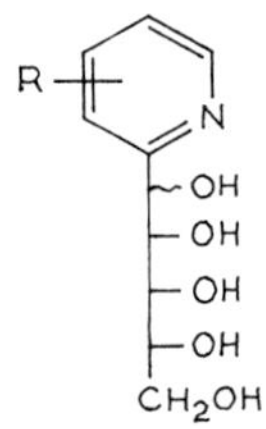

(76) R = H or Me

2-Pyridyl pentitols (76) have been synthesized by addition of metallated pyridines to 2,4:3,5-di-O-isopropylidene-aldehydo-D-ribose.[78] Selective osmylation of 5-vinyl-4,5-dihydroisoxazoles has been investigated as a route to polyols and amino-sugars, optically active compounds being obtained as shown in Scheme 15. exo-Metallation of the nitrile oxide dipolar cycloadduct (77) provided a 1:1 mixture of diastereoisomers from which compound (78) was separated chromatographically, and converted to a 78:22

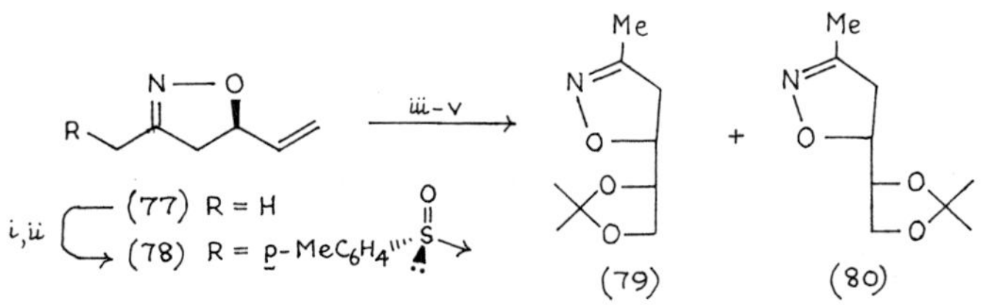

Reagents: i, LDA; ii, Menthyl toluene-p-sulphinate; iii, $OsO_4$-$Me_3NO$; iv, $Me_2CO$-$H^+$; v, Na-Hg-$NaH_2PO_4$-MeOH

Scheme 15

mixture of separable acetonides (79) and (80).[79]

Alditol-1-yl and C-glycofuranosyl pyrazoles, e.g. (81), have been synthesized from aldehydo-sugar derivatives in three steps (i, $Ph_3P=CHCO_2Et$; ii, $CH_2N_2$; and iii, $Br_2$).[80] Alditol-1-yl

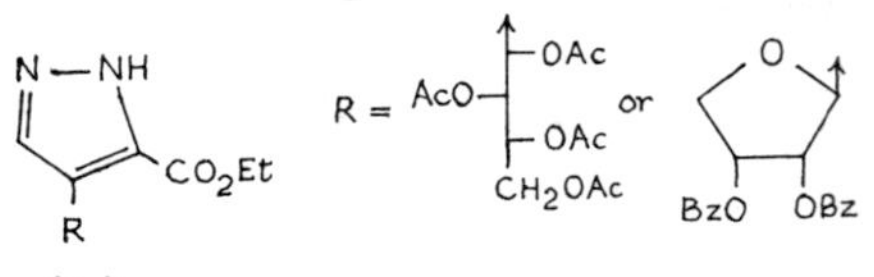

(81)

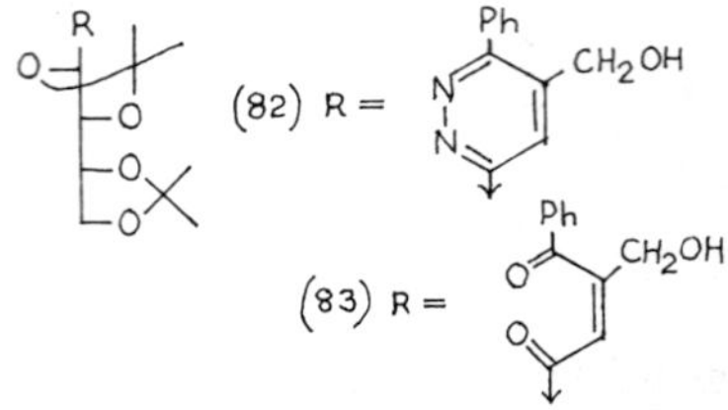

pyridazine derivatives, e.g. (82), have been obtained from photo-oxidation of corresponding alditol-1-yl furans followed by condensation of the α,β-unsaturated γ-ketones so formed, e.g. (83), with hydrazine.[81]

Takagi and co-workers have continued their investigation of the reaction of sugars with o-phenylenediamine (c.f. Vol.17, p.112) by analyzing the quinoxaline derivatives formed from five disaccharides,[82] and from dehydroascorbic acid derivatives (c.f. Vol.18, p.113).[83]

## References


1 J.Kovacs, I.Pinter, A.Messmer, and G.Toth, Carbohydr. Res., 1985, 141, 57.

2 G.Westphal and L.Kroh, Nahrung, 1984, 28, 809 (Chem. Abstr., 1985, 102, 60 884).

3 J.Fernandez-Bolanos, J.Galbis Perez, and F.Zamora Mata, An.Quim., Ser. C, 1983, 79, 206 (Chem. Abstr., 1985, 102, 62 541).

4 H.Booth and K.A.Khedhair, J. Chem. Soc., Chem. Commun., 1985, 467.

5 D.J.Nelson and M.J.Lavin, J. Carbohydr. Chem., 1985, 4, 91.

6 L.A.Abibullaeva and B.N.Stepanenko, Fiz.-Khim. Issled. Sint. Prir. Soedin., 1983, 26 (Chem. Abstr., 1985, 102, 185 370).

7 G.Snatzke and J.Vlahov, Liebigs Ann. Chem., 1985, 439.

8 R.Walczyna and J.Sokolowski, Pol. J. Chem., 1984, 58, 791.

9 Z.Smiatacz and A.Moczulska, Carbohydr. Res., 1985, 135, 219.

10 M.L.Sinnott and W.S.S.Wijesundera, Carbohydr. Res., 1985, 136, 357.

11 M.Sawaki, T.Takeda, Y.Ogihara, and S.Shibata, Chem. Pharm. Bull., 1985, 33, 5134.

12 H.Ishii, Y.Inoue, and R.Chujo, Polym. J. (Tokyo), 1985, 17, 693 (Chem. Abstr., 1985, 103, 123 902).

13 B.Bendiak and D.A.Cumming, Carbohydr. Res., 1985, 144, 1.

14 R.R.Schmidt and J.Michel, J. Carbohydr. Chem., 1985, 4, 141.

15 V.I.Mukhanov, I.L.Kustova, V.N.Gagloev, I.V.Yartseva, V.E.Shevchenko, A.Veveris, B.Spince, and M.N. Preobrazhenskaya, Bioorg. Khim., 1984, 10, 1385 (Chem. Abstr., 1985, 102, 132 393).

16 V.Škarić and J.Matulić-Adamić, J. Chem. Soc., Perkin Trans. 1, 1985, 779.

17 R.Asenjo Asenjo, M.Melgarejo Sampedro, C.Rodriguez Melgarejo, M.Nogueras Montiel, and A.Sanchez Rodrigo, An. Quim., Ser. C, 1983, 79, 417 (Chem. Abstr., 1985, 102, 149 676).

18 A.Sanchez Rodrigo, M.Nogueras Montiel, R.Lopez Garzon, M.D.Gutierrez Valero, and E.Colacio Rodriguez, Thermochim. Acta, 1985, 86, 199 (Chem. Abstr., 1985, 103, 178 548).
19 B.Robillard, M.F.Lhomme, and J.Lhomme, Tetrahedron Lett., 1985, 26, 2659.
20 B.N.Berad, S.P.Deshmukh, and M.G.Paranjge, J. Indian Chem. Soc., 1984, 61, 883 (Chem. Abstr., 1985, 103, 37 682).
21 R.Rico Gomez and M.L.Ruiz Mora, An. Quim., Ser. C, 1984, 80, 208 (Chem. Abstr., 1985, 102, 132 407).
22 L.A.Baltina, R.M.Kondratenko, and G.A.Tolstikov, Zh. Obshch. Khim., 1984, 54, 2573 (Chem. Abstr., 1985, 102, 204 185).
23 J.Fuentes Mota, M.C.Ortiz Mellet, F.Segura Ramos, M.A.Pradera Adrian, and A.Cert Ventula, An. Quim., Ser. C, 1983, 79, 221 (Chem. Abstr., 1985, 102, 46 208).
24 J.Fuentes Mota, M.C.Ortiz Mellet, and M.A.Pradera Adrian, An. Quim., Ser. C, 1984, 80, 48 (Chem. Abstr., 1985, 102, 79 257).
25 I.Yamamoto, K.Fukui, S.Yamamoto, K.Ohta, K.Matsuzaki, Synthesis, 1985, 686.
26 M.Wojtowicz, Acta Pol. Pharm., 1984, 41, 535 (Chem. Abstr., 1985, 102, 221 128).
27 T.Walle, D.D.Christ, U.K.Walle, and M.J.Wilson, J. Chromatogr., 1985, 341, 213.
28 O.Grech-Bélanger and J.Turgeon, J. Chromatogr., 1985, 337, 172.
29 M.G.Peter and W.Vaupel, J. Chem. Soc., Chem. Commun., 1985, 848.
30 J.Gal, J. Chromatogr., 1985, 331, 349.
31 Z.J.Witczak, Tetrahedron, 1985, 41, 4781.
32 F.Hayase and H.Kato, Agric. Biol. Chem., 1985, 49, 467.
33 F.Hayase, S.B.Kim and H.Kato, Agric. Biol. Chem., 1985, 49, 2337.
34 D.L.Ingles and D.Gallimore, Chem. Ind. (London), 1985, 194 (Chem. Abstr., 1985, 103, 71 638).
35 N.Kinae, M.Saito, S.Sawada, M.Yamashita, H.Uto, I.Tomita, and H.Kanamori, Chem. Pharm. Bull., 1985, 33, 2438.
36 M.S.Feather and R.D.Huang, J. Carbohydr. Chem., 1985, 4, 363.
37 Z.Györgydeàk and L.Szilàgyi, Liebigs Ann. Chem., 1985, 103.
38 L.Szilàgyi and Z.Györgydeàk, Carbohydr. Res., 1985, 143, 21.
39 M.Kloosterman, M.P.De Nijs, and J.H. Van Boom, J.R. Neth. Chem. Soc., 1984, 103, 243 (Chem. Abstr., 1985, 102, 6 964).
40 H.Paulsen and J.P.Lorentzen, Carbohydr. Res., 1985, 140, 155.
41 G.Janairo, A.Malik, and W.Voelter, Liebigs Ann. Chem., 1985, 653.
42 A.Guiller, C.H.Gagnieu, and H.Pacheco, Tetrahedron Lett., 1985, 26, 6343.
43 H.Paulsen, J.P.Lorentzen, and W.Kutschker, Carbohydr. Res., 1985, 136, 153.
44 J.Theim and D.Springer, Carbohydr. Res., 1985, 136, 325.
45 G.Kurz, J.Lehmann, and R.Theime, Carbohydr. Res., 1985, 136, 125.
46 J.M.J.Tronchet, K.D.Pallie, and F.Barbalat-Rey, J. Carbohydr. Chem., 1985, 4, 29.
47 J.S.Brimacombe and K.M.M.Rahman, J. Chem. Soc., Perkin Trans. 1, 1985, 1067.
48 J.S.Brimacombe and K.M.M.Rahman, J. Chem. Soc., Perkin Trans. 1, 1985, 1073.
49 J.Dupuis, B.Giese, J.Hartung, M.Leising, H.-G.Korth, and R.Sustmann, J. Am. Chem. Soc., 1985, 107, 4332.
50 A.Gomez-Sanchez, J.Galan, M.Rico, and J.Bellanato, J. Chem. Soc., Perkin Trans. 1, 1985, 2695.
51 K.Satsumabayashi and Y.Nomoto, Nippon Shika Daigaku Kiyo, Ippan Kyoiku-kei, 1983, 12, 131 (Chem. Abstr., 1985, 102, 132 392).
52 R.Meuwly and A.Vasella, Helv. Chim. Acta, 1985, 68, 997.
53 J.Farkaš and I.Frič, Collect. Czech, Chem. Commun., 1985, 50, 1291.
54 J.C.Palacios Albarran, E.Roman Galan, and J.A.Galbis Perez, Carbohydr. Res., 1985, 143, 117.
55 D.Beer and A.Vasella, Helv. Chim. Acta, 1985, 68, 2254.
56 R.Huber, A.Knierzinger, J.-P.Obrecht, and A.Vasella, Helv. Chim. Acta, 1985, 68, 1730.
57 B.Bernet, E.Krawczyk and A.Vasella, Helv. Chim. Acta, 1985, 68, 2299.
58 J.M.J.Tronchet, E.Winter-Mihaly, J.Rupp, F.Barbalat-Rey, and M.Geoffroy, Carbohydr. Res., 1985, 136, 375.

59 T.Tsuchiya, M.Nakano, T.Torii, Y.Suzuki, and S.Umezawa, Carbohydr. Res., 1985, 136, 195.
60 T.Yoshioka, H.Yamada, and T.Uematsu, J. Chem. Soc., Perkin Trans. 1, 1985, 1271.
61 T.Yoshioka and T.Uematsu, Carbohydr. Res., 1985, 143, 282.
62 T.Yoshioka and T.Uematsu, J. Chem. Soc., Perkin Trans. 1, 1985, 1261.
63 T.Yoshioka and T.Uematsu, J. Chem. Soc., Perkin Trans. 2, 1985, 1377.
64 K.Katano, P.-I.Chang, A.Millar, V.Pozsgay, D.K.Minster, T.Ohgi, and S.M.Hecht, J. Org. Chem., 1985, 50, 5807.
65 N.A.Chursina, B.K.Sadybakasov, I.F.Strel'tsova, V.A.Afanas'ev, M.Yu. Antipin, and Yu.T.Struchkov, Zh. Obshch. Khim., 1984, 54, 706 (Chem. Abstr., 1984, 101, 55 463).
66 B.K.Sadybakasov, V.A.Afanas'ev, M.Yu.Antipin, and Yu.T.Struchkov, Zh. Strukt. Khim., 1984, 25, 104 (Chem. Abstr., 1985, 102, 185 378).
67 P.Keil, H.B.Mortensen, and C.Christophersen, Acta Chem. Scand. Ser. B., 1985, 39, 191.
68 K.Atta, S.Hauptmann, and H.Wilde, Carbohydr. Res., 1985, 145, 37.
69 E.R.Dilanyan, T.R.Ovsepyan, F.G.Arsenyan, G.M.Stepanyan, E.Mironov, and M.E.Vol'pin, Khim.-Farm. Zh., 1984, 18, 927 (Chem. Abstr., 1985, 102, 132 366).
70 L.Somogyi, Carbohydr. Res., 1985, 142, 315.
71 L.Somogyi, Carbohydr. Res., 1985, 145, 156.
72 L.Somogyi, Carbohydr. Res., 1985, 144, 71.
73 M.Gomez Guillen, L.M.Vazquez de Miguel, and J.Valazquez Jimenez, An. Quim., Ser. C, 1983, 79, 152 (Chem. Abstr., 1985, 102, 62 522).
74 J.A.Galbis Perez, J.L.Jimenez Requejo, J.C.Palacios Albarran, and M.Alvarez Gonzalez, Carbohydr. Res., 1985, 138, 153.
75 J.Fernandez-Bolanos, J.Fuentes Mota, and I.Robina Ramirez, An. Quim., Ser. C, 1983, 79, 317.
76 J.Fernandez-Bolanos, M.Trujillo Perez-Lanzac, J.Fuentes Mota, and A.Cert Ventula, Carbohydr. Res., 1985, 143, 260.
77 J.Fernandez-Bolanos, R.Ruiz Contreras, and F.Zamora Mata, An. Quim., Ser. C 1984, 80, 195 (Chem. Abstr., 1985, 102, 149 675).
78 M.Belmans, E.Esmans, R.Dommisse, J.Lepoivre, F.Alderweireldt, J.Balzarini, and E.de Clercq, Nucleosides Nucleotides, 1985, 4, 523.
79 R.Annunziata, M.Cinquini, F.Cozzi, and L.Raimondi, J. Chem. Soc., Chem. Commun., 1985, 403.
80 J.Molina Molina,A.Sanz Tejedor, and J.A.Lopez Sastre, An. Quim., Ser. C, 1984, 80, 111 (Chem. Abstr., 1985, 102, 149 698).
81 F.J.Lopez Aparicio, M.T.Plaza Lopez-Espinosa, and R.Robles Diaz, An. Quim., Ser. C, 1984, 80, 156 (Chem. Abstr., 1985, 102, 149 674).
82 N.Morita, K.Inoue, and M.Takagi, Agric. Biol. Chem., 1985, 49, 3279.
83 Y.Tsujimoto, M.Ohmori, and M.Takagi, Carbohydr. Res., 1985, 138, 148.

# 11
# Thio- and Seleno-sugars

The crystal structure by direct methods of isobutyl 2,3,4-tri-O-acetyl-1-thio-β-D-xylopyranoside shows that it exists in the $^{4}C_{1}$ (D) conformation.[1] Reaction of the 1-thioauric-glucose derivative (1), an auric antiarthritic drug, with hydrogen chloride in 95% aqueous methanol yields 2,3,4,6-tetra-O-acetyl-1-thio-D-glucose, triethylphosphinylgold(I) chloride and the thionium sugar (2). With nitric acid the gold complex (3) is formed but triethylphosphinylgold(I) nitrate is not.[2] The range of (1-thio-glucosyl) triazines

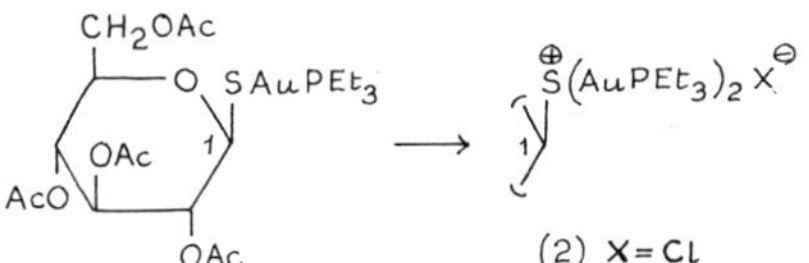

(1)

(2) X = Cl
(3) X = $NO_3$

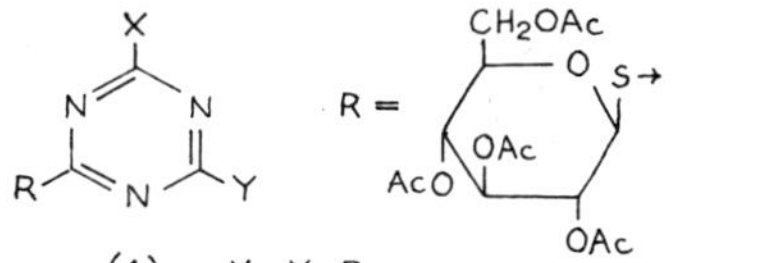

(4) X = Y = R
X = R, Y = ←$OCH_2$-ferrocene
X = R or Cl, Y =

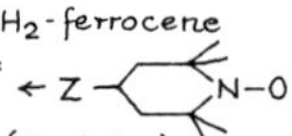

(Z = O, NH)

(4) have been synthesized as spectroscopic probes.[3] Thiourea

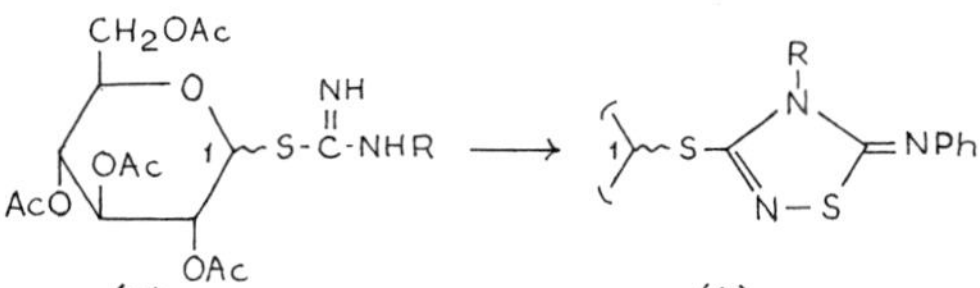

(5) (6)

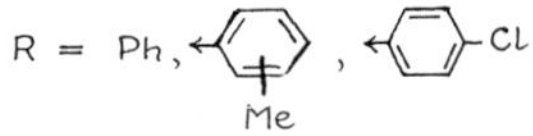


derivatives of 2,3,4,6-tetra-O-acetyl-D-glucose (5) condense with ClSCCl=NPh to yield the corresponding thiazolidines (6).[4] The glucosyl thioureides (7) have been synthesized from the corresponding isodithiobiurets (8) and $PhN{=}CCl_2$.[5] Phase-transfer

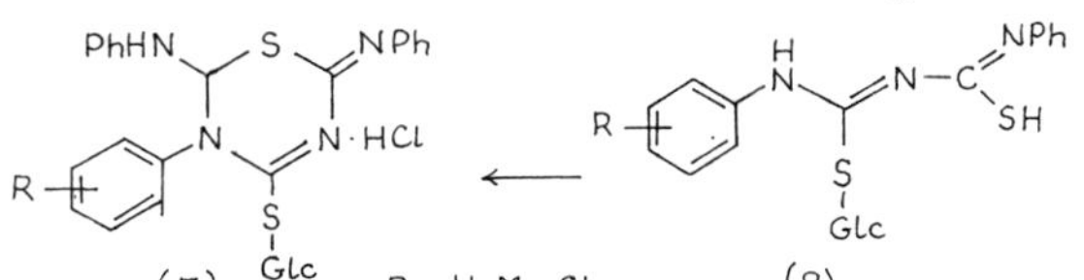

(7) R = H, Me, Cl (8)

Glc =

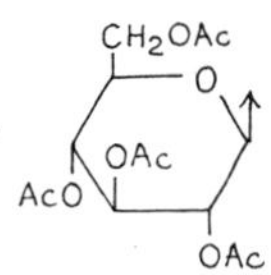

catalysis has been used to synthesize the sulphides (9) in high yield from the corresponding α-glycosylbromides and sodium sulphide.[6]

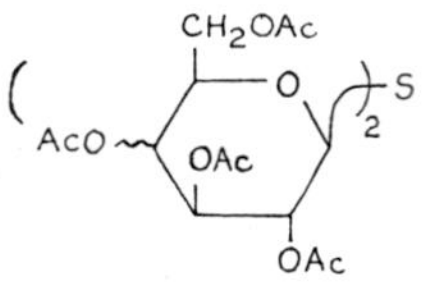


(9) (Glc or Gal)

Heating of the 2-thio-arabinose cyclonucleoside (10) with 80% acetic acid produced the thiazolidine (11).[7]

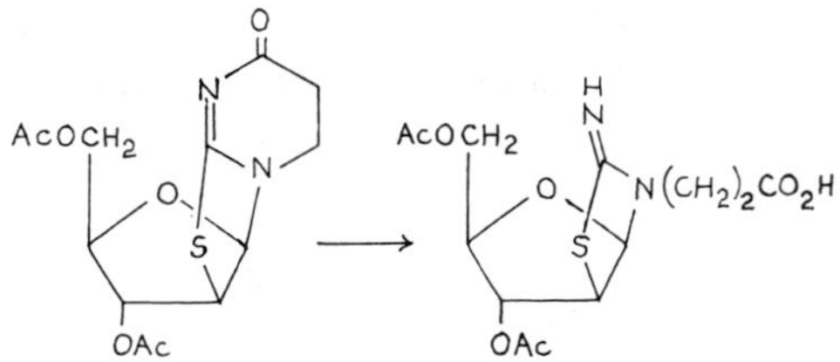


(10) (11)

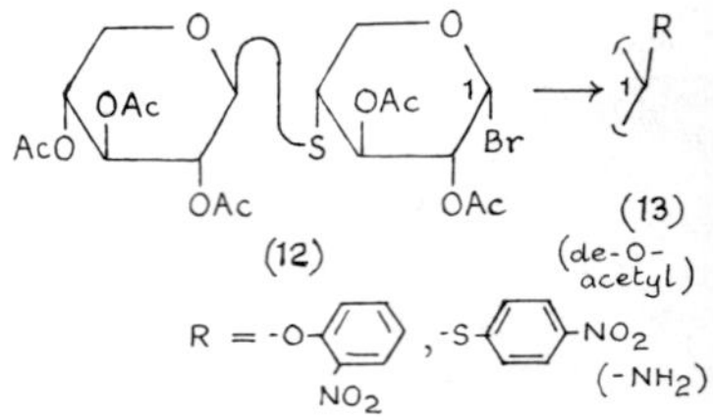


(12) (13) (de-O-acetyl)

Reaction of penta-O-acyl-4-thio-β-D-xylobiosyl α-bromide (12) with o-nitrophenol, p-nitrobenzenethiol or p-aminobenzenethiol in acetone in the presence of potassium carbonate, followed by deacylation, gave the corresponding β-glycosides (13) in excellent yield. The o-nitrophenyl bioside was a good chromogenic substrate for xylanases A and B obtained from Sporotrichum dimorphosporum while the p-aminophenyl 1,4-dithiobioside was a competitive inhibitor for xylanase B and a rate enhancer for xylanase A in its hydrolysis of D-xylan. The results suggest the possibility of a regulatory mechanism for D-xylan hydrolysis by xylanase, which may involve extracellular oligosaccharides. Attachment of p-aminophenyl 1,4-dithiobioside to Sepharose 4B gave an immobilized agent which retained its activity towards xylanase.[8]

The syntheses of 5-thio-D-arabinose and -lyxose have been achieved conventionally by thiocarboxylate displacements on 5-sulphonate ester derivatives of these sugars, and their methyl glycosides were also prepared, both anomers being obtained.[9] Methyl 5-thiopentopyranosides (D-*arabino*, L-*lyxo*, D-*ribo*, and D-*lyxo* derivatives) have also been prepared from methyl 5-thio-3-O-tosyl-α-D-xylopyranoside as outlined in Scheme 1.[10] The syntheses of S-(5'-deoxy-5'-adenosyl)-(+)-2-methylhomocysteine (14) has been achieved in two steps from (+)-2-methyl-homocysteine.[11] Methyl 2-O-tosyl-4,6-thioanhydro-α-

D-gulopyranoside (15) was obtained on alkali treatment of the thio-acetate (16). The reaction was thought to proceed as shown in Scheme 2. The product, which was unstable at room temperature, was

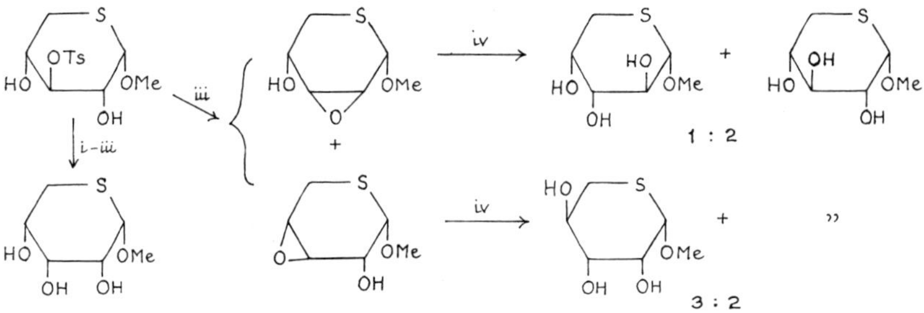

Reagents: i, BzCl-Py; ii, NaOBz-DMF-$H_2O$; iii, NaOMe-MeOH; iv, $OH^-$

Scheme 1

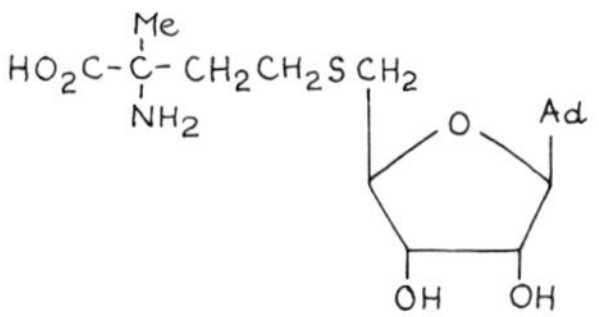

(14)

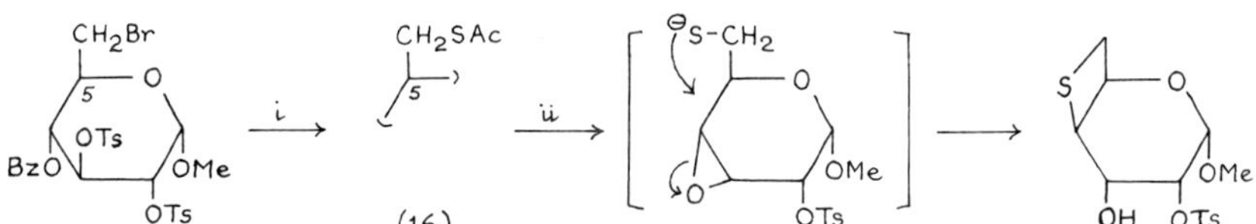

(16) (15)

Reagents: i, KSAc; ii, MeONa-MeOH

Scheme 2

characterized as its stable sulphone and as its 2,3-ditosylate.[12]

Reference to thiosugar-containing cephalosporins and anthracycline antibiotics will be found in Chapter 19, while the synthesis of 4-thioxylobiose is described in Chapter 4.

In a paper describing the merits of the odourless, stable, available selenylating reagent N-phenylselenophthalimide[13] the two carbohydrate examples shown in Scheme 3 were given.

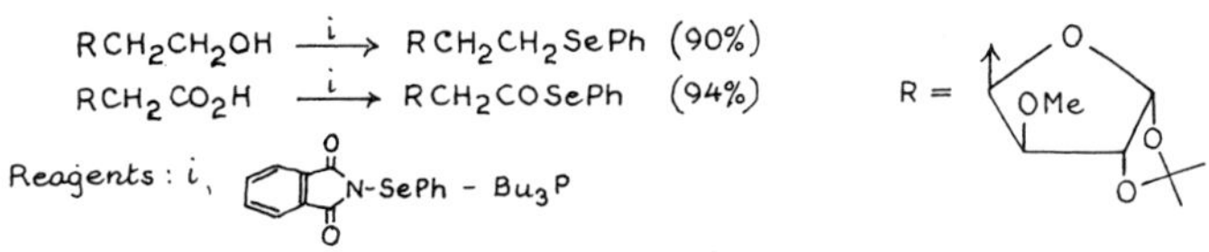

Scheme 3

## References

1 G.A.Jeffrey, J.R.Ruble, and B.Sepehrnia, Carbohydr. Res., 1985, 144, 197.
2 J.Hempel and Y.Mikuriya, Bioinorg. Chem. Gold Co-ord. Compd., Proc. Symp., 1981 (Chem. Abstr., 1985, 102, 32 022).
3 M.J.Adam and L.D.Hall, Carbohydr. Res., 1985, 144, 23.
4 I.D.Bedekar, A.H.Lokhande, and M.G.Paranjpe, J. Indian Chem. Soc., 1984, 61, 432 (Chem. Abstr., 1985, 102, 24 948).
5 I.Bedekar and M.G.Paranjpe, J. Indian Chem. Soc., 1984, 61, 781 (Chem. Abstr., 1985, 103, 37 663).
6 J.Bogusiak and W.Szeja, Carbohydr. Res., 1985, 141, 165.
7 V.Skaric and J.Matulic-Adamic, J. Chem. Soc., Perkin Trans. I, 1985, 779.
8 J.Comtat, J.Defaye, H.Driguez, and E.Ohleyer, Carbohydr. Res., 1985, 144, 33.
9 N.A.Hughes and N.M.Munkombwe, Carbohydr. Res., 1985, 136, 397.
10 N.A.Hughes and N.M.Munkombwe, Carbohydr. Res., 1985, 136, 411.
11 K.Ramalingam and R.W.Woodard, Tetrahedron Lett., 1985, 26, 1135.
12 D.Miljkovic, V.Popsavin, and J.Harangi, Tetrahedron, 1985, 41, 2737.
13 K.C.Nicolaou, N.A.Petasis, and D.A.Claremon, Tetrahedron, 1985, 41, 4835.

# 12
# Deoxy-sugars

Potent new antibiotics isolated from Actinomadura sp. fermentation have released 3'- and 4'-N-(2-methoxypropenoyl)-4,5-dimethoxy-anthranilate esters of methyl 2,6-dideoxy-α,β-L-lyxo-hexopyranoside (i.e. a 2-deoxy-L-fucoside) on methanolysis.[1] Four oligoglycosides named cynatratosides B-E have been isolated from the Chinese crude drug "Pai-Wei" and shown to have variously linked L-cymarose, D-digitoxose, D-oleandrose, and D-glucose moieties; only the digitoxose and glucose moieties have free hydroxy groups.[2] Full details have appeared on the steroidal oligoglycosides isolated from Cynanchum wilfordi (Asclepiadaceae) which contain both D- and L-cymarose moieties, as well as other deoxy- and O-methyl-deoxy-sugar units (c.f. Vol.18, p.120).[3,4] The occurrence of a 4,6-dideoxy-3-O-methyl-D-erythro-hexos-2-ulose moiety in a cardenolide glycoside is covered in Chapter 15. Macrolide antibiotics of the complex mycinamicin, which contain components with 6-deoxy-D-allose moieties and their 2-mono- and 2,3-di-O-methyl ether analogues, are covered in Chapter 19.

The thermodynamic properties of 2-deoxy-sugars in water have been measured by microcalorimetric and isopiestic methods and compared to those of their parent monosaccharides.[5]

6-Deoxy-1,2:3,4-di-O-isopropylidene-α-D-galactopyranose has been obtained in 85% yield by irradiation of the corresponding 6-O-acetate.[6] Photolytic deoxygenations of the 2-6-dipivaloates (1) have provided the 2,6-dideoxyhexosides (2) in ∿40% yields;

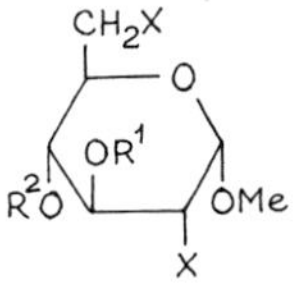


(1) X = $O_2CBu^t$

(2) X = H

$R^1, R^2 = SiMe_2Bu^t$, H

these products provided routes to 3- and 4-O-substituted 2,6-dideoxy-D-arabino-hexoses, including D-oleandrose (2,6-dideoxy-3-O-methyl-D-arabino-hexose).[7]

Following their earlier work on the lithium triethylborohydride reductions of tosylate derivatives of methyl 4,6-*O*-benzylidene-α-D-glucopyranoside (Vol.16, p.122), Baer and co-workers have studied the reduction of the 2-mono-, 2,2'-di-, 2,3,2'-tri-, and 2,3,2',3'-tetra-tosylate derivatives of 4,6:4',6'-di-*O*-benzylidene-α,α-trehalose. The reactions proceed *via* epoxide intermediates which yield expected 2- and/or 3-mono-and/or di-deoxy analogues.[8] 4-Substituted allyl 6-deoxy-α-D-mannopyranosides have been synthesized following reductive de-tosylation of a 2,3-*O*-iso-propylidene-6-*O*-tosyl-precursor.[9] 1,2-*O*-Cyclohexylidene-5-deoxy-3-*O*-methyl-α-D-xylofuranose has been synthesized using a similar sulphonate reduction.[10]

Barton's radical deoxygenation procedure has been applied to methyl xanthate derivatives of sugars to prepare methyl 2-deoxy-β-D-*erythro*-pentopyranoside,[11] the 3-acetamido-2-deoxy-D-*arabino*-hexoside (3)[12] and 2-acetamido-3-deoxy-D-*ribo*-hexoside (4),[13] and

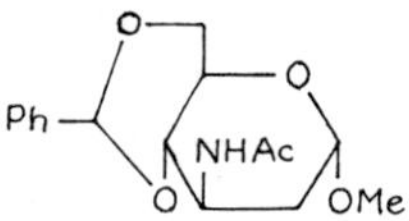

(3)

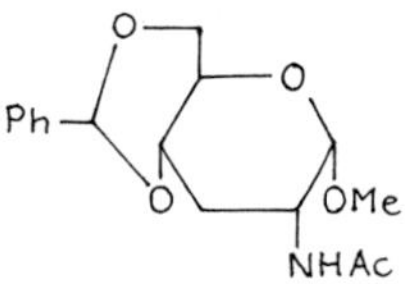

(4)

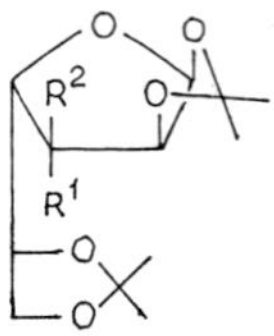

(5) $R^1 = R^2 = H$
(6) $R^1 = D, R^2 = H$
(7) $R^1 = H, R^2 = D$

1,2:5,6-di-*O*-isopropylidene-3-deoxy-D-*lyxo*-hexofuranose (5) and its 3-*C*-deuterio derivatives (6) and (7) (*via* β-D-idose, β-D-talose, and (3-$^2$H)-β-D-talose 3-xanthate precursors). These latter deuterated products were obtained in the ratio 65:35, respectively, on reduction of the D-idose 3-xanthate with tributyltin deuteride.[13] 3'-Deoxy-tiazofurin has been synthesized in four steps from the *C*-nucleoside tiazofurin [*i.e.* 2-(β-D-ribofuranosyl)-thiazole-4-carboxamide] *via* its 2',5'-di-*O*-tert-butyldimethylsilyl-3'-*O*-imidazolethiocarbonyl derivative.[14] Related syntheses of branched-chain deoxy-sugars involving radical intermediates are covered in Chapters 13 and 14.

Deoxy-sugars have also been synthesized from halogeno-sugars. A convenient synthesis of 6-deoxy-D-glucopyranose from D-glucose (three steps, 54% overall yield) involved the catalytic hydrogenation of 6-bromo-6-deoxy-1,2,3,4-tetra-*O*-(trimethylsilyl)-D-gluco-

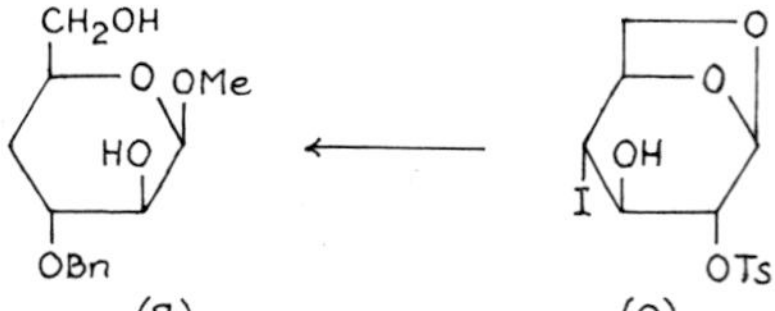

(8) (9)

pyranose.[15] Methyl 3-O-benzyl-4-deoxy-β-D-arabino-hexopyranoside (8) and its 3-O-methyl analogue have been synthesized from the levoglucosan-derived iodo-tosylate (9) in four steps.[16] Methyl α-L-digitoxose (10) and the isomeric L-lyxo-hexoside (11) (i.e. a

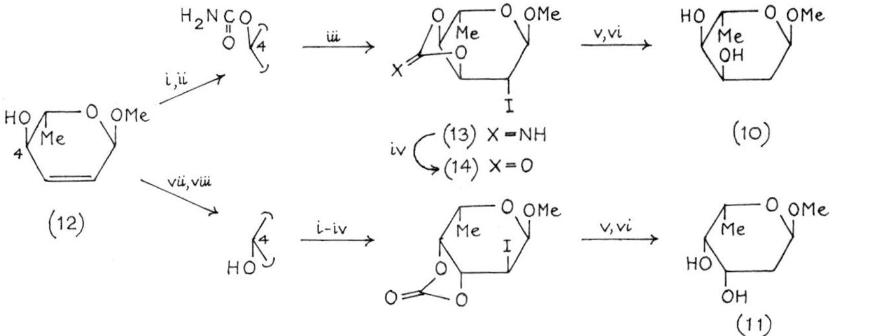

Reagents: i, $Cl_3CONCO$; ii, $K_2CO_3$-MeOH; iii, $I(Collidine)_2ClO_4$; iv, $H_2O$; v, $Bu_3SnH$-AIBN, vi, $Ba(OH)_2$; vii, DEAD-TPP-BzOH; viii, MeONa-MeOH

Scheme 1

2-deoxy-L-fucoside) have been synthesized from the α-L-erythro-hex-2-enoside (12) as shown in Scheme 1. Intramolecular iodocyclization of allylic urethanes gave iminocarbonates [e.g. (13)] which yielded carbonates [e.g. (14)] on hydrolysis.[17]

The synthesis of 2-acetamido-1,6-anhydro-2,3-dideoxy-β-D-ribo-hexopyranose from a 3-thio-sugar by reductive desulphurization is covered in Chapter 9. Reduction ($NaBH_4$) of secondary mesylates with a vicinally related trans-diallylamino group proceeds to give deoxygenated products via aziridinium ion intermediates. In consequence, sometimes the product is that of amino migration;[18] examples are detailed in Chapter 9.

Deoxy-sugar phosphates have been synthesized by catalytic hydrogenation of various enol phosphates. Thus the alditol derivatives (15) and (16) were obtained from 1,2- and 2,3-unsaturated pre-

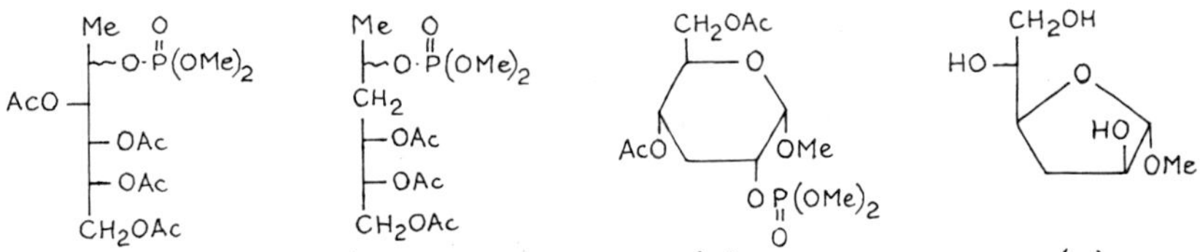

(15) (16) (17) (18)

cursors, while the 3-deoxy-2-O-phosphate (17) was obtained from methyl 4,6-O-benzylidene-3-deoxy-2-O-(dimethylphosphoryl)-α-D-glucopyranoside.[19] Methyl 3-deoxy-α-D-arabino-hexofuranoside (18) has been synthesized from D-glucono-1,5-lactone via the known 3-deoxy-

D-arabino-hexono-1,5-lactone tribenzoate. This 1,5-lactone was converted to the corresponding 1,4-lactone by debenzoylation-rebenzoylation, and reduced at C-1 with disiamylborane.[20] Racemic diginofuranosides (19) and rhodinofuranosides (20) were obtained in

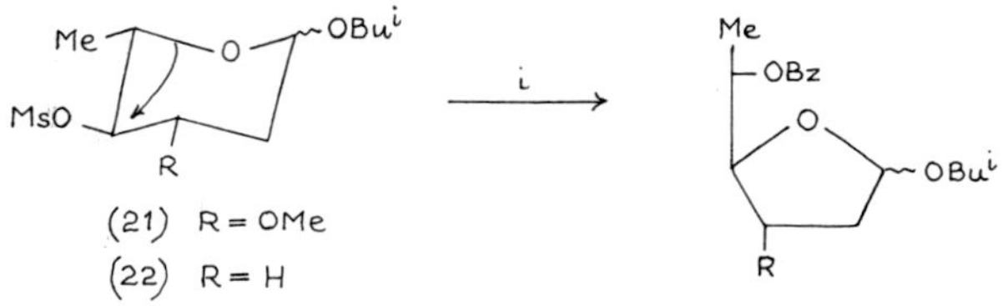


(19) R = OMe
(20) R = H
(21) R = OMe
(22) R = H

Reagents: i, NaOBz-DMF

Scheme 2

substantial amounts due to ring-contraction during the attempted nucleophilic substitution of the racemic oleandropyranoside and amicetopyranoside 4-mesylates, (21) and (22) respectively (Scheme 2) These products result from inversion at C-4 but not at C-5.[21]

The radiochemical 2-deoxy-D-[1-$^{11}$C]glucose has been synthesized from [$^{11}$C]hydrogen cyanide and 1-deoxy-1-iodo-2,3:4,5-di-O-isopropylidene-D-arabinitol in 20% yield (in 50-55 min.).[22] The four 4-deoxy-D-heptosides (23) (isomers at C-2 and C-3) have been obtained from the corresponding 2,3,4-trideoxy-hept-2-enoside by cis-hydroxylation (the catalytic $OsO_4$ method), or by epoxidation-hydrolysis.[23]

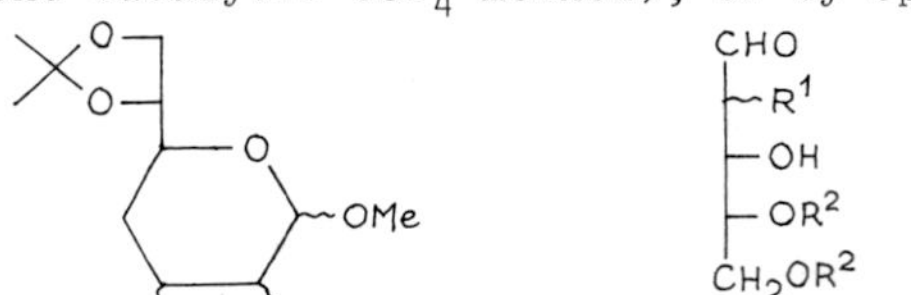


(23)

(24) $R^1 = R^2 = H$
(25) $R^1$ = NHCHO, $R^2,R^2 = CMe_2$

Deoxy-sugars have been elaborated from chiral three-carbon starting materials. 2-Deoxy-D-ribose (24) has been obtained in eight steps from the 2-formylamino derivatives (25), the synthesis of which from 2,3-O-isopropylidene-D-glyceraldehyde is covered in Chapter 9.[24] Alternatively, this same overall conversion to either enantiomer could be achieved using the diastereoselective aldol condensation of 2,3-O-isopropylidene-D- or-L-glyceraldehyde

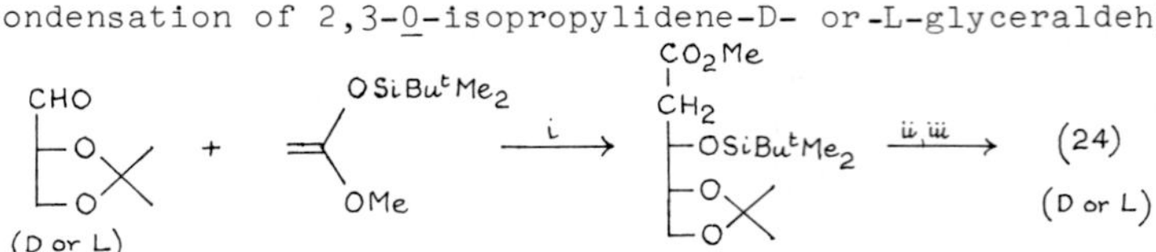


(26) (27)

Reagents: i, $ZnI_2$-MeCN; ii, $CF_3CO_2H$; iii, Disiamylborane

Scheme 3

with ketene acetal (26) (Scheme 3), a reaction which yields predominantly the <u>erythro</u> isomer (<u>e.g.</u> 27) with silyl transfer.[25] The 2,6-dideoxy-L-<u>arabino</u>-hexoside (28) and the 4,6-dideoxy-3-<u>O</u>-methyl-D-<u>xylo</u>-hexoside (29) have been synthesized from the major isomers generated by coupling α- and β-alkoxy-aldehydes (30) and

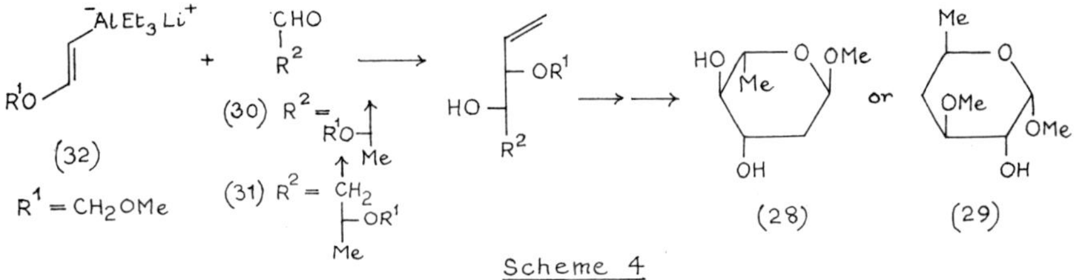


Scheme 4

(31) respectively with the allylic aluminium 'ate' complex (32), as shown in Scheme 4. The diastereofacial selectivity in the construction of three chiral centres on addition of such allylic organometallic reagents to chiral aldehydes is influenced by the nature of the metal, and can be predicted from the results of an extensive study of simpler systems.[26] The chiral epoxysulphide (33) has been converted into the 2,3,6-trideoxy-sugars L-rhodinose (34) and D-amicetose (35) by alkylation of the common aldehyde (36) with or without chelation control respectively, as shown in Scheme 5.[27]

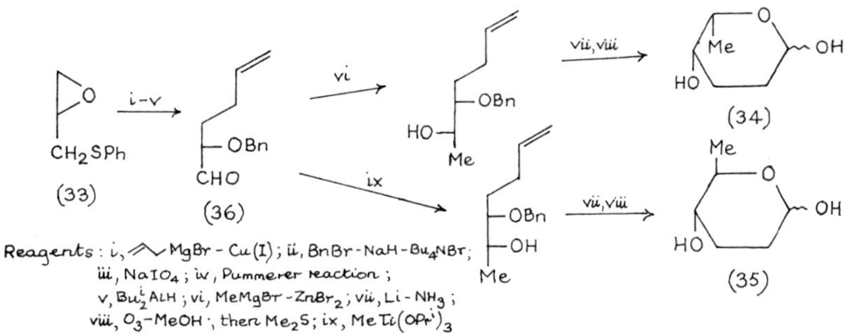


Scheme 5

A number of deoxy-sugar derivatives have been prepared from the chiral diols obtained by reduction of α-acetoxyketones (37)-(39)

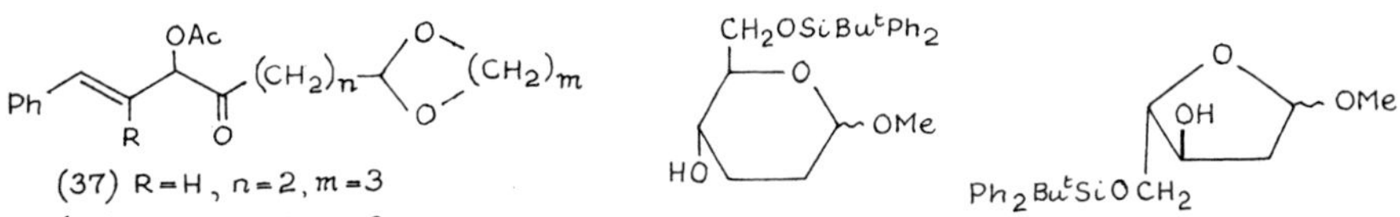

(37) R=H, n=2, m=3

(38) R=H, n=1, m=2

(39) R=Me, n=1, m=2

(40)

(41)

with Baker's yeast. In this way the 2,3-dideoxy-D-*erythro*-hexosides (40), the free sugar of the 2-deoxy-L-fucoside (11), 2-deoxy-D-*erythro*-pentosides (41), and D-digitoxose [*c.f.* compound (10), the methyl glycoside of the L-enantiomer] were prepared.[28]

Racemic deoxy-sugars have been synthesized by hetero-Diels-Alder strategies. The DL-fucose derivative (42) was obtained *via* the racemic dihydropyrone (43) (Scheme 6).[29] The racemic 2,4,6-tri-

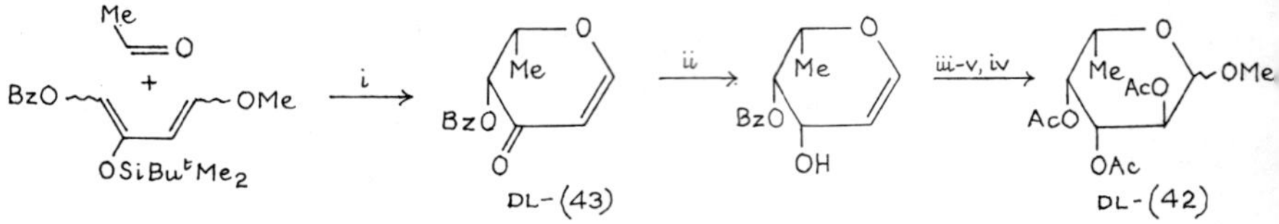

Reagents: i, $BF_3.OEt_2$; ii, $NaBH_4$-$CeCl_3$; iii, $K_2CO_3$-MeOH; iv, $Ac_2O$-$NEt_3$-DMAP; v, MCPBA-MeOH

Scheme 6

deoxy- and 2,4-dideoxy-sugars (44) and (45) (*i.e.* DL-olivose) were obtained *via* the common intermediate (46) (Scheme 7); by using α-methoxyphenylacetate derivatives, a separable 2:1 mixture of L- and D-sugar diastereomers (*e.g.* 47) was obtained, and the major isomer was converted to the L-enantiomers of (44) and (45).[30]

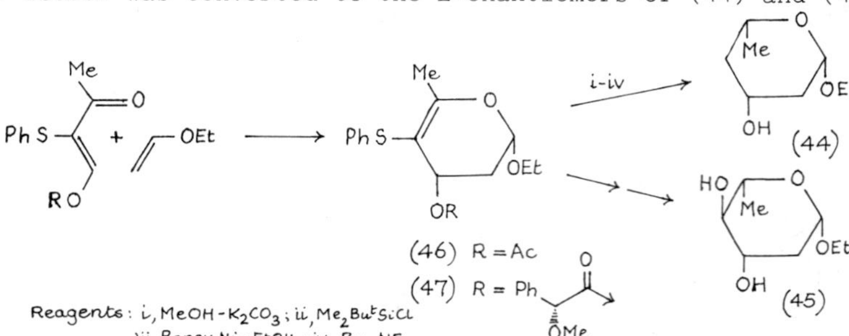

Reagents: i, MeOH-$K_2CO_3$; ii, $Me_2Bu^tSiCl$ iii, Raney Ni-EtOH; iv, $Bu_4NF$

Scheme 7

The racemic 2-deoxy-pentonic acid lactones (48) and (49) have been obtained in a 4.3:1 ratio, in one step, by oxylactonization (MCPBA) of alkene (50).[31] Application of the Sharpless asymmetric

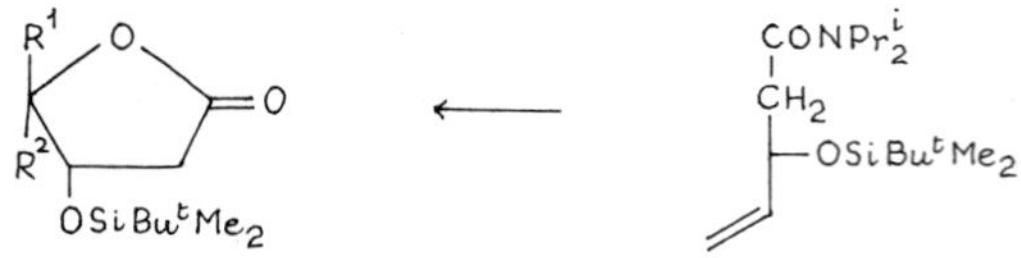

DL-(48) $R^1$=$CH_2OH$, $R^2$=H

DL-(49) $R^1$=H, $R^2$=$CH_2OH$

DL-(50)

epoxidation procedure to the racemic alkene (51) yielded a mixture containing the L-*trans* isomer of the starting alkene, and the *erythro*- (52) and *threo*-D-epoxides in a 96:4 ratio. Hydrolysis (base then acid) of epoxide (52) yielded 2,6-dideoxy-D-*ribo*-hexose (53).[32]

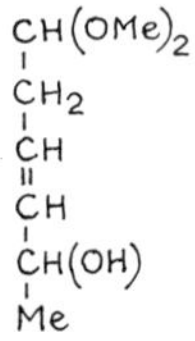
(51)

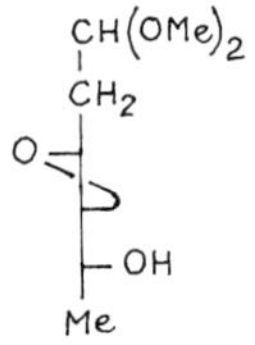
(52)

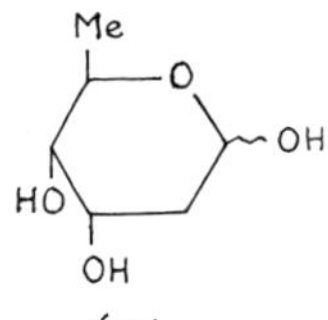
(53)

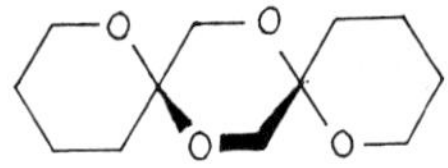
(54)

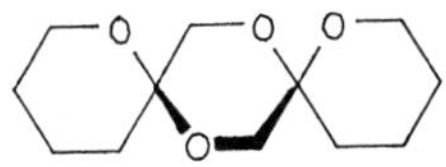
(55)

1,6-Dihydroxy-2-hexanone, a 3,4,5-trideoxy-ketohexose, has been synthesized, and shown to exist in the acyclic form in aqueous solution to the extent of 60%. On standing, it slowly forms a hemiacetal dimer, and with acid one or both of the possible anhydrides (54) and (55) by dehydrative dimerization.[33]

Other examples of deoxy-sugar synthesis can be found in Chapter 9.

## References


1 J.H.Wilton, G.C.Hokanson, and J.C.French, J. Chem. Soc., Chem. Commun., 1985, 919.
2 Z.-X.Zhang, J.Zhou, K.Hayashi, and H.Mitsuhashi, Chem. Pharm. Bull., 1985, 33, 1507.
3 S.Tsukamoto, K.Hayashi, and H.Mitsuhashi, Tetrahedron, 1985, 41, 927.
4 S.Tsukamoto, K.Hayashi, and H.Mitsuhashi, Chem. Pharm. Bull., 1985, 33, 2294.
5 G.Barone, G.Castronuovo, V.Elia, C.A.Mattia, R.Puliti, and R.Rizzo, Thermochim. Acta, 1985, 85, 447 (Chem. Abstr., 1985, 103, 160 773).
6 J.P.Pete, C.Portella, and D.Scholler, J. Photochem., 1984, 27, 128 (Chem. Abstr., 1985, 102, 79 217).
7 J.Dornhagen, A.Klausener, J.Runsink, and H.-D.Scharf, Liebigs Ann. Chem., 1985, 1838.
8 H.H.Baer, M.Mekarska, and F.Boucher, Carbohydr. Res., 1985, 136, 335.
9 J.Gigg, R.Gigg, S.Payne, and R.Conant, Carbohydr. Res., 1985, 141, 91.
10 D.A.Miljkovic, N.S.Vukojevic, and M.M.Popsavin, Glas. Hem. Drus., Beograd, 1984, 49, 309 (Chem. Abstr., 1985, 102, 113 800).
11 S.Inokawa, T.Mitsuyoshi, H.Kawamoto, H.Yamamoto, and M.Yamashita, Carbohydr. Res., 1985, 142, 321.
12 J.J.Patroni and R.V.Stick, Aust. J. Chem., 1985, 38, 947.
13 R.J.Conway, J.P.Nagel, R.V.Stick, and D.M.G.Tilbrook, Aust. J. Chem., 1985, 38, 939.
14 R.H.Baur and D.C.Baker, Nucleosides Nucleotides, 1984, 3, 77 (Chem. Abstr., 1985, 102, 185 422).
15 R.Csuk, Carbohydr. Res., 1985, 140, 167.
16 A.Banaszek, Bull. Pol. Acad. Sci., Chem., 1984, 32, 99 (Chem. Abstr., 1985, 102, 149 663).
17 H.W.Pauls and B.Fraser-Reid, J. Carbohydr. Chem., 1985, 4, 1.
18 D.Picq, G.Carret, D.Anker, and M.Abou-Assali, Tetrahedron Lett., 1985, 26, 1863.
19 J.Thiem and D.Rasch, Liebigs Ann. Chem., 1985, 536.
20 C.Du Mortier and R.M.De Lederkremer, Carbohydr. Res., 1985, 139, 47.
21 G.Berti, G.Catelani, F.Colonna, M.Ferretti, and L.Monti, Gazz. Chim. Ital., 1985, 115, 85 (Chem. Abstr., 1985, 102, 204 188).

22 D.Van Haver, N.A.Rabi, M.Vandewalle, P.Goethals, and C.Vandecasteele, J. Labelled Compd. Radiopharm., 1985, 22, 657 (Chem. Abstr., 1985, 103, 215 663).

23 J.Jurczak, T.Bauer, and J.Kihlberg, J. Carbohydr. Chem., 1985, 4, 447.

24 Y.Yamamoto, M.Kirihata, I.Ichimoto, and H.Ueda, Agric. Biol. Chem., 1985, 49, 1435.

25 Y.Kita, H.Yasuda, O.Tamura, F.Itoh, Y.Y.Ke, and Y.Tamura, Tetrahedron Lett., 1985, 26, 5777.

26 Y.Yamamoto, T.Komatsu, and K.Maruyama, J. Organomet. Chem., 1985, 285, 31 (Chem. Abstr., 1985, 103, 215 660).

27 T.Itoh, A.Yoshinaka, T.Sato, and T.Fujisawa, Chem. Lett., 1985, 1679.

28 G.Fronza, C.Fuganti, P.Grasselli, and S.Servi, Tetrahedron Lett., 1985, 4961.

29 S.J.Danishefsky and C.J.Maring, J. Am. Chem. Soc., 1985, 107, 1269.

30 R.R.Schmidt and M.Maier, Tetrahedron Lett., 1985, 26, 2065.

31 Y. Ichikawa, T.Miwa, and K.Narasaka, Bull. Chem. Soc. Jpn., 1985, 58, 3309.

32 Yu.E.Raifel'd, L.L.Zilberg, B.M.Arshava, and S.M.Makin, Zh. Org. Khim., 1984, 20, 1330 (Chem. Abstr., 1985, 102, 24 928).

33 W.A.Szarek, O.R.Martin, R.J.Rafka, and T.S.Cameron, Can. J. Chem., 1985, 63, 1222.

# 13
# Unsaturated Derivatives

## 1 Glycals

A range of differently substituted derivatives of the glycal of D-ribofuranose have been produced from 2,3-O-isopropylidene-D-ribono-γ-lactone by substitution at O-5 followed by reduction, chlorination at C-1, and then elimination (Scheme 1).[1] Racemic

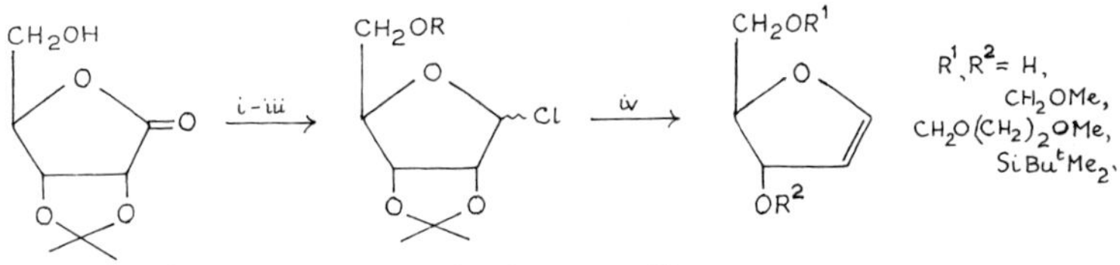


Reagents: i, RCl ; ii, $Bu^i_2AlH$ ; iii, $CCl_4$-$P(NMe_2)_3$ ; iv, Li-$NH_3$(l)

Scheme 1

xylo-furanoid glycals of aminodeoxy-aldoses ($R^1$ = H, $R^2$ = Me, $CH_2OBu^t$), -dialdoses [$R^1$ = H, $R^2$ = $CH(OEt)_2$] and deoxyketoses ($R^1$ = Me) (1) have been produced in two steps (yields 41-67%) from the furoisoxazolines (2) obtained by dipolar cycloaddition of aliphatic nitrile oxides to furan or 2-methylfuran (Scheme 2).[2] 5-Amino-5-deoxy-ido-furanosides are also obtainable from (2).

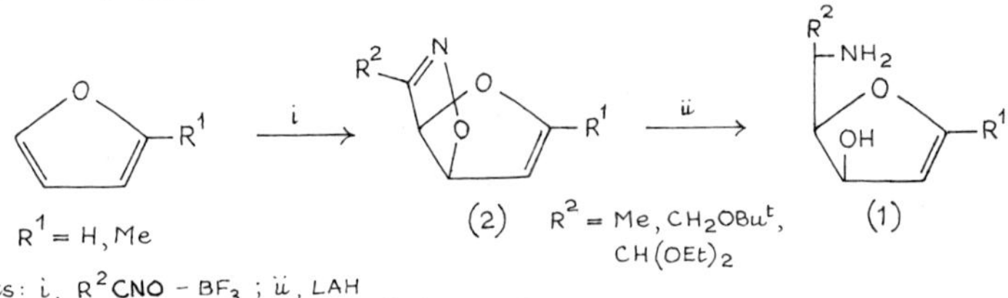


Reagents: i, $R^2CNO$ - $BF_3$ ; ii, LAH

Scheme 2

The conformational equilibria of the peracetylated glycals derived from four hexoses and two pentoses have been calculated by use of a molecular mechanics programme and compared with those determined by X-ray crystallography and n.m.r. spectroscopy. For the pentose derivatives there are inconsistencies between the calculated and experimental results.[3]

Two complementary cycloaddition approaches have provided access to racemic pyranoid glycal derivatives. The first, involving the use of aldehydes and oxygenated butadienes, led to compounds related to glucal (Scheme 3),[4] and it has also been extended to

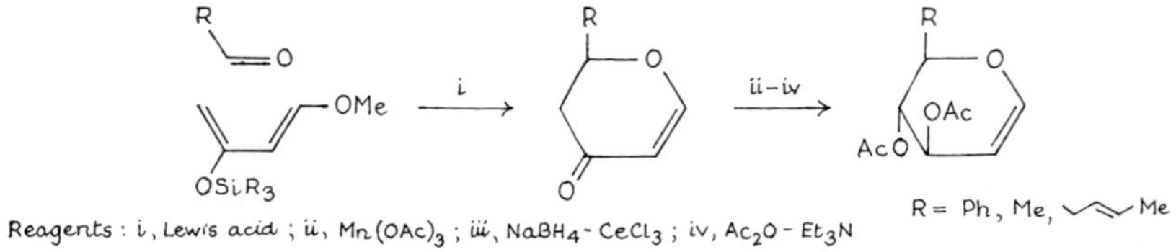

Scheme 3

give galactal and fucal analogues.[5] The alternative procedure used α,β-unsaturated carbonyl compounds and various enol ethers (Scheme 4). A range of 5-substituted glycals, e.g. compound (3), was obtained, and the reduction and conformation of an extensive series of such compounds were described.[6]

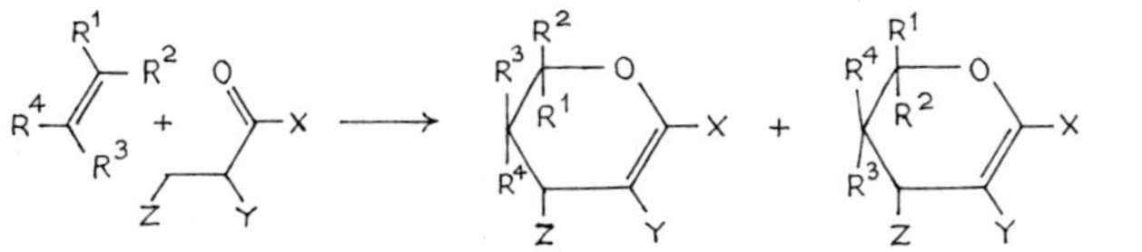

Scheme 4

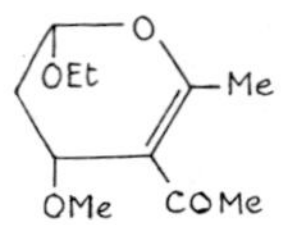

(3)

An unusual reaction of some 1,6-anhydroaldohexopyranose derivatives was uncovered on treatment of the levoglucosenone derivative (4) with sodium in liquid ammonia, which is indicated in Scheme 5; likewise compounds (5) (ribo or arabino) gave the glycal (6) in good yield after acetylation. By contrast, a 1,6-anhydro-β-D-galactopyranose analogue gave only poor yields.[7]

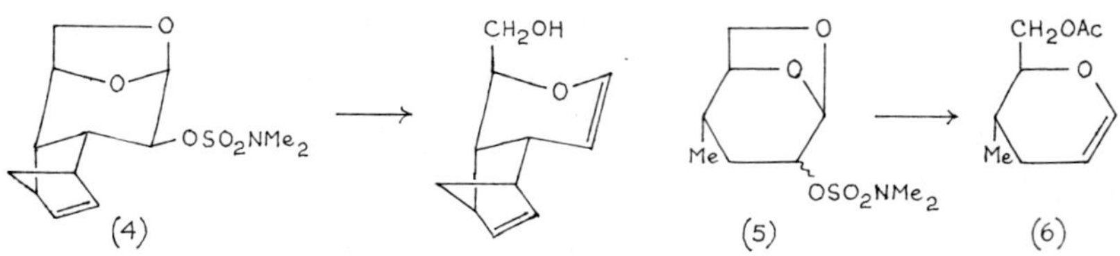

Scheme 5

Substitutions of the allylic hydroxy groups of L-rhamnal and L-fucal were favoured when acetyl chloride, N-acetylimidazole, benzoyl chloride and N-benzoylimidazole were used, and 40-60% yields of the 3-esters were obtainable. With acetic anhydride in pyridine the homoallylic groups (O-4) were preferentially substituted. 3-O-Acetyl-L-fucal underwent O-3 to O-4 ester migration.[8] A similar study of silylation reactions of these glycals was then undertaken. Several reagents gave 3-ethers selectively, the best selectivity being obtained with t-butyldimethylsilyl chloride which allowed 89% and 70% yields of the 3-O-silyl ethers of L-rhamnal and L-fucal, respectively.[9]

Several studies of the synthesis of 2-deoxy-2-fluoro-hexoses from glycals, including $^{18}F$ labelled compounds, are reported in Chapter 8.

An extended study of the [2+2]cycloaddition of p-toluenesulphonyl isocyanate to glycal derivatives (M. Chmielewski et al., Tetrahedron Lett., 1984, 25, 4797) has led to the following generalizations: (i) in the case of 3-O-substituted glycals addition occurs trans to the substituent; mixtures are obtained from 3-deoxy compounds; (ii) the additions are reversible, retroaddition occurring on standing at room temperature; (iii) on heating, the lactams rearrange to α,β-unsaturated amides (e.g. Scheme 6); (iv) the adducts react with methanol at C-1, and with

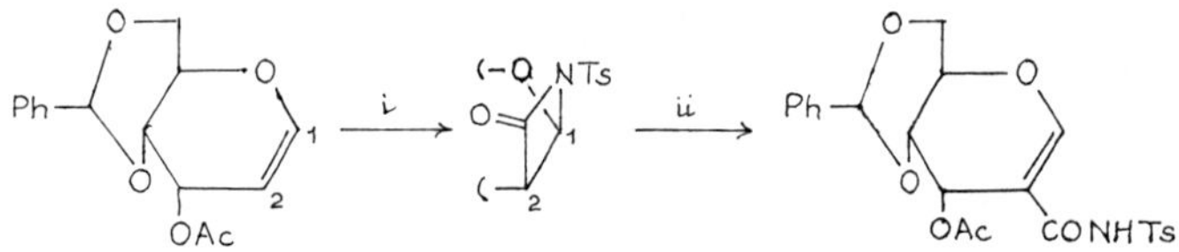

Reagents: i, TsNCO ; ii, Δ

Scheme 6

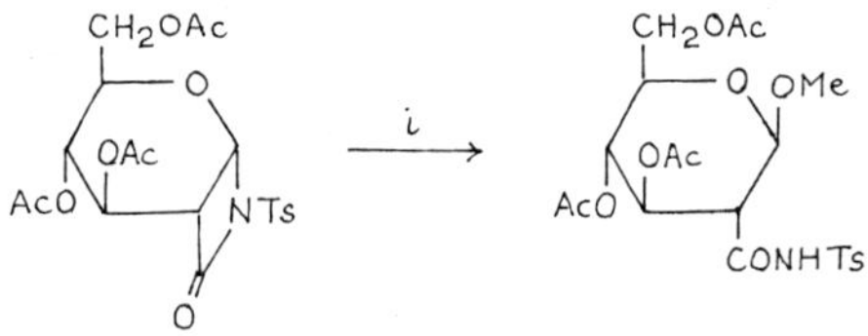

Reagents: i, MeOH

Scheme 7

inversion of configuration (e.g. Scheme 7).[10] [4+2]Cycloaddition to tri-O-acetyl-D-glucal of 1,2,4,6-tetrazine-3,6-dimethyl dicarboxylate gives initially compound (7) which rearranges mainly to give the acyclic product (8) together with small amounts of the oxidized product (9) (Scheme 8).[11]

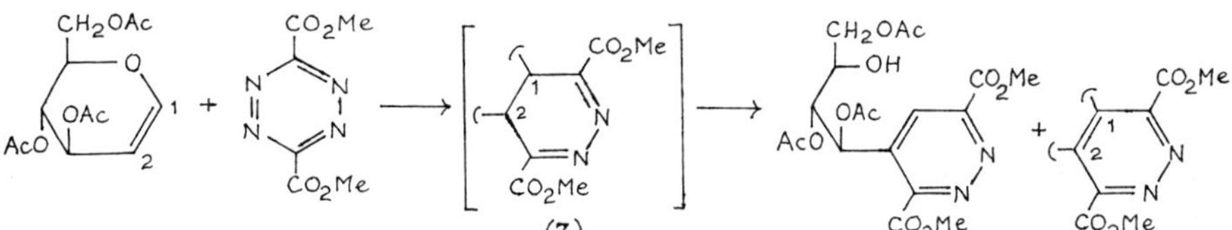

(7) (8) (9)

Scheme 8

As is the case with other acylated glycals, 3,4-di-O-acetyl-L-rhamnal reacts with alcohols in the presence of tin(IV) chloride to give 2,3-unsaturated glycosides. Various alcohols gave mainly α-linked products, the selectivity increasing in going from primary to tertiary alcohols.[12] By an addition, elimination process, tri-O-acetyl-D-galactal can be converted into the 2,3-unsaturated phenyl C-glycoside (10) by use of benzene and palladium acetate in the presence of acetic acid.[13] Treatment of the allylic alcohol (11) with 1-fluoro-1-methoxy-2-phenylsulphinylpropene gave an intermediate allylic ether which underwent Claisen rearrangement and subsequent acid elimination to afford the acrylic acid C-glycoside (12) in 67% yield.[14]

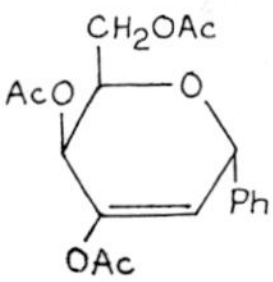


(10)

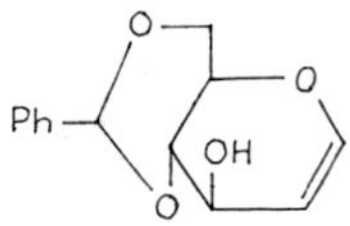


(11)

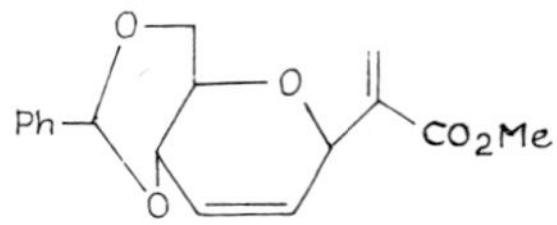


(12)

## 2 Other Unsaturated Derivatives

Fraser-Reid has provided a sequel to his 10 year old review of some unsaturated derivatives, and has stressed the usefulness of 2,3-unsaturated sugars, especially as a source of chiral natural products.[15]

The unsaturated lactones (13) are obtainable directly from the corresponding 2,3-unsaturated ethyl α-glycosides by treatment with pyridinium chlorochromate in the presence of acid resin or with Jones reagent.[16] Treatment of 2,4,6-tri-O-benzyl-3-O-tosyl-D-mannose with sodium hydride or other base gave compound (14) rather than the 1,3-anhydride,[17] and the 2-deoxy analogue (15) was produced as a by-product of the treatment of methyl 3-O-benzyl-4,6-O-

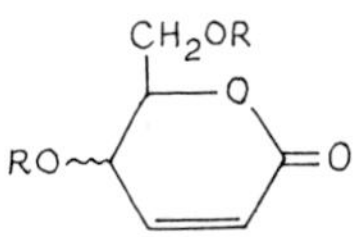


(13) R = Ac, Bz

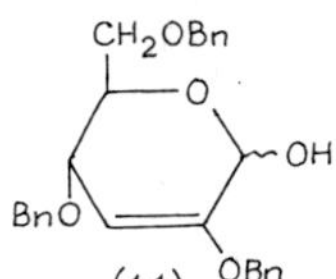


(14)

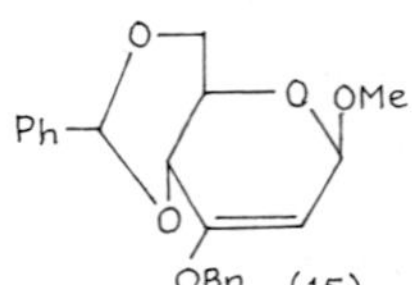


(15)

benzylidene-2-O-trifluoromethanesulphonyl-β-D-mannopyranoside with tetraethylammonium fluoride.[18]

Methyl 2,3,4-trideoxy-6,7-O-isopropylidene-D-*erythro*-hept-2-enopyranoside was converted into 4-deoxy-D-*allo*-, -D-*manno*-,

-D-*gluco*- and -D-*altro*-heptopyranosides by hydroxylation procedures.[19] 1,3-Dipolar cycloaddition of various nitrones to different 2,3-unsaturated lactones and 4-ulose derivatives gave compounds exemplified by (16) and (17),[20] and Diels-Alder addition

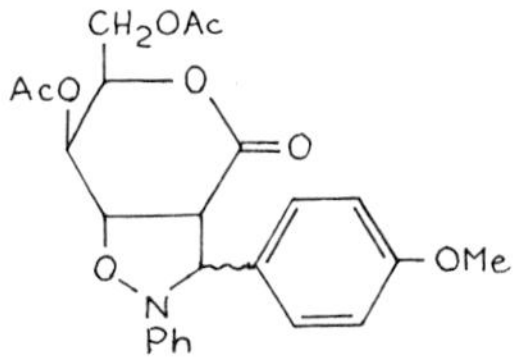

(16) (two isomers)

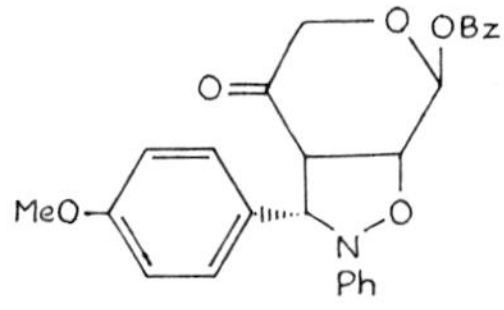

(17) (one isomer)

to the latter type of enone afforded access to "second generation enones" (Scheme 9).[21] The use of an allyl 2,3-unsaturated *C*-

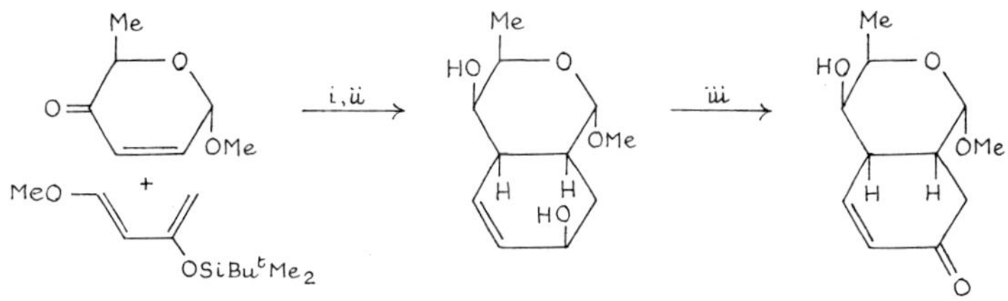

Reagents: i, Δ ; ii, LAH ; iii, $MnO_2$

Scheme 9

glycoside in a natural-product synthesis is noted in Chapter 24.

1,2:5,6-Di-*O*-isopropylidene-α-D-glucose 3-triflate gives the corresponding 3-deoxy-3-enose on treatment with the fluorinating agent TASF, whereas the D-allose isomer affords the product of substitution (see Chapter 8). A nucleoside containing a 3-deoxy-3-enofuranose ring is noted in Chapter 20, and the *X*-ray crystal structure of a 3,4-dideoxy-5-fluorohex-3-enopyranoside derivative is recorded in Chapter 22.

A cycloaddition reaction to levoglucosenone, a hex-3-enopyranosid-2-ulose derivative, has been used to give a tricyclic product (Chapter 24), and some nucleophilic addition reactions of related enones are indicated in Scheme 10. The *X*-ray crystal structure of the 4,5-unsaturated compound (19), the enantiomer of (18), was recorded.[22] Not surprisingly, some methyl 2,3,4-tri-*O*-acetyl-β-D-glucopyranosyl uronate nucleosides underwent elimination on treatment with DBU to give 4',5'-unsaturated products.[23] An antibiotic containing a hex-4-enofuranosyl unit is noted in Chapter 19.

In the area of 6-deoxyhexopyranosyl derivatives, *N*-(phenylselenyl)phthalimide has been used to effect elimination from the

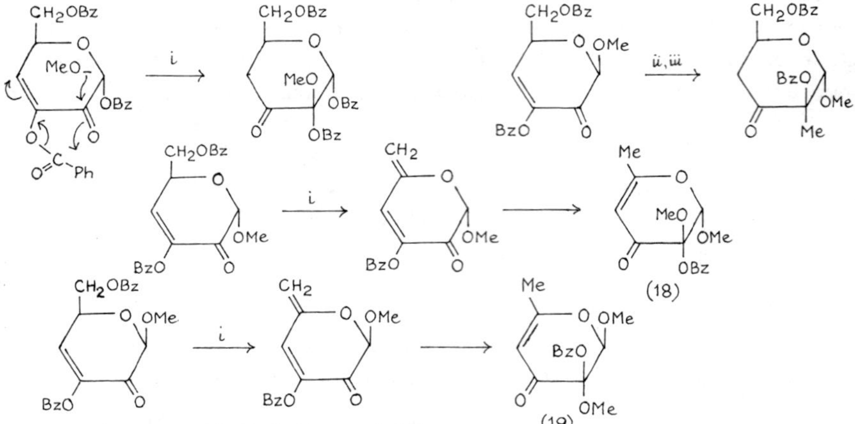

(18) (19)

Reagents: i, MeOH - $OH^-$(cat.); ii, MeMgI; iii, $OH^-$

Scheme 10

6-hydroxy compound to give alkene (20), and the same reagent allowed access to the related unsaturated sugar (21) from the acyclic alkene (22) (obtained using a Wittig reaction) by an

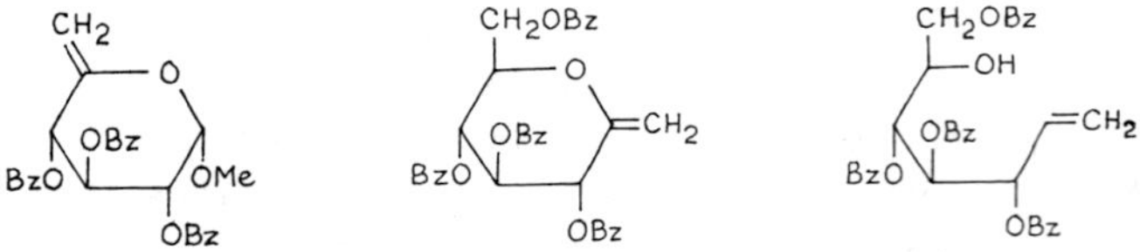

(20) (21) (22)

addition-elimination procedure.[24] Treatment of compounds such as (20) with mercury salts in aqueous acetone is known to give deoxyinososes. It has now been shown that mercury-containing intermediates are involved (Scheme 11).[25] Applications of this

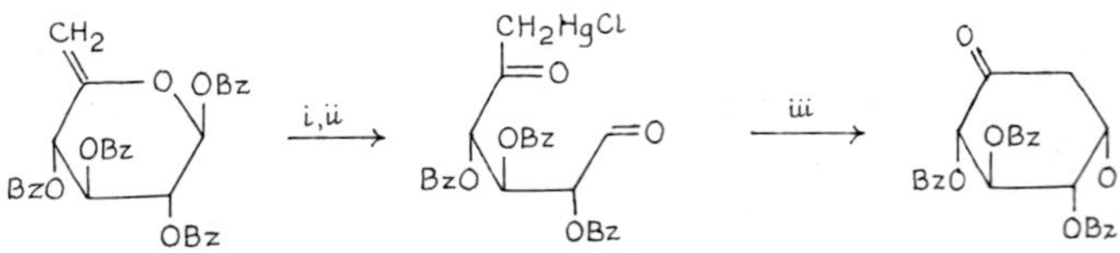

Reagents: i, $HgCl_2$-$H_2O$; ii, remove $H_2O$; iii, continue i or use $H_2S$

Scheme 11

reaction to give aminocyclohexane derivatives are recorded in Chapter 18.

Methylated 6-deoxy-6-iodohexopyranosides carrying β-D-xylopyranosyl groups at O-2, O-3, O-4 and O-3 and O-4 have been converted into 5,6-dideoxyhex-5-enose derivatives by treatment with zinc in model studies required for the development of a specific method of polysaccharide chain cleavage. The procedure was then applied to reduced, permethylated tragacanthic acid.[26]

As is indicated in Scheme 12, reaction of the anion derived from

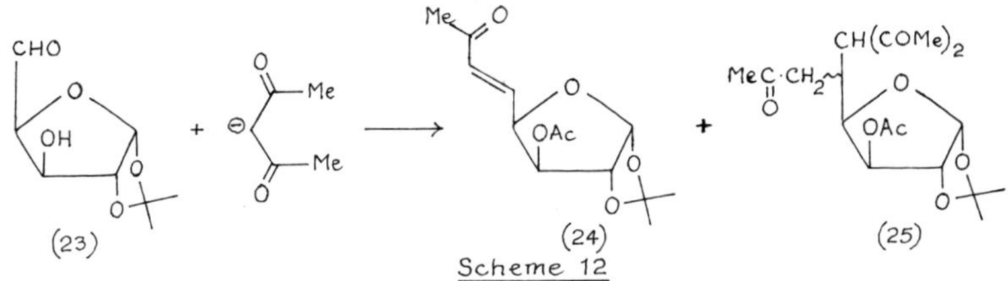

Scheme 12

pentane-2,4-dione with aldehyde (23) gives a product which undergoes acetyl group migration to afford the 5-ene (24), together with the product of Michael addition (25).[27] Cycloaddition of trimethylenemethane to the unsaturated uronate (26) gave the cyclopentane derivative (27) (Scheme 13).[28]

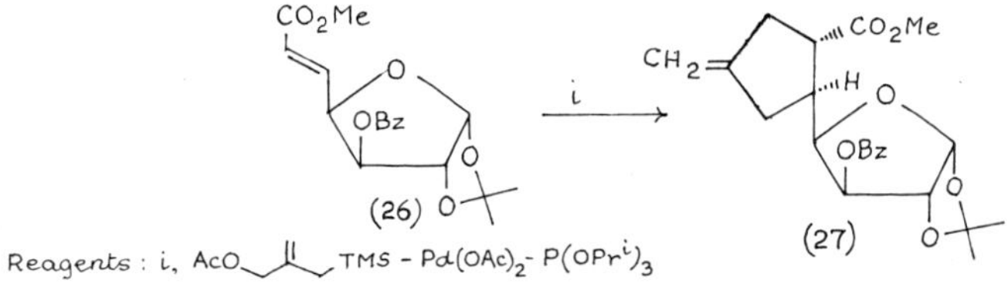

Scheme 13

The isomer ratio of 6,7- and 7,8-alkenes produced by reaction of furanoid and pyranoid aldehydes with allyltrimethylsilane depends greatly on the Lewis acid catalyst used (Scheme 14).[29] The

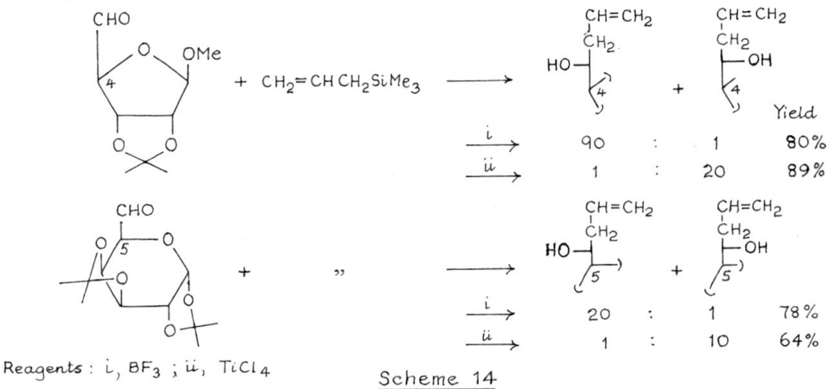

Scheme 14

elaborated 6-ene (28) has been prepared from a 3-O-benzyl-6-deoxy-6-iodo-glycoside by use of a sulphoxide-based carbanion. It is a potent inhibitor of 3-hydroxy-3-methyl glutaryl coenzyme A reductase.[30]

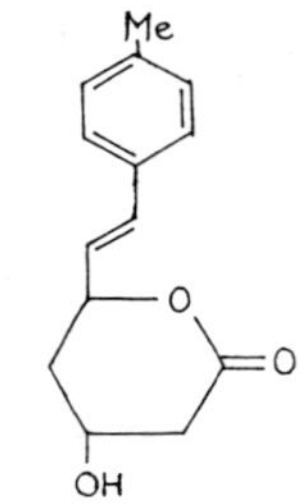

(28)

Interesting and novel carbon-carbon bond-forming radical reactions have been applied to halogenated derivatives, xanthate esters and related compounds. One paper describes the use of vinylstannanes which involves radical addition, elimination processes (Scheme 15);[31] another uses allylstannanes (Scheme 16).[32]

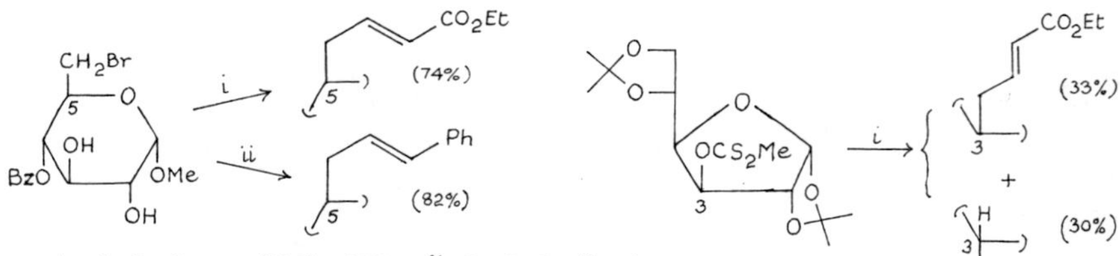

Reagents: i, $Bu_3Sn$–CH=CH–$CO_2Et$ - AIBN ; ii, $Bu_3Sn$–CH=CH–Ph - AIBN

Scheme 15

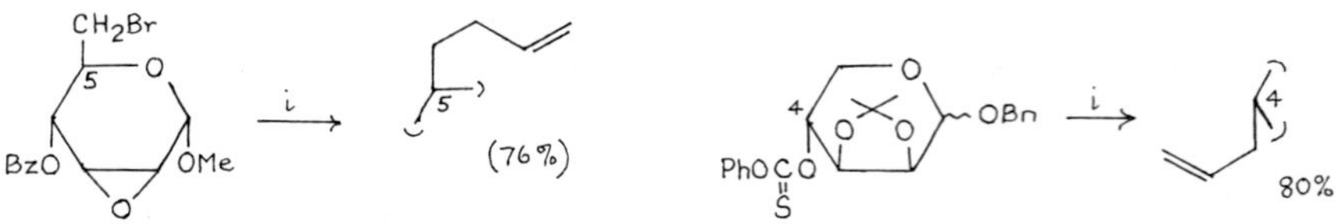

Reagents: i, CH$_2$=CHCH$_2$SnBu$_3$ - AIBN or hν

Scheme 16

Both approaches offer means of obtaining extended-chain unsaturated sugars, branched-chain unsaturated compounds, or C-glycosides.

Acetylide condensation with an aldehyde has afforded two stereoisomers of the 7,8'-yne (29). The X-ray structure of one, as its acetate, was determined.[33]

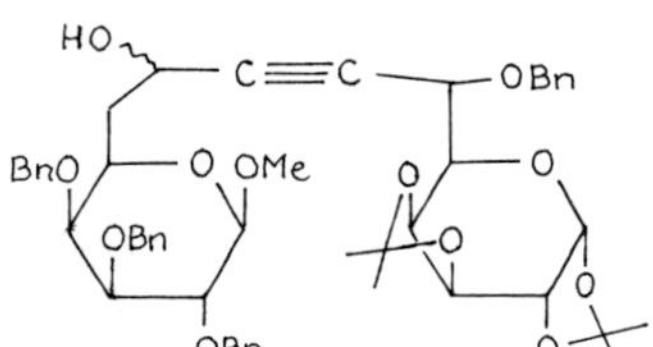

(29)

Some useful acyclic alkenes have been produced by a ring-opening reaction involving aldehydic intermediates (Scheme 17).[34] An

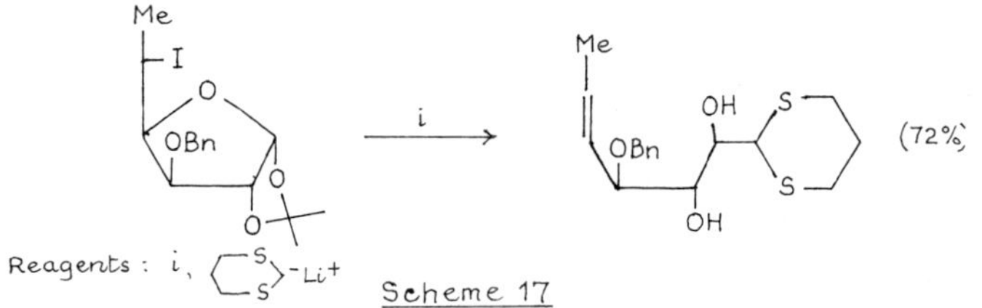

Scheme 17

octulosonic acid disaccharide synthesis involving an acyclic intermediate made by a Wittig condensation is referred to in Chapter 2.

A set of acyclic dienes (30) have been prepared from the unsaturated lactone (31) and photolysed in attempts to obtain [2+2]-

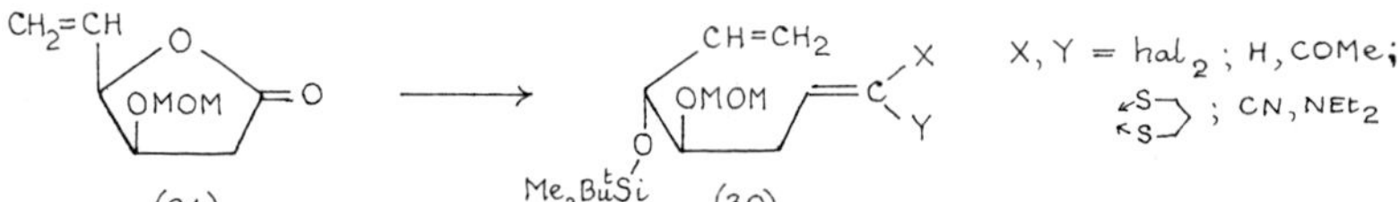

(31) (30)

cycloaddition products. Earlier work with a related compound led to a new approach to the prostaglandins, but compound (30) failed to undergo the required reaction. In some cases (e.g. X, Y = H, COMe) double-bond migration occurred.[35]

The branched-chain unsaturated compound (32) has been synthesized from 2,3-O-isopropylidene-D-glyceraldehyde (Scheme 18).[36]

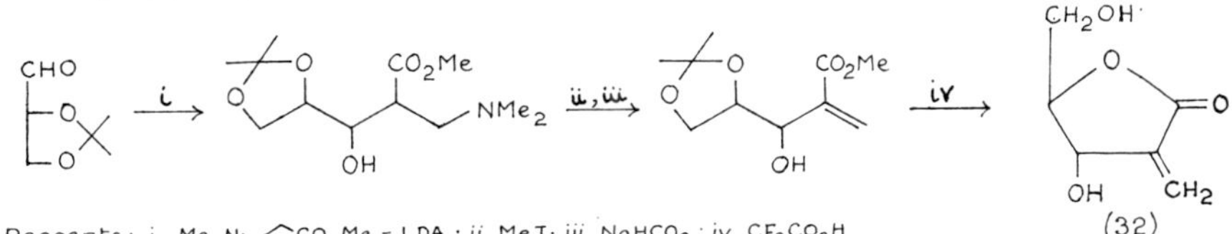

(32)

Scheme 18

## References


1 J.C.-Y.Cheng, U.Hacksell, and G.D.Daves, J. Org. Chem., 1985, 50, 2778.
2 V.Jager and I.Muller, Tetrahedron, 1985, 41, 3519.
3 W.Korytnyk and O.Dodson-Simmons, Prog. Clin. Biol. Res., 1985, 172 (Chem. Abstr., 1985, 102, 185 364).
4 S.J. Danishefsky and M.Bednarski, Tetrahedron Lett., 1985, 26, 3411.
5 S.J.Danishefsky and C.J.Maring, J. Am. Chem. Soc., 1985, 107, 1269.
6 M.Maier and R.R.Schmidt, Liebigs Ann. Chem., 1985, 2261.
7 P.Bhate and D.Horton, Carbohydr. Res., 1985, 139, 191.
8 D.Horton, W.Priebe, and O.Varela, Carbohydr. Res., 1985, 144, 317.
9 D.Horton, W.Priebe, and O.Varela, Carbohydr. Res., 1985, 144, 325.
10 M.Chmielewski, Z.Kaluza, C.Belzecki, P.Salanski, J.Jurczak,and H.Adamowicz, Tetrahedron, 1985, 41, 2441.

11 H.Gnichtel and C.Gumprecht, Liebigs Ann. Chem., 1985, 628.
12 P.Bhate, D.Horton, and W.Priebe, Carbohydr. Res., 1985, 144, 331.
13 H.Gillier-Pandraud, R.Brahmi, V.Bellosta-Dechavanne, and S.Czernecki, Can. J. Chem., 1985, 63, 491.
14 J.-M.Vatele, Carbohydr. Res., 1985, 136, 177.
15 B.Fraser-Reid, Acc. Chem. Res., 1985, 18, 347.
16 A.Hernandez, M.Martin-Lomas, R.M.Rabanal, and S.Valverde, An. Quim., Ser.C, 1985, 81, 76 (Chem. Abstr., 1985, 103, 215 689).
17 F.Kong, B.Su, and C.Schuerch, Carbohydr. Res., 1985, 142, 152.
18 T.Haradahira, M.Maeda, Y.Kai, H.Omae, and M.Kojima, Chem. Pharm. Bull., 1985, 33, 165.
19 J.Jurczak, T.Bauer, and J.Kihlberg, J. Carbohydr. Chem., 1985, 4, 447.
20 I.Panfil and M.Chmielewski, Tetrahedron, 1985, 41, 4713.
21 M.A.Rahman, D.R.Kelly, R.M.Srivastava, and B.Fraser-Reid, Carbohydr. Res., 1985, 136, 91.
22 F.W.Lichtenthaler, S.Nishiyama, P.Kohler, and H.J.Lindner, Carbohydr. Res., 1985, 136, 13.
23 V.A.Timoshchuk and L.N.Kulinkovich, Carbohydr. Res., 1985, 145, 123.
24 J.-M.Lancelin, J.-R.Paugny, and P.Sinay, Carbohydr. Res., 1985, 136, 369.
25 R.Blattner, R.J.Ferrier, and S.R.Haines, J. Chem. Soc., Perkin Trans. 1, 1985, 2413.
26 G.O.Aspinall, D.Chatterjee, and L.Khondo, Can. J. Chem., 1984, 62, 2728.
27 F.J.Lopez Aparicio, F.Zorrilla Benitez, P.Garcia Mendoza, and F.Santoyo Gonzalez, Carbohydr. Res., 1985, 135, 303.
28 B.M.Trost, J.Lynch, and P.Renaut, Tetrahedron Lett., 1985, 26, 6315.
29 S.Danishefsky and M.DeNinno, Tetrahedron Lett., 1985, 26, 823.
30 J.D.Prugh, C.S.Rooney, A.A.Deana, and H.G.Ramjit, Tetrahedron Lett., 1985, 26, 2947.
31 J.E.Baldwin and D.R.Kelly, J. Chem. Soc., Chem. Commun., 1985, 682.
32 G.E.Keck, E.J.Enholm, J.B.Yates, and M.R.Wiley, Tetrahedron, 1985, 41, 4079.
33 J.W.Krajewski, P.Gluzinski, S.Jarosz, A.Zamojski, J.Bleidelis, A.Mishnyov, and A.Kemme, Carbohydr. Res., 1985, 144, 183.
34 H.Redlich, J.B.Lenfers, and W.Bruns, Liebigs Ann. Chem., 1985, 1570.
35 R.J.Ferrier and P.C.Tyler, Carbohydr. Res., 1985, 136, 249.
36 C.Papageorgiou and C.Benezra, J. Org. Chem., 1985, 50, 157.

# 14
# Branched-chain Sugars

## 1 Compounds with an R-C-O Branch

The occurrence of 6-deoxy-3-C-methyl-β-D-altropyranose as a C-aryl glycoside in the antibiotic virenomycin is referred to in Chapter 19.

Standard Grignard reactions with glycosidulose derivatives have been employed in several syntheses. Methyl α-D-mycaroside (1) has been prepared from mannose in a 10-step sequence via the 6-deoxy intermediate (2) in 22% overall yield (Scheme 1), and L-rhamnose

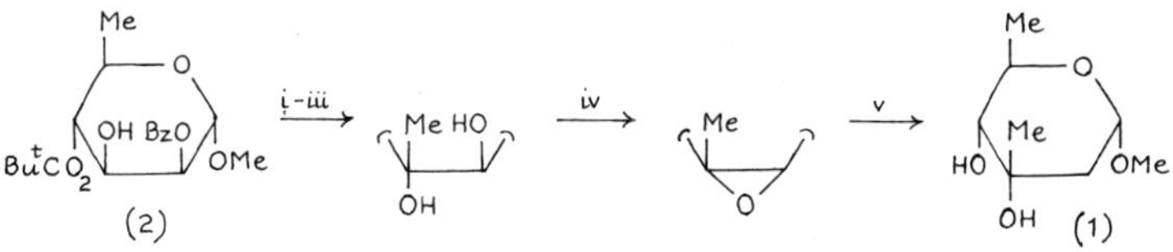

Reagents: i, PCC-NaOAc-Mol.sieve; ii, MeMgI; iii, Resin(OH⁻)-MeOH; iv, $Tf_2O$; v, $LiEt_3BH$

Scheme 1

similarly furnished the L-enantiomer. LTBH treatment of the 2-O-triflate of the minor manno epimer from the Grignard reaction did not give the corresponding 2-deoxy-arabino sugar (olivomycoside) but instead methyl migration occurred to give the 2,6-dideoxy-2-C-methyl-allopyranoside (3), with the suggested mechanism shown in Scheme 2.[1] The doubly branched-chain sugar (4) has been prepared

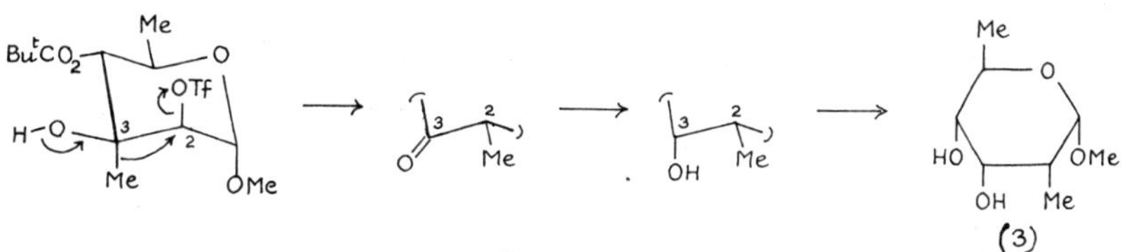

Scheme 2

from 4-O-allyl-1,6;2,3-dianhydro-D-mannopyranose using dimethylmagnesium to open the oxiran ring and methyl magnesium bromide to introduce the 4-C-methyl group on the osulose (5). C-3 and C-4 epimers of (4) were also prepared by the alternative Wittig procedure on (5), with or without preliminary base-catalysed C-3 epimerization (Scheme 3). The 4-O-mesylate of (4) on solvolysis

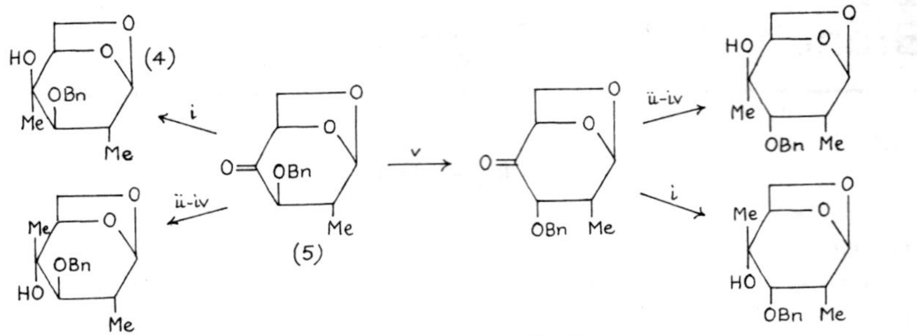

Reagents: i, MeMgBr; ii, $CH_2$=$PPh_3$; iii, MCPBA; iv, LAH; v, $MeO^-$

Scheme 3

rearranged to the furanoside (6)(previously assigned a different structure by the same group).[2] A Grignard - osulose reaction has also been employed to prepare the 3-C-methyl-6-deoxy-D-gulose derivative (7) from D-glucose.[3] The isomeric sugar L-vinelose (8)

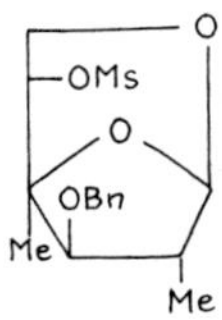

(6)

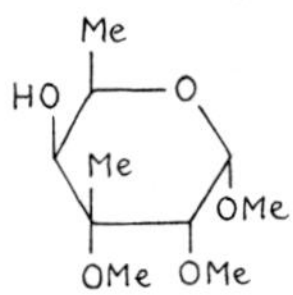

(7)

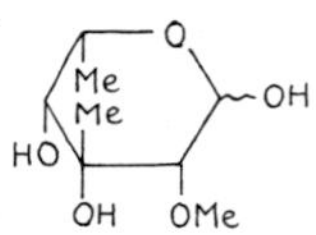

(8)

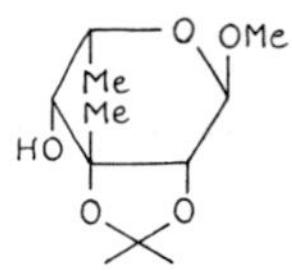

(9)

has been synthesized from rhamnose via a standard glycosid-4-ulose by enolate methylation followed by borohydride reduction; the resulting intermediate (9) on deacetonation - reacetonation gave the 3,4-O-isopropylidene isomer leading to the required 2-O-methyl ether product.[4]

Non-carbohydrate precursors have been used in stereo-controlled syntheses of 3-C-methyl hexoses. The aldol condensation shown in Scheme 4 gave a mixture of stereoisomers leading to the diastereo-

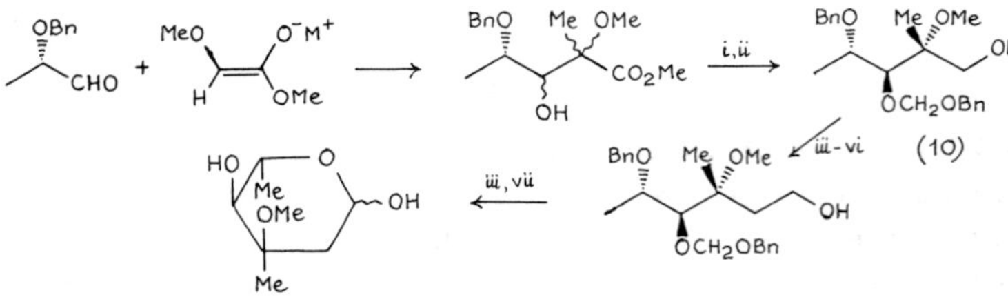

Reagents: i, $BnOCH_2Cl$; ii, LAH; iii, $(COCl)_2$-DMSO; iv, $Ph_3P$=$CH_2$; v, 9-BBN; vi, $H_2O_2$-$OH^-$; vii, $H_2$-Pd

Scheme 4

isomer (10) after chromatography, which was converted to L-cladinose (11) as indicated in the Scheme.[5] Syntheses of racemic mycarose (12) and <u>epi</u>-axenose (13) resulted from pent-4-en-1-yne as outlined in Scheme 5.[6] Lewis acid catalysed condensation of 2-benzyloxy-ethanal with the furan (14) led to the racemic glycenonolactone (15), which was then converted to the trifluoromethyl branched-chain sugar derivatives (16) and (17)(Scheme 6).[7]

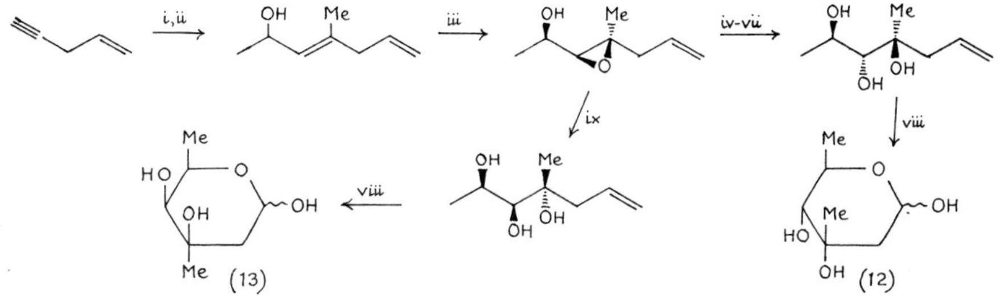


Reagents: i, $Me_3Al-(Cp)_2ZrCl_2$; ii, MeCHO; iii, $Ti(OPr^i)_4-Bu^tO_2H$; iv, PhNCO; v, $Et_2AlCl$; vi, $H_3O^+$; vii, $MeO^-$; viii, $O_3-Me_2S$; ix, $H_2SO_4$-THF-$H_2O$

Scheme 5

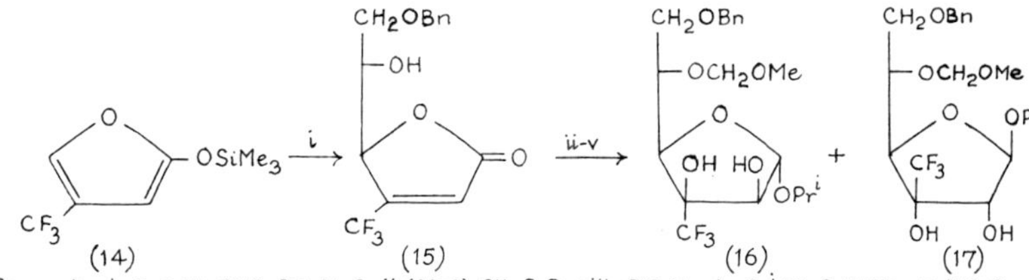


Reagents: i, $BnOCH_2CHO-BF_3.Et_2O$; ii, $(MeO)_2CH-P_2O_5$; iii, DIBAL; iv, $Pr^iOH$-PyHOTS; v, $KMnO_4$

Scheme 6

A new synthesis of L-apiose uses the photochemical reaction of the (-)-8-phenylmenthyl ester of phenylglyoxylic acid with 2,2-dimethyl-1,3-dioxalen to generate the oxetan (18) with 96% d.e., leading in turn to the branched-chain sugar (19) and the apioside (20)(Scheme 7).[8]

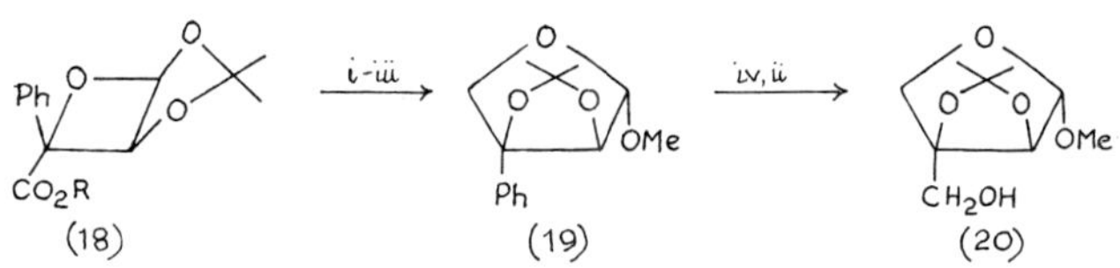


Reagents: i, $H^+$-MeOH; ii, LAH; iii, $Me_2CO-H^+$; iv, $RuO_4$

Scheme 7

A new Claisen-type rearrangement for making $\alpha$-substituted acrylic esters from allyl alcohols has been applied to carbohydrate allylic alcohols, glycenoside (21) giving the C-5 substituted heptenuronate (22).[9]

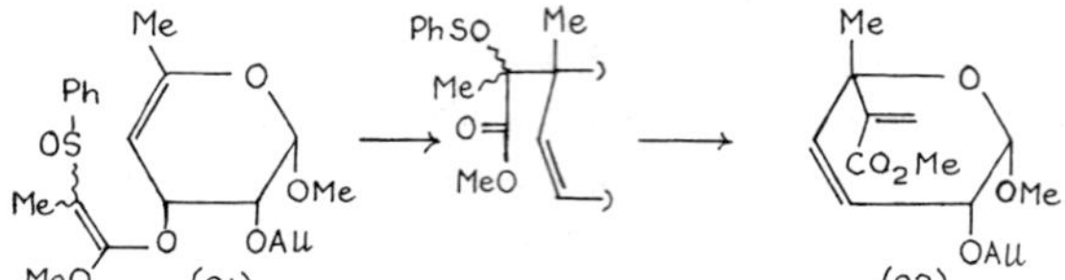

(21) (22)

Whereas the cyclic carbonate (23) reacts with tributylstannane to give only the expected 3-deoxy-3-C-hydroxymethyl product (24), the corresponding α-hydroxyethyl derivative (25) gave both the deoxy product (26) and the 3-C-ethyl branched-chain isomer (27), the ratio depending on the concentration of the reactants.[10]

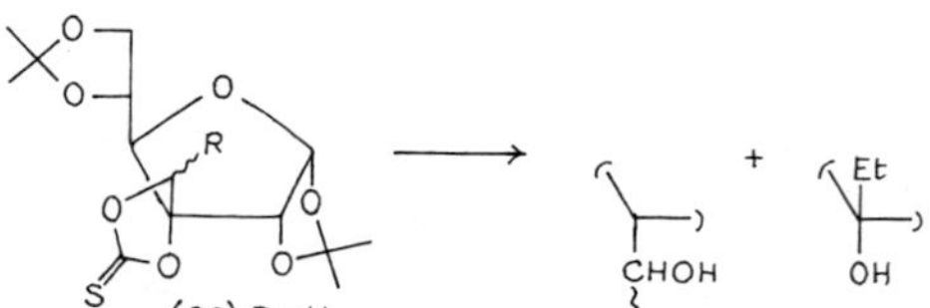

(23) R = H
(25) R = Me
(24) R = H
(26) R = Me
(27)

The formation of branched-chain amino-sugars from enaminals and from branched-chain glycosiduloses is referred to in Chapter 9.

## 2 Compounds with an R-C-N Branch

A note has been published on the conversion of an acetamido branched-chain sugar derivative to the corresponding nitro analogue (L-decilonitrose) by sequential N-deacetylation using calcium in liquid ammonia followed by m-chloroperbenzoic acid oxidation.[11] Syntheses of D- and L-rubranitrose by Brimacombe's group are mentioned in Chapter 10, and the synthesis of 3'-C-methyldaunorubicin (containing vancosamine) is covered in Chapter 19.

## 3 Compounds with an R-C-H or R-C-R Branch

A review on syntheses involving C-C bond formation via radicals from organotin and organomercury compounds includes examples of branched-chain sugars.[12] Thermally or photochemically induced free-radical C-C bond formation between allyl- or methallyl-tributylstannane with carbohydrates with suitably labile halogen, oxygen, or selenium substituents can yield deoxy branched-chain sugars in good yield, as illustrated in Scheme 8; the stereochemistry results from steric approach control.[13] A similar reaction of a vinyl-stannane with the 3-O-methyl xanthate of diacetone glucose gave a mixture of the branched-chain sugar (28) and the reduced 3-deoxy sugar (Scheme 9).[14]

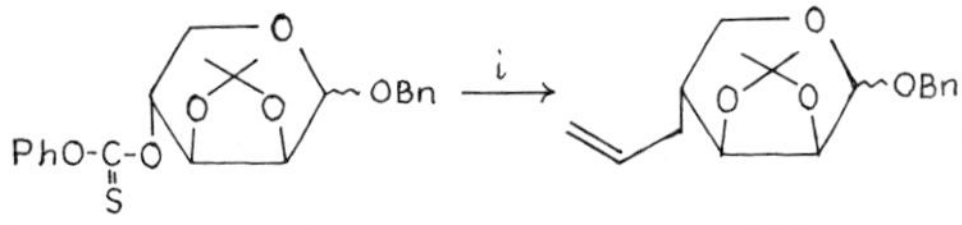

Reagent: i, $CH_2{=}CHCH_2SnBu_3$ - hν

Scheme 8

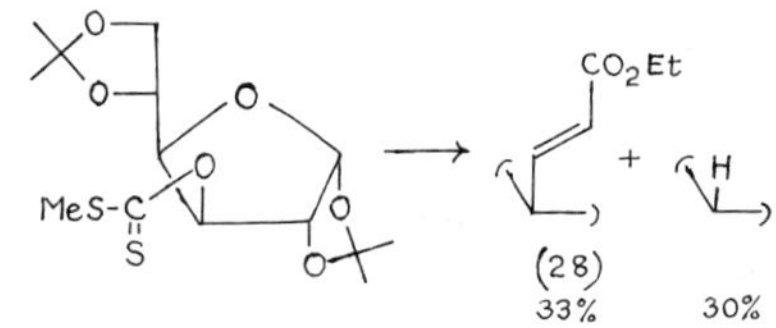

Reagents: i, $Bu_3SnCH{=}CHCO_2Et$ - AIBN

Scheme 9

Cycloaddition reactions leading to branched-chain sugars have been further exploited. Addition of toluene-p-sulphonyl isocyanate to glycals under high pressure (see Vol.18,p.142) leads to β-lactam derivatives of 2-C-carboxylated sugars; further glycal examples suggest that (i) addition occurs on the face opposite to the C-3 substituent, (ii) addition is reversible, retro-addition occurring on standing at room temperature or on heating, (iii) on heating under pressure lactams rearrange to α,β-unsaturated amides, e.g. (29)→(30), and (iv) methyl glycosides arise by methanol substitution at C-1 with inversion, e.g. (31)→(32).[15] 1,3-Dipolar

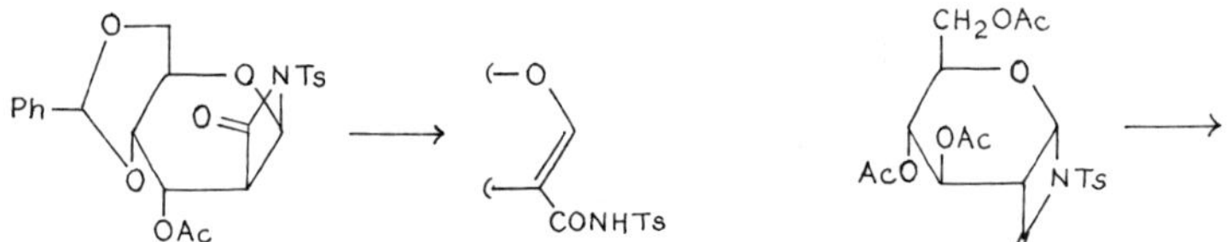

(29) (30) (31) (32)

cycloadditions of nitrones to sugar α,β-unsaturated lactones and enones have been found to be regiospecific, with addition preferentially trans to existing ring substituents; examples are given in Scheme 10.[16] Another paper reports a similar observation on an

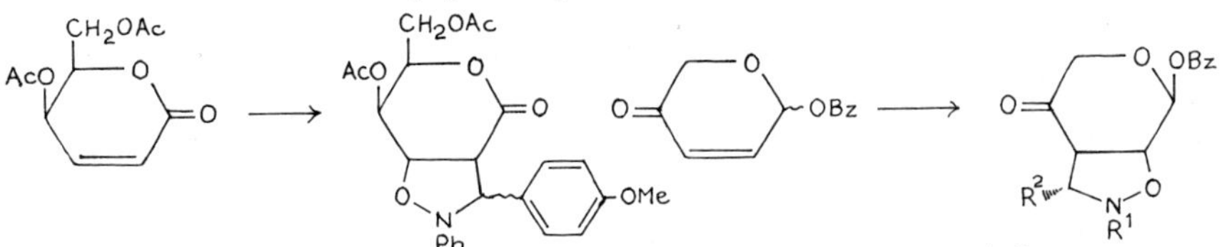

endo : exo 7:3

$R^1,R^2$ = Ph, Bn, p-MeOC$_6$H$_4$

Scheme 10

enone and notes that nitrone addition to a hexenoside was less regiospecific, as indicated in Scheme 11.[17]

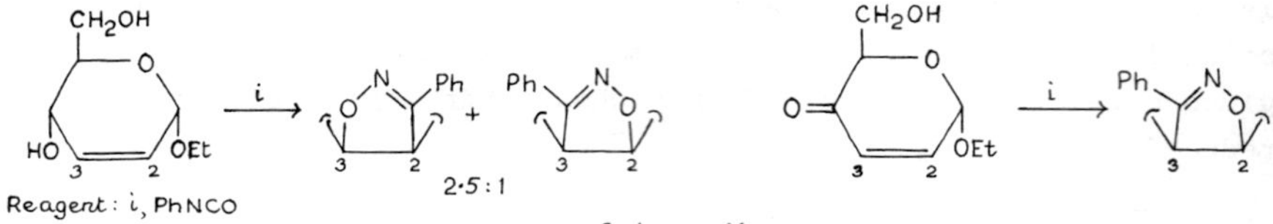

Scheme 11

Reaction of organocuprates with the hexenoside (33) yields 2-C-alkyl branched-chain sugars (34) by allylic displacement. Analogously, the terminal alkene (35) gave the chain-extended sugar (36).[18]

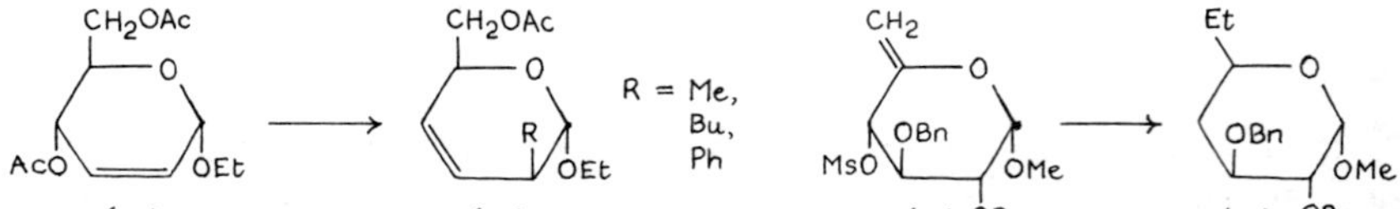

(33) (34) (35) (36)

Several methyl 6-deoxy-hexopyranoside 2- and 4-tosylates undergo rearrangement to branched-chain sugars on treatment with lithium triethylborohydride, e.g. (37)→(38) and (39)→(40).[19] Methyl 6-deoxy-2,3-di-O-tosyl-α-D-galactopyranoside yielded both a 3-deoxy-

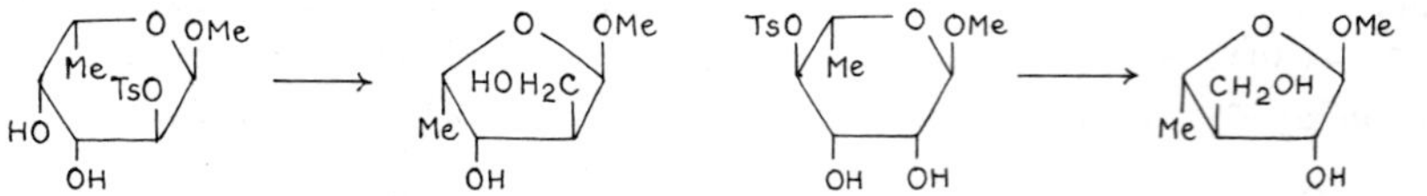

(37) (38) (39) (40)

3-C-hydroxymethyl-xylofuranoside and a mixture of 3,6-dideoxy-hexosides, epimeric at C-4, as outlined in Scheme 12.[20]

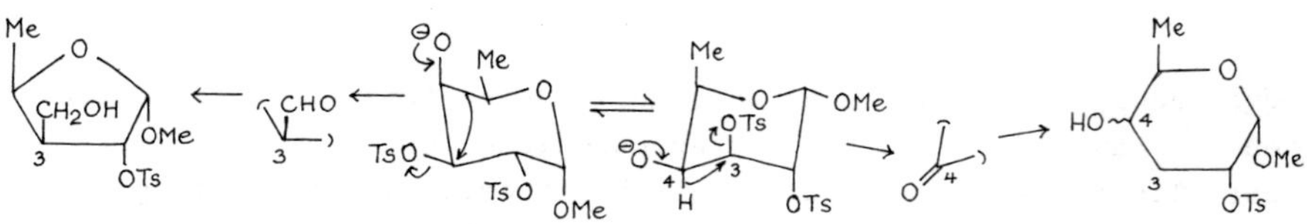

Scheme 12

A study on the Knoevenagel reaction of some aldehydo dialdose derivatives with pentane-2,4-dione revealed that acetyl transfer can occur, with further Michael addition of the reagent to the resulting enone (Scheme 13).[21]

The absolute configurations of 1,3-dideoxy-5,6-O-isopropylidene-3-C-methyl-D-ribo- and D-lyxo-hex-2-ulose have been reported.[22]

An orthoester derived from the exocyclic unsaturated branched-chain sugar (41) underwent Claisen rearrangement to form the <u>gem</u>-disubstituted branched-chain sugar (42)(Scheme 14); the stereochemistry of (42) was deduced from the fact that the derived C-formyl sugar (43) did not form a hemiacetal. The 5-epimer of (41) behaved similarly.[23]

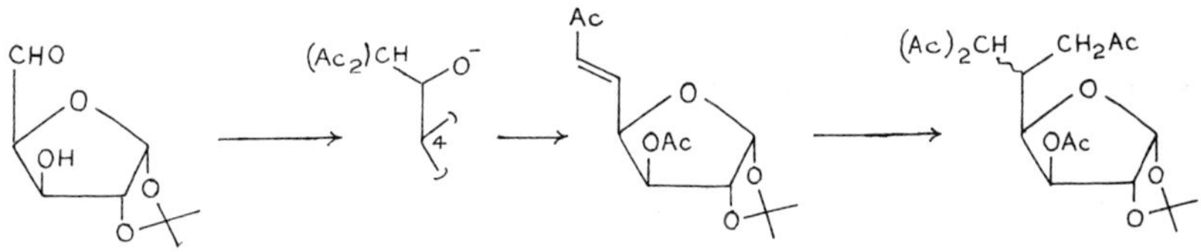

Scheme 13

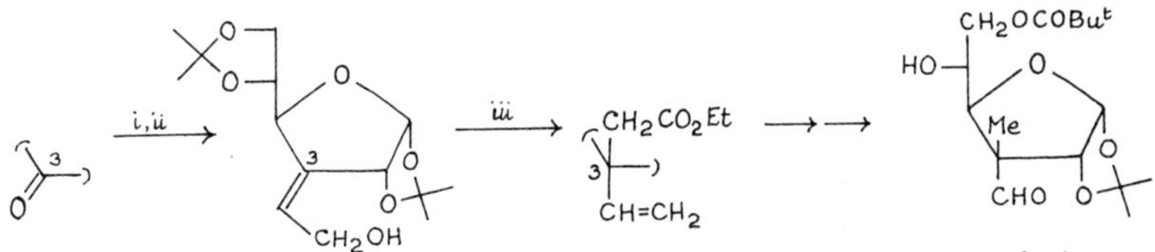

(41) (42) (43)

Reagents: i, $Ph_3P{=}CHCO_2Et$; ii, DIBAL; iii, $MeC(OEt)_3$-$EtCO_2H$ (135°)

Scheme 14

Methods for preparing geminal di-methyl branched-chain sugars have been described, using either dichlorocarbene addition to a methylene-substituted derivative (satisfactory for the 3-substituted pyranoside (44) derived from glucose) or by methylcuprate addition to the 4-<u>C</u>-methyl-hex-3-en-2-one (45) leading to the 3-deoxyhexosid-2-ulose (46) or the <u>xylo</u>-hexopyranoside (47) (Scheme 15). A similar sequence on a 2-<u>C</u>-methyl-hex-2-en-4-one

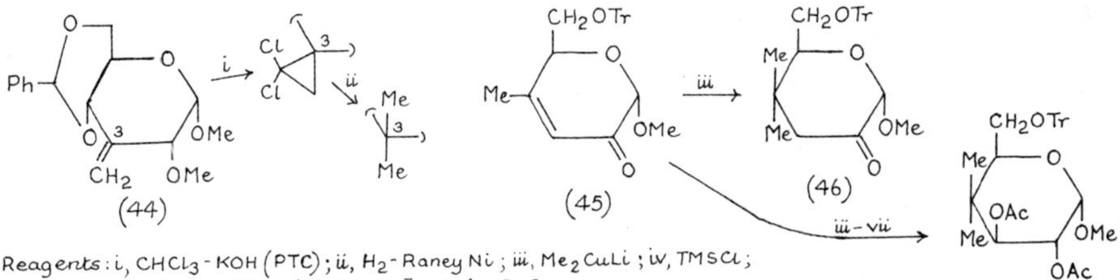

(44) (45) (46) (47)

[Xyl : Ara = 2:1]

Reagents: i, $CHCl_3$-KOH (PTC); ii, $H_2$-Raney Ni; iii, $Me_2CuLi$; iv, TMSCl; v, $NaBH_4$-$CoCl_2$; vi, $H_2O_2$-$OH^-$; vii, $Ac_2O$-Py

Scheme 15

gave the corresponding 2,2-di-<u>C</u>-methyl branched-chain sugar.[24]

The same dichlorocarbene approach for preparing a 4-di-C-methyl derivative from mannose is mentioned in Chapter 24.

Reaction of 2,3-O-isopropylidene-D-glyceraldehyde with a silylated enacetal leads to the 2-deoxy-2,2-di-C-methyl-*erythro*-pentonolactone (48).[25] Full details of the Diels-Alder reaction of a glycosenone mentioned last year (Vol.18,p.143,ref.41) have been reported.[26] Wittig reaction on a nucleoside glycosulose derivative led to the branched-chain sugar analogue (49), which was cyclized to the lactone (50) and dehydro derivative (51).[27] Other branched-

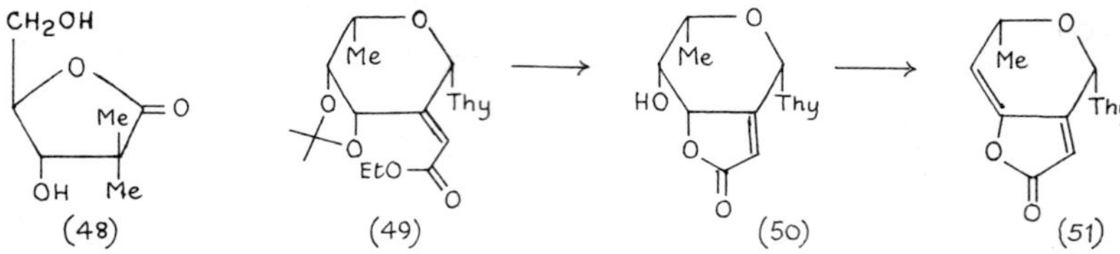

chain sugar nucleosides are mentioned in Chapter 20.

Reaction of a fused-ring branched-chain sugar is referred to in Chapter 13, and other branched-chain sugar derivatives are mentioned in Chapter 24.

## References


1 R.W.Binkley, J. Carbohydr. Chem., 1985, 4, 227.
2 A.F.Sviridov, M.S.Ermolenko, D.Y.Yashunsky, A.S.Shashov, and N.K.Kochetkov, Carbohydr. Res., 1985, 136, 101.
3 A.U.Rahman, N.Hong, and J.Yoshimura, J. Chim. Chem. Soc. (Taipei), 1984, 31, 287 (Chem. Abstr., 1985, 102, 79 223).
4 A.Klemer and H.Stegt, J. Carbohydr. Chem., 1985, 4, 205.
5 C.H.Heathcock and S.H.Montgomery, Tetrahedron Lett., 1985, 26, 1001.
6 W.R.Roush and S.M.Hagadorn, Carbohydr. Res., 1985, 136, 187.
7 K.Kawada, O.Kitagawa, T.Taguchi, Y.Hanzawa, Y.Kobayashi, and Y.Iitaka, Chem. Pharm. Bull., 1985, 33, 4216.
8 A.Nehrings, H.-D.Scharf, and J.Runsink, Angew. Chem. Int. Ed. Engl., 1985, 24, 877.
9 J.-M.Vatèle, Carbohydr. Res., 1985, 136, 177.
10 H.Redlich, W.Sudau, and H.Paulsen, Tetrahedron, 1985, 41, 4253.
11 J.S.Brimacombe and K.M.M.Rahman, Carbohydr. Res., 1985, 140, 163.
12 B.Giese, Angew. Chem. Int. Ed. Engl., 1985, 24, 553.
13 G.E.Keck, E.J.Enholm, J.B.Yates, and M.R.Wiley, Tetrahedron, 1985, 41, 4079.
14 J.E.Baldwin and D.R.Kelly, J.Chem. Soc., Chem. Commun., 1985, 682.
15 M.Chmielewski, Z.Kaluza, C.Belzecki, P.Salanski, J.Jurczak, and H.Adamowicz., Tetrahedron, 1985, 41, 2441.
16 I.Panfil and M.Chmielewski, Tetrahedron, 1985, 41, 4713.
17 H.Gnichtel and L.Autenrieth-Ansorge, Liebig's Ann. Chem., 1985, 2217.
18 Y.Chapleur and Y.Grapsas, Carbohydr. Res., 1985, 141, 153.
19 H.H.Baer, D.J.Astles, H.-C. Chin, and L.Siemsen, Can. J.Chem., 1985, 63, 432.
20 R.W.Binkley, J. Org. Chem., 1985, 50, 5646.
21 F.J.Lopez Aparicio, F.Zorrilla Benitez, P.Garcia Mendoza, and F.Santoyo Gonzalez, Carbohydr. Res., 1985, 135, 303.
22 F.J.Lopez Aparicio, I.Izquierdo Cubero, and M.D.Portal Olea, An. Quim., Ser. C, 1983, 79, 295 (Chem. Abstr., 1985, 102, 62 525).

23 K.Tadano, Y.Idogaki, H.Yamada, and T.Suami, Chem. Lett., 1985, 1925.
24 H.Hashimoto, N.Kawauchi, and J.Yoshimura, Chem. Lett., 1985, 965.
25 Y.Kita, H.Yasuda, O.Tamura, F.Itoh, Y.Y.Ke, and Y.Tamura, Tetrahedron Lett., 1985, 26, 5777.
26 M.A.Rahman, D.R.Kelly, R.M.Srivastava, and B.Fraser-Reid, Carbohydr. Res., 1985, 136, 91.
27 J.Herscovici, M.-J.Egron, and K.Antonakis, Carbohydr. Res., 1985, 142, 324.

# 15
# Aldosuloses and Dialdoses

## 1 Aldosuloses

A review - "Regioselective Manipulation of Hydroxy Groups via Organotin Derivatives" - refers to the preparation of aldosulose compounds.[1] (See examples in Scheme 1).

Glycosid-2-ulose derivatives mesylated at O-4 give access to enolones, e.g. (1), from which glycosid-3-ulose compounds, e.g. (2), can be obtained. Such compounds give good access to specifically deuterated analogues.[2] Some reactions of compound (1) and closely related enones are detailed in Chapter 13.

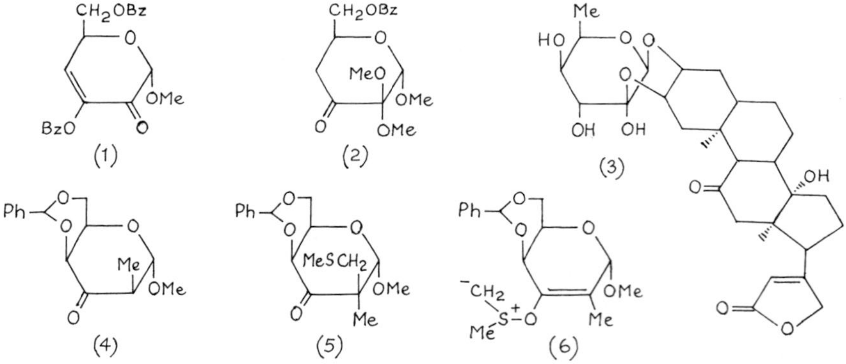


Several cardiac glycosides isolated from plants are glycosides of aldos-2-uloses, e.g. (3), which is based on 6-deoxy-D-xylo-hexos-2-ulose.[3] Others are based on the 3-methyl ether of this sugar,[3] on 4,6-dideoxy-3-O-methyl-D-erythro-hexos-2-ulose,[4] and on 4-deoxy-3-O-methyl-D- and -L-glycero-pentos-2-ulose.[5]

The glycosid-3-ulose derivative (4) was obtained from the corresponding D-ido-alcohol by use of acetic anhydride - DMSO for 4 h. When, however, this was extended to 4 days the doubly branched product (5) was obtained, by sigmatropic rearrangement of the intermediate (6).[6]

The potentially valuable observation has been made that treatment with bromine of per-O-tributylstannyl derivatives of methyl glycopyranosides with equatorial aglycones usually gives 3-ulosides in greater than 90% yield, whereas reaction of those with axial aglycones normally afford the 4-ketones in 70-80% yield, although, exceptionally, methyl β-D-galactopyranoside and methyl α-D-mannopyranoside showed no high regioselectivity.[7] These generalizations are illustrated in Scheme 1.

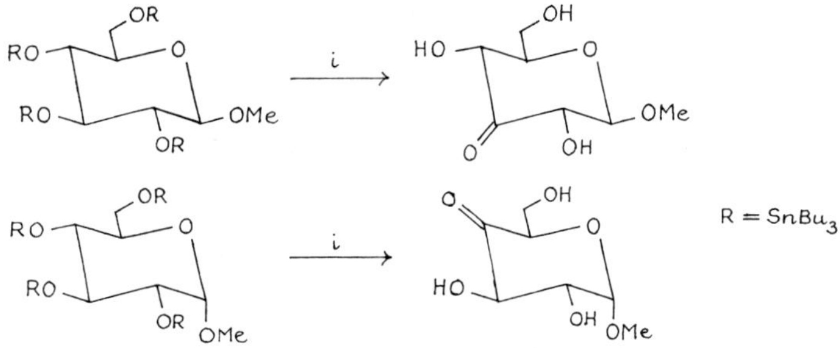


Reagent : i, $Br_2$

Scheme 1

The oxidation of 1,2:5,6-di-O-isopropylidene-α-D-glucofuranose with acetic anhydride-DMSO is improved when the reagents are premixed to allow formation of the reactive complex.[8]

Collins and coworkers have continued their studies of photochemical reactions of aldosulose derivatives. Pyranos-3-uloses

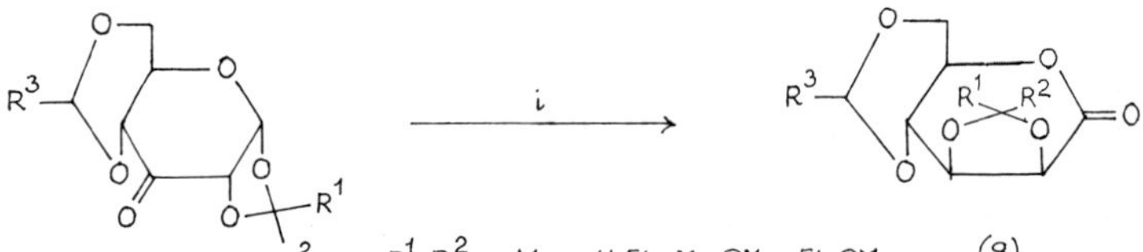

(7) $R^1, R^2$ = $Me_2$; H,Et; Me,OMe; Et,OMe (8)

$R^3$ = Me, Ph

Reagent : i, hν

Scheme 2

(7) photoisomerized to the D-mannono-1,5-lactones (8) by a

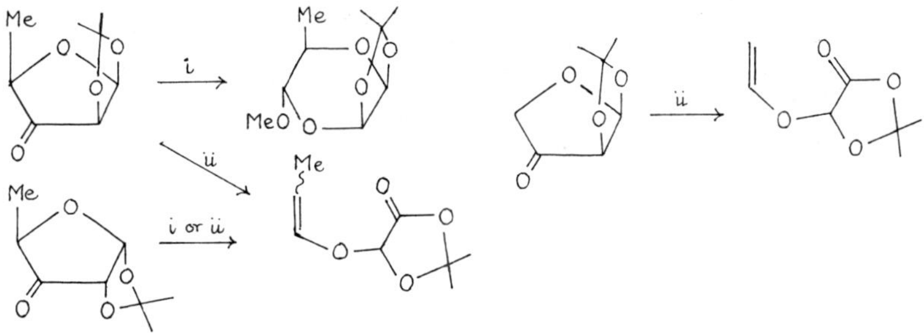


Reagents: i, hν - MeOH ; ii, hν - $C_6H_6$

Scheme 3

mechanism which presumably involves hydrogen transfer from C-1 to C-3 (Scheme 2),[9] whereas furanos-3-ulose derivatives reacted quite differently (Scheme 3), and these differences were discussed in detail.[10]

Synthesis of the racemic octosid-6-ulose derivative (9) has been reported from the uronate (10) in connection with lincomycin

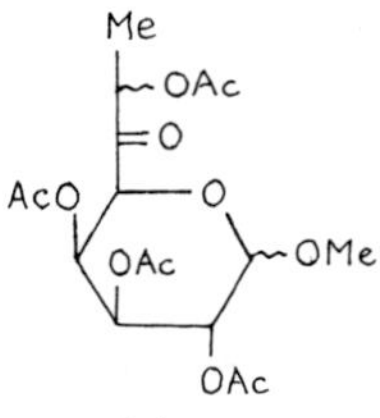


(9)

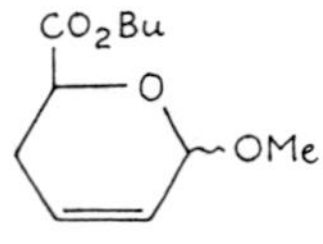


(10)

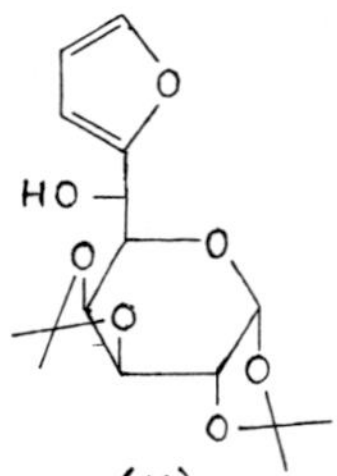


(11)

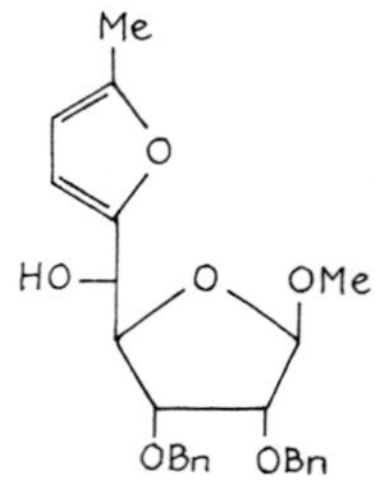


(12)

studies.[11]

## 2 Dialdoses

Reactions of several dialdose derivatives with furan or 2-methylfuran gave, in the presence of chloroacetic acid, condensation products with the (S) configuration, e.g. (11) and (12).[12] Other chain-extension reactions of such aldehydes are noted in Chapter 13.

## References


1 S.David and S.Hanessian, Tetrahedron, 1985, 41, 643.
2 A.Klausener, J.Runsink, and H.-D.Scharf, Carbohydr. Res., 1985, 138, 335.
3 F.Abe and T.Yamauchi, Chem. Pharm. Bull., 1985, 33, 3662.
4 F.Abe and T.Yamauchi, Chem. Pharm. Bull., 1985, 33, 847.
5 K.Shimada, T.Kyuno, T.Nambara,and I.Uchida, Phytochemistry, 1985, 24, 1345.
6 S.Nagarajan and K.L.Rinehart, J. Org. Chem., 1985, 50, 380.
7 Y.Tsuda, N.Matsuhira, and K.Kanemitsu, Chem. Pharm. Bull., 1985, 33, 4095.
8 A.Mazur, B.E.Tropp, and R.Engel, Tetrahedron, 1984, 40, 3949.
9 P.M.Collins, F.Farnia, R.N.Munasinghe, N.N.Oparaeche, and F.Siavoshy, J. Chem. Soc., Chem. Commun., 1985, 1038.
10 P.M.Collins and F.Farnia, J. Chem. Soc., Perkin Trans. 1, 1985, 575.
11 M.Chmielewski, B.Doboszewski, O.Achmatowicz, Jr., and A.Zamojski, Bull. Pol. Acad. Sci., Chem., 1984, 32, 423 (Chem. Abstr., 1985, 103, 88 136).
12 B.Grzeszczyk, K.Dziewiszek, S.Jarosz, and A.Zamojski, Carbohydr. Res., 1985, 145, 145.

# 16
# Sugar Acids and Lactones

## 1 Aldonic Acids

The rate and efficiency of crystallization of calcium gluconate from solution following the electrode oxidation of glucose have been investigated,[1] and this reaction at a platinum electrode is markedly catalysed by submonolayers of heavy metals (e.g. thallium, palladium, bismuth) deposited at under-potentials. This has been interpreted in terms of reduced electrode poisoning.[2] Treatment of 3-O-methyl-D-glucose with lead hydroxide in water gave the isomeric 2-deoxy-3-O-methyl-D-*arabino*-hexonic acid; in the presence of sodium hydroxide, however, the products were α- and β-gluco-metasaccharinic acid.[3]

Cellobiose, maltose and 4-O-methyl-D-glucose with anthraquinone 2-sulphonic acid in dilute sodium hydroxide solution degrade to give a series of substituted aldonic acids:

| | Cellobiose | Maltose | 4-O-Methyl-D-glucose |
|---|---|---|---|
| 2-O-substituted-D-erythronic acid | 19 | 32 | 20 |
| 3-O-substituted-D-arabonic acid | 2 | 5.5 | 17 (arabonic + ribonic) |
| 3-O-substituted-D-ribonic acid | 0.5 | 2 | |
| 4-O-substituted-D-gluconic acid | 2 | 5 | 25 (gluconic + mannonic) |
| 4-O-substituted-D-mannonic acid | 2.5 | 7 | |
| 3-O-substituted pentose acid (1) | 8 | 8 | 32 |

Mechanisms for the reactions were formulated based on these products and non-carbohydrate products, the main pathways being thought to involve 2-uloses and 2,3-diuloses. Compounds (1) are believed to be formed by benzilic acid rearrangement reactions

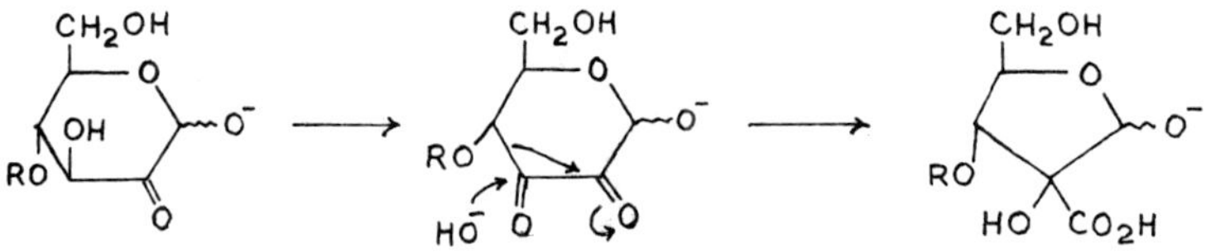


(1)

Scheme 1

(Scheme 1).[4]

An interesting photochemical rearrangement of 1,2-O-alkylidene-hexos-3-ulose derivatives to 2,3-O-alkylidenehexono-1,5-lactones is described in Chapter 15.

A stereoselective synthesis of methyl 2-amino-2-deoxy-D-gluconate from 2,3-O-isopropylidene-D-glyceraldehyde is outlined in Scheme 2.[5]

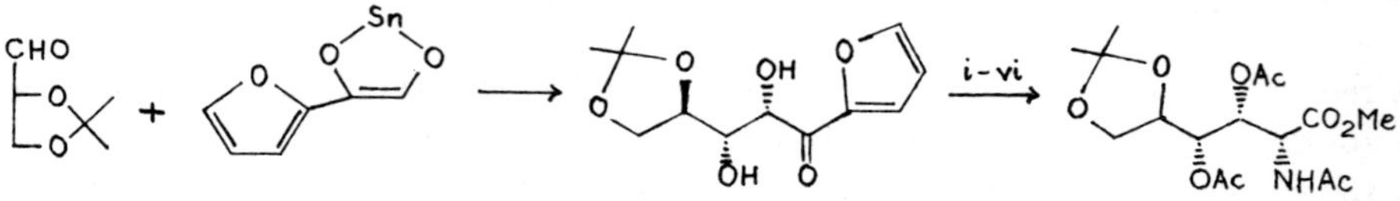


*Reagents*: i, $(Im)_2CO$; ii, $NH_2OBz$; iii, LAH; iv, $Ac_2O$-Py; v, $RuO_2$-$IO_4^-$; vi, $CH_2N_2$

Scheme 2

The preparation of several aldonolactone-based orthoester-linked disaccharide derivatives, e.g. (2), which has a similar structural

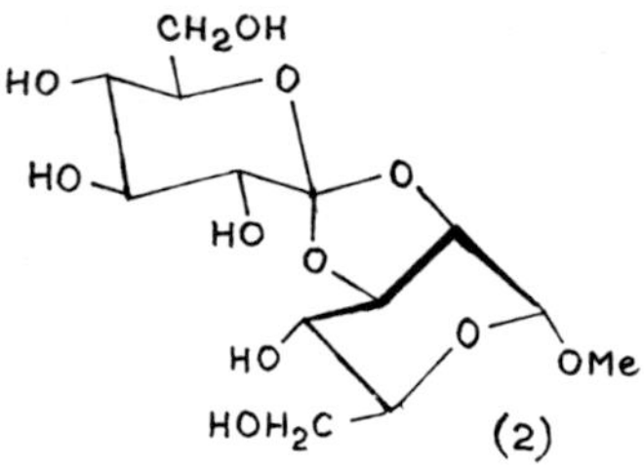


(2)

feature to that of the orthosomycin family of antibiotics, has been reported (c.f. Carbohydr. Res., 1983, 121, 175, 187). Distereo-isomers were formed and were structurally assigned from optical rotational and $^{13}C$ n.m.r. data using an X-ray crystallographic structure for reference (Chapter 22).[6]

Reaction of D-ribono-1,4-lactone with benzaldehyde in the presence of acid gives a 3,4-O-benzylidene derivative of the 1,5-lactone (Chapter 6); borate esters of aldonic and aldaric acids are referred to in Chapter 7; 2,3-unsaturated aldonolactones are noted in Chapter 13. Other syntheses and reactions of aldonic acid derivatives are referred to in Chapters 12 and 16.

## 2 Saccharinic Acids

A study has been reported of the alkaline degradation of alginates to various carboxylic acids. High alkalinities favour the production of glucoisosaccharinic, anhydroisosaccharinic and 2-deoxy-3-C-methyltetraric acids, whereas 2,3-dideoxypentaric acid was the major product formed at low concentrations. Calcium ions promoted the production of α-glucoisosaccharinic acid, 3,4-dideoxyhexaric and 2-hydroxybutanoic acids. The products suggest that the polymer chains were degraded from the reducing ends or from these ends of polymer fragments.[7]

Acid-catalysed reaction of α-D-glucoisosaccharinic acid (3) in an autoclave gave the C-2 epimer of (3) and the anhydrides (4)-(9).[8]

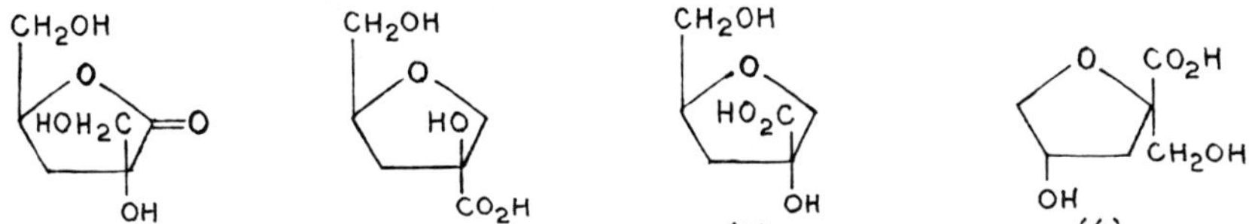

(3) (4) (5) (6)

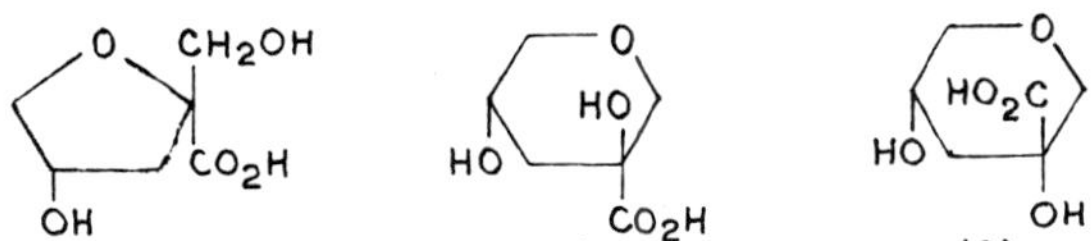

(7) (8) (9)

## 3 Ulosonic Acids

Optimum conditions for the oxidation of the primary hydroxy group of 2,3:4,6-di-O-isopropylidene-L-sorbose at a chromium-nickel steel grid anode were developed to give yields in excess of 70%.[9]

Danishefsky and coworkers have developed the route outlined in Scheme 3 for the synthesis of 3-deoxy-DL-manno-2-octulopyranosonic

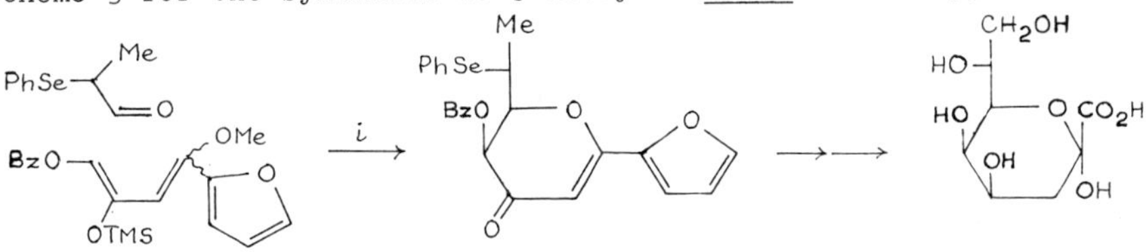

Reagent: *i*, $BF_3$

Scheme 3

acid (KDO).[10] Aldol condensation of 2-acetamido-2-deoxy-D-glyceraldehyde with di-tert-butyl oxaloacetate at pH 12 gave the branched-

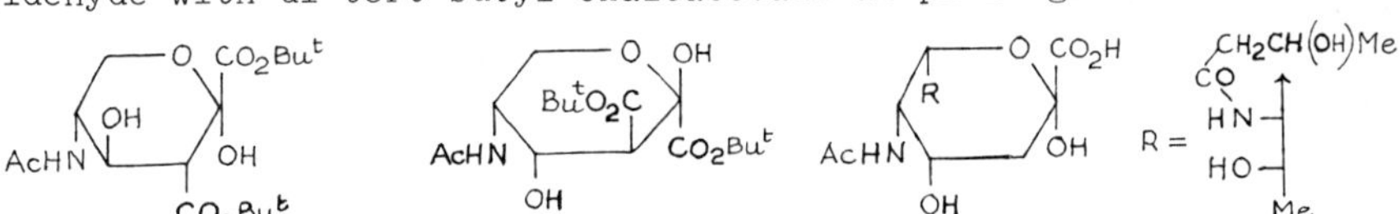

(10) (11) (12)

chain ulosonic acids (10) and (11) in modest yields in the ratio 3:1. The work was repeated in the L-series.[11]

The nonulosonic acid (12) has been identified as a component of some O-specific *Shigella* and *Pseudomonas* polysaccharides,[12] and 5-acetamido-9-*O*-acetyl-3,5-dideoxy-D-*glycero*-D-*galacto*-2-nonulopyranos-1-onic acid (*N*-acetyl-9-*O*-acetylneuraminic acid) has been made by treatment of 2-acetamido-6-*O*-acetyl-2-deoxy-D-mannose with pyruvic acid in the presence of immobilized acylneuraminate pyruvate ligase. The product has been reported to be present in embryonic and tumour cells. The starting sugar was prepared from 2-acetamido-2-deoxy-D-mannose via partially silylated intermediates.[13]

## 4 Uronic Acids

D-Glucuronic acid, isolated as the sodium salt, has been prepared following palladium-catalysed air oxidation of 1,2-*O*-isopropylidene-α-D-glucofuranose.[14] The biosynthetic precursors of the L-guluronic acid of alginic acid are D-glucitol and its 6-phosphate,[15] and the trihydroxypiperidine (13), a derivative of 1-deoxynojirimycin, has been isolated from seeds of *Bapia racemosa* and found to be an inhibitor of human liver β-D-glucuronidase.[16]

Products from oxidative and non-oxidative alkaline treatment of D-galacturonic acid were analysed using glc-mass spectrometry and found to contain 13 hydroxymonocarboxylic acids and 26 dicarboxylic acids. In the absence of oxygen the main products were 3-deoxy-

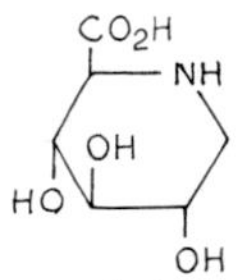

(13)

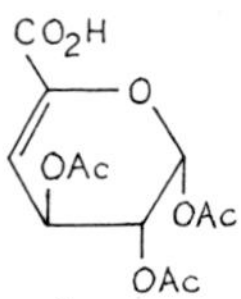

(14)

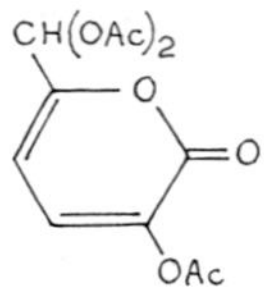

(15)

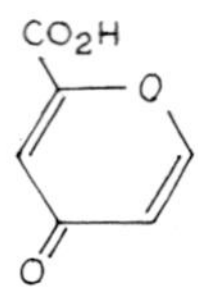

(16)

*lyxo*-hexaric, 3-deoxy-*xylo*-hexaric, malic, tartronic, *C*-methyltartronic, lactic and 3-deoxytetronic acids. In the presence of oxygen they were arabinonic, threonic, malic, tartronic and glycolic acids.[17] Degradation of D-galacturonic acid, or its tetra-*O*-acetate, in the presence of various bases, gives compounds (14)-(16); conditions were found for obtaining good yields of either (14) or (16).[18]

Compounds (17), required for carcinogenicity studies, were prepared from either D-glucuronic acid triethylammonium salt or D-glucurono-6,3-lactone by treatment with the corresponding aryl

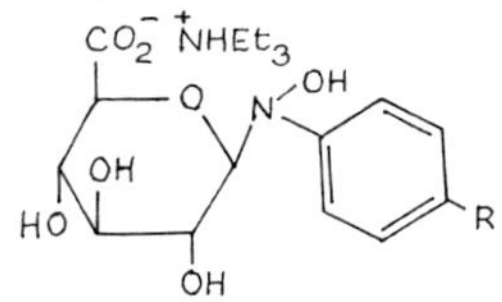

(17)

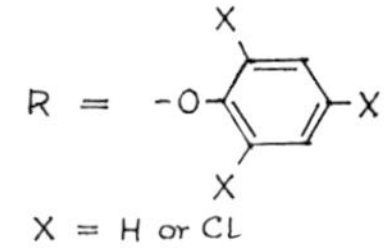

hydroxylamines.[19]

Some unsaturated uronic acid derivatives are referred to in Chapter 13.

In the area of furanoid uronic acids, a sterol conjugated as a D-riburonofuranoside has been isolated from an Argentinian plant Bauhinia candicans,[20] and 6-amino-6-deoxy-D- and -L-*glycero*-D-*gluco*-hepturonic acid have been made following condensation between 3-*O*-benzyl-1,2-*O*-isopropylidene-D-*xylo*-pentodialdose and ethyl nitroacetate.[21]

Some interesting reactions of compounds of this class have been reported. D-Glucurono-6,3-lactone (18) has given rise to a series

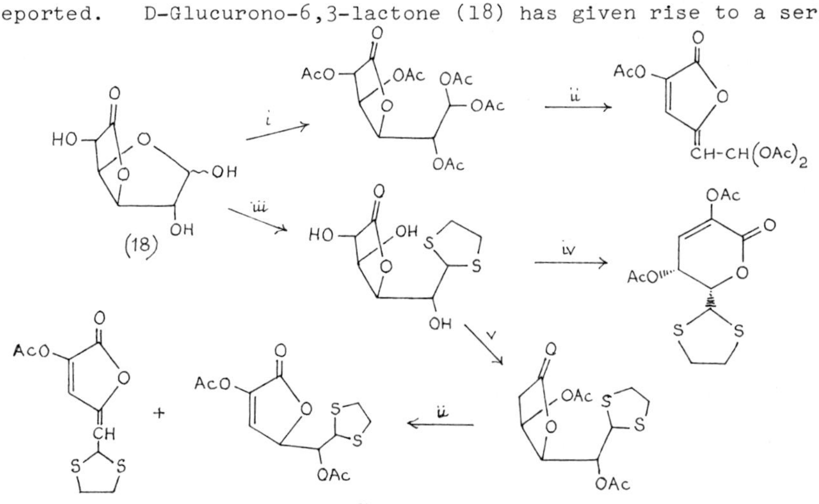

Reagents: i, $Ac_2O-BF_3$; ii, $Et_3N$; iii, $HS(CH_2)_2SH-H^+$; iv, $Ac_2O-Et_3N$; v, $Ac_2O$

Scheme 4

of unsaturated compounds (Scheme 4),[22] and compound (19) can be

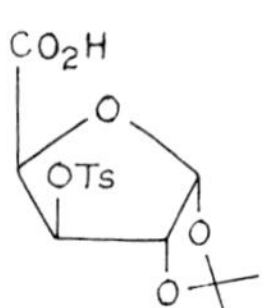

(19)

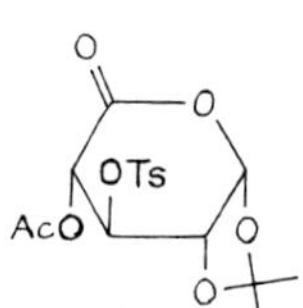

(20)

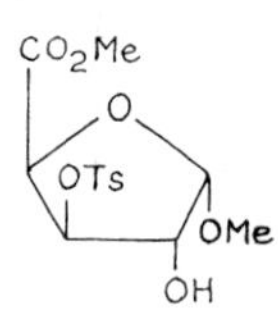

(21)

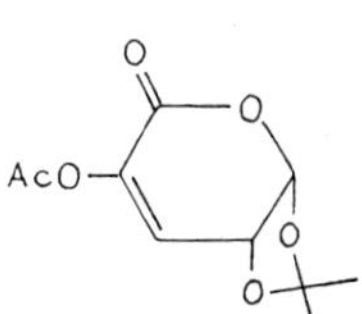

(22)

converted in moderate yield into the unusual compound (20) and thence into (21) and (22).[23]

## 5 Ascorbic Acids

An extensive review of L-ascorbic acid covering synthesis, substitution reactions and oxidation has appeared.[23a]

Reaction of ascorbic acid with superoxide ion ($O_2^{\cdot-}$) gives the derived radical anion.[24] The CIDEP technique has been used to detect the neutral ascorbate radical in photochemical reactions between ascorbic acid 6-palmitate and benzophenone,[25] and other carbonyl compounds,[26] and e.s.r. was employed to show the presence of ascorbate radical derived photochemically from sodium ascorbate and methylene blue. In the presence of oxygen, hydroxy radicals are produced, suggesting that it could initiate radical damage in this photodynamic system.[27]

A review has been published in Japanese on the crystal structure of dehydro-L-ascorbic acid and its dimer,[28] and the *X*-ray analysis of a *cis*-diaminocyclohexane-platinum-L-ascorbic acid complex has shown that the acid is bonded to the metal at C-2 and O-5 (23).[29] The iron-chelating properties of L-ascorbic acid and the structure of iron(II) ascorbate have been examined in relation to iron absorption. The acid forms 1:1 cationic complexes with $Ca^{2+}$, $Mn^{2+}$, $Co^{2+}$, $Cu^{2+}$, $Zn^{2+}$ and $Cd^{2+}$.[30]

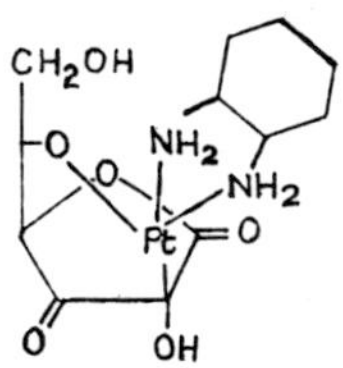


(23)

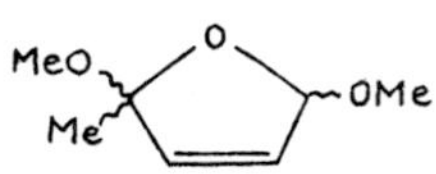


(24)

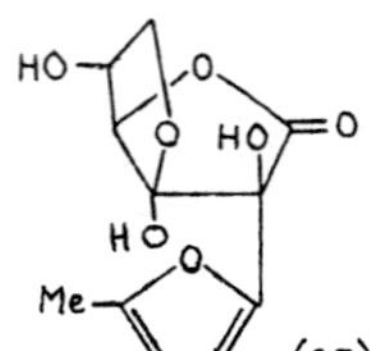


(25)

Dimolar allylation of L-ascorbic acid gives the 2,3-diether as major product,[31] and reaction with the dihydrofuran (24) gives the branched lactone (25) probably by way of pent-2-enal-4-one.[32] Some hydrazine products are noted in Chapter 10.

Various aspects of the oxidation of L-ascorbic acid still attract attention. An immobilized pumpkin oxidase has been used for quantitative analysis of the vitamin in foods; the immobilization stabilizes the enzyme significantly.[33] The acid acts synergistically with α-tocopherol and 6-hydroxy-2,5,7,8-tetramethylchroman-2-carboxylate as an antioxidant in linoleic acid autoxidations in micelles.[34]

A review (in Japanese) on oxidations with heavy-metal catalysts has appeared,[35] and specific studies with the following metal ions have been reported: palladium black,[36] copper(II)-EDTA (in Chinese),[37] and chitosan-copper(II) and -iron(III) complexes (in Korean).[38] Bromate-cerium-sulphuric acid oxidation leads to oscillations of the Belousov-Zhabolinskii type,[39] and aerobic oxidation studies have indicated that further oxidation products can be formed without the initial formation of dehydroascorbic acid.[40]

## References


1 B.W.Krol, Process Technol. Pri., 1984, 2, 425 (Chem. Abstr., 1985, 102, 54 115).
2 N.Xonoglou and G.Kokkinidis, Bioelectrochem. Bioeng., 1984, 12, 485 (Chem. Abstr., 1985, 102, 149 654).
3 R.A.Gakhokidze and A.A.Surmava, Soobshch. Akad. Nauk Gruz. SSR, 1984, 115, 297 (Chem. Abstr., 1985, 102, 185 360).
4 T.Vuorinen, Carbohydr. Res., 1985, 141, 307.
5 T.Mukaiyama, R.Tsuzuki, and J.Kato, Chem. Lett., 1985, 837.
6 K.Asano, S.Horito, J.Yoshimura, T.Nakazawa, Z.-I.Ohya, and T.Watanabe, Carbohydr. Res., 1985, 138, 325.
7 K.Niemela and E.Sjostrom, Carbohydr. Res., 1985, 144, 241.
8 R.Alen, Carbohydr. Res., 1985, 144, 163.
9 I.Pesheva, M.Dolapchieva, and D.Merdzhanova, Farmatsiya (Sofia), 1985, 35,1 (Chem. Abstr., 1985, 103, 37 661).
10 S.J.Danishefsky, W.H.Pearson, and B.E.Segmuller, J. Am. Chem. Soc., 1985, 107, 1280.
11 M.Miljkovic and P.Hagel, Carbohydr. Res., 1985, 141, 213.
12 Y.A.Knirel, E.V.Vinogradov, A.S.Shashkov, N.K.Kochetkov, V.L'Vov, and B.A.Dmitriev, Carbohydr. Res., 1985, 141, Cl.
13 C.Auge, S.David, C.Gautheron, and A.Veyrieres, Tetrahedron Lett., 1985, 26, 2439.
14 D.P.Fel'dman, A.D.Voitenko, M.V.Shimanskaya, and M.Lidaks, Khim.-Farm.Zh., 1984, 18, 1356 (Chem. Abstr., 1985, 102, 185 399).
15 M.Quillet and G.de Lestang-Bremond, Phytochemistry, 1985, 24, 43.
16 K.S.Manning, D.G.Lynn, J.Shabanowitz, L.E.Fellows, M.Singh, and B.D.Schrire, J. Chem. Soc., Chem. Commun., 1985, 127.
17 K.Niemela and E.Sjostrom, Carbohydr. Res., 1985, 144, 93.
18 K.Tajima, Chem. Lett., 1985, 49.
19 T.Yoshioka, H.Yamada, and T.Uematsu, J. Chem. Soc., Perkin Trans. 1, 1985, 1271.
20 A.M.Iribarren and A.B.Pomilio, Phytochemistry, 1985, 24, 360.
21 H.Paulsen, M.Brieden, and V.Sinnwell, Liebigs Ann. Chem., 1985, 113.
22 H.Itoh, K.Aoki, and T.Matsumoto, Bull. Chem. Soc. Jpn., 1985, 58, 777.
23 V.A.Timoshchuk, Carbohydr. Res., 1985, 145, 131.
23a P.A.Seib, Int. J. Vitam. Nutr. Res., Suppl., 1985, 27 (Vitamins), 259 (Chem. Abstr., 1985, 103, 196 328).
24 D.T.Sawyer, T.S.Calderwood, C.L.Johlman, and C.L.Wilkins, J. Org. Chem., 1985, 50, 1409.
25 M.T.Craw, M.C.Depew, and J.K.S.Wan, J. Am. Chem. Soc., 1985, 107, 1084.
26 M.T.Craw, M.C.Depew, and J.K.S.Wan, Photobiochem. Photobiophys., 1985, 9, 109 (Chem. Abstr., 1985, 103, 177 878).
27 G.R.Buettner, T.P.Doherty, and T.D.Bannister, Radiat. Environ. Biophys., 1984, 23, 235 (Chem. Abstr., 1985, 102, 32 055).

28 M.Ohmori, Seikatsu Eisei, 1985, 29, 105 (Chem. Abstr., 1985, 102, 209 200).
29 L.S.Hollis, A.R.Amundsen, and E.W.Stern, J. Am. Chem. Soc., 1985, 107, 274.
30 C.M.Plug, D.Dekker, and A.Bult, Pharm. Weekbl., Sci. Ed., 1984, 6, 245 (Chem. Abstr., 1985, 102, 119 517).
31 K.S.Kim, H.S.Lee, S.C.Kim, and C.S.Hahn, Bull. Korean Chem. Soc., 1984, 5, 264 (Chem. Abstr., 1985, 102, 149 684).
32 G.Fodor, K.Sussangkarn, H.Mathelier, R.Arnold, I.Karle, and C.George, J. Org. Chem., 1984, 49, 5064.
33 M.Esaka, K.Suzuki, and K.Kubota, Agric. Biol. Chem., 1985, 49, 2955.
34 L.R.C.Barclay, S.J.Locke, and J.M.MacNeil, Can. J. Chem., 1985, 63, 366.
35 N.Shinohara, Yokohama-shiritsu Diagaku Ronso, Shizen Kagaku Keiretsu, 1984, 35, 45 (Chem. Abstr., 1985, 103, 6 625).
36 H.Sakurai and K.Ishii, Nihon Daigaku No-juigakuba Gakujutsu Kenkyu Hokoku, 1985, 44 (Chem. Abstr., 1985, 103, 6 650).
37 Y.Pang and J.Xu, Yaoxue Xuebao, 1984, 19, 374 (Chem. Abstr., 1985, 103, 178 552).
38 H.Lee and Y.Shin, Pollimo, 1985, 9, 125 (Chem. Abstr., 1985, 103, 196 352).
39 R.Blume and H.J.Bader, Z. Naturforsch., B: Anorg. Chem., Org. Chem., 1984, 39B, 1795 (Chem. Abstr., 1985, 102, 149 686).
40 O.E.Hughes, Anal. Chem., 1985, 57, 555 (Chem. Abstr., 1985, 102, 50 806).

# 17
# Inorganic Derivatives

## 1 Carbon-bonded Phosphorus Derivatives

The structure of compound (1) has been determined by *X*-ray crystallography, and its solution conformation and those of several analogues by $^{1}H$ n.m.r. spectroscopy.[1] The phenylphosphinyl compound (2)[2] and the hydroxyphosphinyl analogue (3)[3] have been made from D-xylofuranose derivatives carrying phosphorus-containing substituents at C-5, and the D-*gluco* analogue of (3) has also been reported.[3]

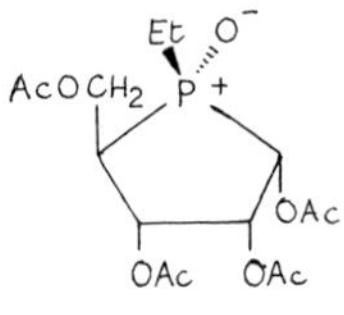

(1)

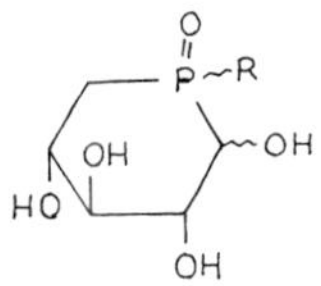

(2) R = Ph

(3) R = OH

An isosteric analogue of D-fructose 1-phosphate has been produced as indicated in Scheme 1, and analogues of D-ribulose 1-

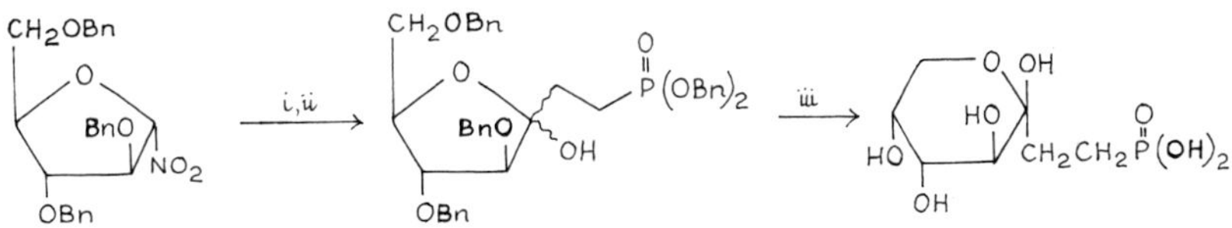

Reagents: i, $CH_2=CHP(O)(OBn)_2-Bu_4NF$; ii, $H_2O$; iii, $H_2$-Pd

Scheme 1

phosphate and D-sedoheptulose 1,7-diphosphate have been synthesized similarly.[4] In related work, the nitroglycosylsulphone (4) was converted mainly into the nitrophosphonate (5) [with some phosphate (6)] on treatment with diethyl phosphite anions. On the other hand, the nitroglycosyl halide (7) gave the phosphate (8) (Scheme 2). Reductive denitration of the nitrophosphonate (5) gave the glycosyl phosphonate (9).[5]

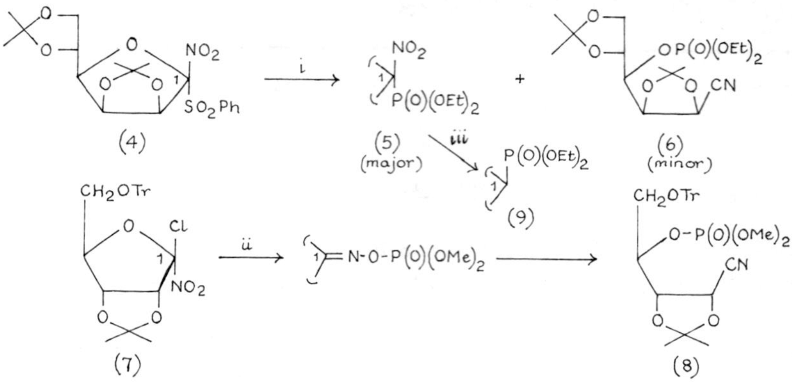

Reagents: i, $NaP(O)(OMe)_2$ ; ii, $KP(O)(OEt)_2$ ; iii, $Bu_3SnH$ - AIBN

Scheme 2

Sugar aldehydes give phosphonates on treatment with dialkyl phosphites[6] which have been used to prepare cyclic hydroxylamine phosphonates and related compounds (Scheme 3).[7]

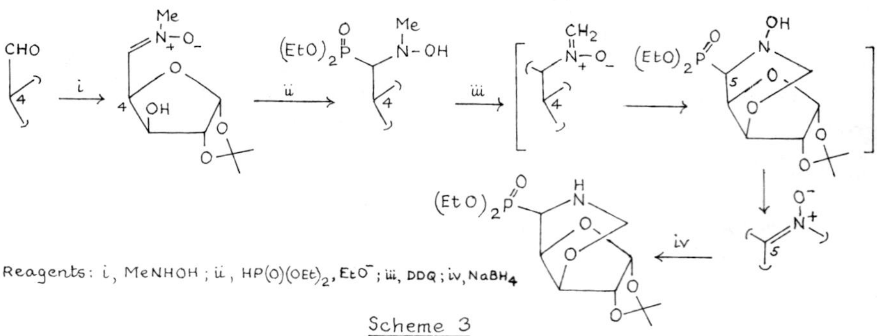

Reagents: i, MeNHOH ; ii, $HP(O)(OEt)_2$, $EtO^-$ ; iii, DDQ ; iv, $NaBH_4$

Scheme 3

A branched-chain phosphonate analogue (10) of a nucleotide has been synthesized from the corresponding bromide and used to prepare a trinucleotide analogue.[8]

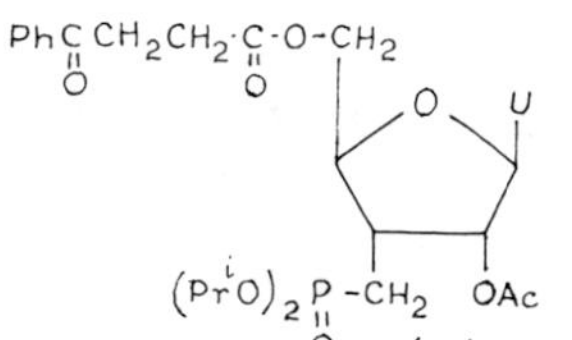

(10)

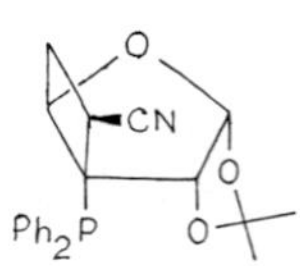

(11)

The X-ray crystal structure of compound (11) has indicated that the configuration at the nitrile-bearing centre is as shown (c.f. Vol. 14, p.98).[9]

## 2 Other Carbon-bonded Compounds

Several glycosyl-metal derivatives have been reported. Tin compounds have been prepared and used to obtain lithium analogues and hence C-glycosides, all with good steric control (Scheme 4).[10]

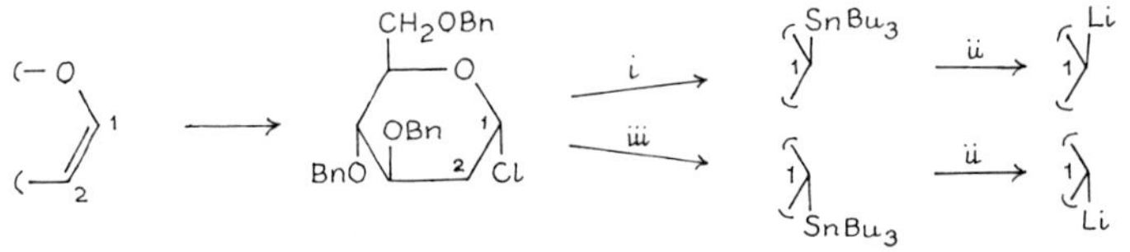


Reagents: i, $Bu_3SnLi$; ii, BuLi; iii, Li naphthalemide; iv, $Bu_3SnCl$

Scheme 4

The first record of tellurocarbohydrates has appeared (Scheme 5).[11] A glycosyl manganese compound has been made in similar fashion and used to prepare a C-glycoside (Scheme 6).[12]

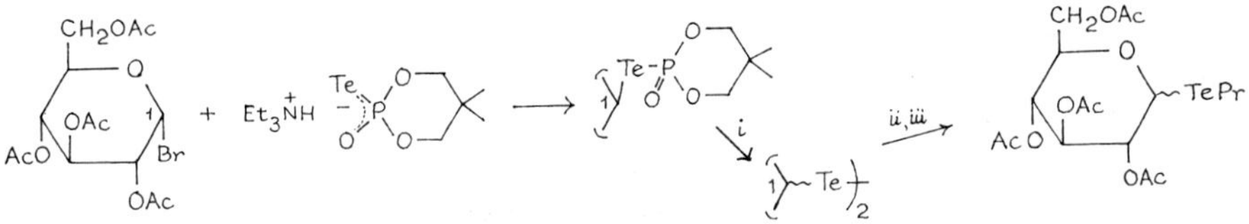


Reagents: i, $MeOH-O_2$; ii, $NaBH_4$; iii, PrBr

Scheme 5

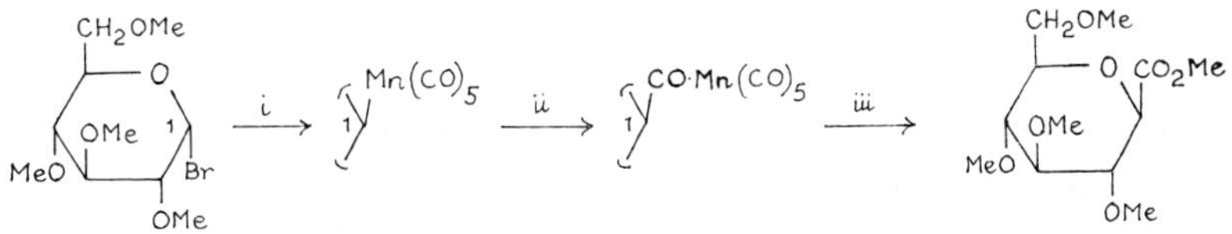


Reagents: i, $NaMn(CO)_5$; ii, CO; iii, $MeOH-Na_2CO_3$

Scheme 6

Reaction of tetra-O-methyl-β-D-glucopyranosyl ($\eta^5$-cyclopentadienyl)dicarbonyl iron with triphenylphosphine in benzene causes ligand exchange leading to the analogue with triphenylphosphine replacing a carbonyl group (85% yield). The process occurs with partial chirality transfer to iron and gives a 70:30 mixture of diastereoisomers.[13]

Fast atom bombardment mass spectrometry has been used to analyse organometallic mixtures produced in solution on addition of heterocyclic palladium compounds to furanoid and pyranoid glycals. The products were nucleoside analogues bearing palladium at C-2 and further products of their reaction.[14] Compound (12) was examined by n.m.r. and m.s. methods.[15]

Mercury-containing intermediates have been isolated from the reaction of 6-deoxyhex-5-enopyranose derivatives with mercury salts in aqueous media, which leads to deoxyinososes (Chapter 13), and a C-2-platinum derivative of ascorbic acid is noted in Chapter 16.

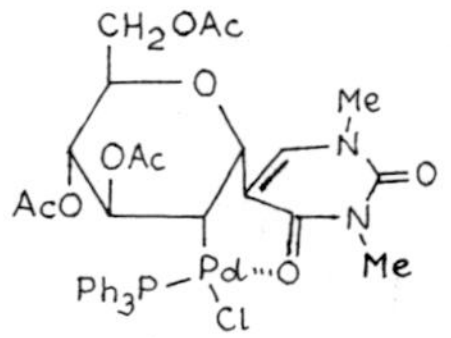


(12)

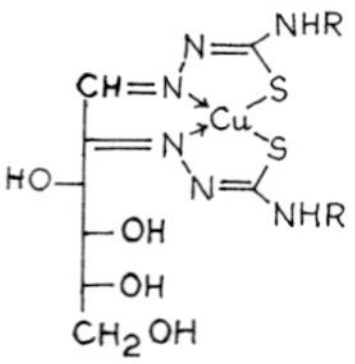


(13) R = H, aliphatic, aromatic.

## 3 Oxygen-bonded Compounds

A useful review has appeared on the application of tin reagents in the selective substitution and oxidation of carbohydrate derivatives.[16] Mannofuranosyl mannofuranosides have been made using 2,3:5,6-di-O-ethylboranediyl-α-D-mannofuranoyl bromide and 1-O-tributylstannyl and 1-O-triethylborate derivatives of the sugar.[17] 2',3'-Dialkylstannylene derivatives of nucleosides have been made by use of dialkyltin oxides and characterized by infrared and $^{119}$Sn Mössbauer spectroscopy.[18] Selective alkylation of glycosides by use of tin intermediates is referred to in Chapter 5.

Copper complexes of D-glucosone bisthiosemicarbazones (13) have antitumour activity.[19]

Considerable work continues in the field of metal complexing with sugar derivatives. Complexes between univalent cations and cyclogentiotetraose peracetate have been shown by FAB-MS to be 1:1 crown-like species.[20]

Complexes of sugars have been reported as follows: L-arabinose with magnesium halides;[21] sugars, polyols and glycosides with calcium and lanthanum ions (t.l.c. study, alditols with the threo-threo sequence complex more strongly with lanthanum than with calcium);[22] mannose, glucose, xylose, arabinose with boric acid ($^{13}$C and equilibria study);[23] mono- and disaccharides with Sn(IV), Sb(V), and Te(VI) hydroxyanions;[24] mannose and nickel N,N'-dimethyl-ethylenediamine (a crystalline complex containing two β-pyranose and one β-furanose units is formed).[25,26] Related studies were carried out with D-glucose and D-fructose, the latter giving the crystalline compound (14), which is a glycosylamine derivative, on reaction with nickel tris(ethylenediamine).[27]

Complexes formed between <u>epi</u>-inositol 1,6-anhydro-β-D-hexopyranoses and ytterbium(III), thulium(III) and erbium(III) have been studied by $^{13}C$ n.m.r. methods.[28] Similar studies were carried out on the natural product muellitol [1,3,5-tri(3-methylbut-2-enyl)-<u>scyllo</u>-inositol] in the presence of several lanthanide shift reagents.[29]

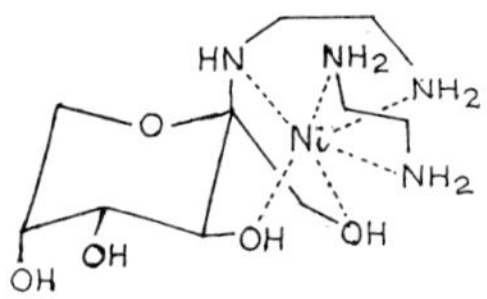


(14)

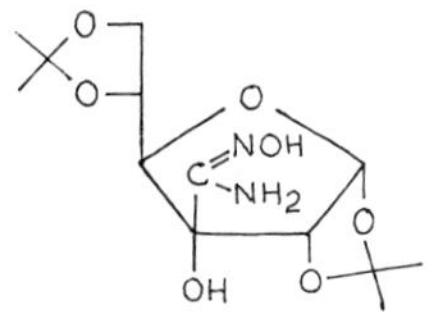


(15)

The structures of the complexes formed between phytic acid (<u>myo</u>-inositol hexaphosphate) and zinc ions have been studied by <u>X</u>-ray diffraction and thermal analyses.[30]

H.p.l.c. retention times have been used to study the binding of metal ions ($Mg^{2+}$, $Ni^{2+}$, $Zn^{2+}$, $Cu^{2+}$) to nucleosides and nucleotides.[31] The reaction of nucleosides with <u>cis</u>-dichlorobis(dimethylsulphoxide) platinum(II) has been studied by 360 MHz $^{1}H$ n.m.r. and $^{13}C$ n.m.r. spectroscopy and shown to be dependent on substitution of DMSO at the metal.[32] In the presence of oxygen, <u>cis</u>-diamminedichloroplatinum(II) and the D-allofuranose aldoxime (15) give a blue paramagnetic species which is believed (e.s.r., n.m.r. studies) to involve coordination between the metal and the nitrogen atoms of the carbohydrate.[33]

## References

1 P.Luger, E.Müller, H.Yamamoto, and S.Inokawa, Carbohydr. Res., 1985, 145, 25.

2 K.Seo and M.Yamashita, Carbohydr. Res., 1985, 141, 335.

3 H.Yamamoto, T.Hanaya, H.Kawamoto, S.Inokawa, M.Yamashita, M.-A.Armour, and T.T.Nakashima, J. Org. Chem., 1985, 50, 3516.

4 R.Julina and A.Vasella, Helv. Chim. Acta, 1985, 68, 819.

5 R.Meawly and A.Vasella, Helv. Chim. Acta, 1985, 68, 997.

6 Yu.A.Zhdanov, Z.I.Glebova, G.K.Kist'yan, V.A.Polenov, and L.A.Uzlova, Zh. Obshch. Khim., 1984, 54, 1660 (Chem. Abstr., 1985, 102, 167 031).

7 J.M.J.Tronchet, E.Winter-Mihaly, J.Rupp, F.Barbalat-Rey, and M.Geoffroy, Carbohydr. Res., 1985, 136, 375.

8 A.Mazur, B.E.Tropp, and R.Engel, Tetrahedron, 1984, 40, 3949.

9 S.Saito, T.Kato, Y.Morita, and H.Miyamae, Chem. Pharm. Bull., 1985, 33, 3092.

10 P.Lesimple, J.-M.Beau, and P.Sinaÿ, J. Chem. Soc., Chem. Commun., 1985, 894.

11 J.Czyzewska-Chlebny and M.Michalska, J. Chem. Soc., Chem. Commun., 1985, 693.

12 P.DeShong, G.A.Slough, V.Elango, and G.L.Trainor, J. Am. Chem. Soc., 1985, 107, 7788.

13 G.L.Trainor, J. Organomet. Chem., 1985, 282, C43 (Chem. Abstr., 1985, 103, 142 277).

14 H.T.Kalinoski, U.Hacksell, D.F.Barofsky, E.Barofsky and G.D.Daves, J. Am. Chem. Soc., 1985, 107, 6476.

15 U.Hacksell, H.T.Kalinkoski, D.F.Barofsky, and G.D.Daves, Acta Chem. Scand., Ser. B, 1985, 39, 469.

16 S.David and S.Hanessian, Tetrahedron, 1985, 41, 643.

17 W.V.Dahlhoff and A.Geisheimer, Z. Naturforsch., B: Anorg. Chem., Org. Chem., 1985, 40B, 141 (Chem. Abstr., 1985, 103, 88 138).

18 G.Ruisi, M.T.Lo Giudice, and L.Pellerito, Inorg. Chim. Acta, 1984, 93, 161 (Chem. Abstr., 1985, 102, 185 404).

19 E.R.Dilanyan, T.R.Ovsepyan, F.G.Arsenyan, G.M.Stepanyan, E.Mironov, and M.E.Vol'pin, Khim.-Farm. Zh., 1984, 18, 927 (Chem. Abstr., 1985, 102, 132 366).

20 C.Bosso, G.Excoffier, M.Vignon, and J.Ulrich, Spectrosc.: Int. J., 1984, 3, 72 (Chem. Abstr., 1985, 102, 149 665).

21 H.A.Tajmir-Riahi, J. Inorg. Biochem., 1984, 22, 55 (Chem. Abstr., 1985, 102, 95 913).

22 S.J.Angyal and J.A.Mills, Aust. J. Chem., 1985, 38, 1279.

23 J.Liu and L.Jiang, Youji Huaxue, 1984, 428 (Chem. Abstr., 1985, 102, 149 645).

24 J.Mbabazi, Carbohydr. Res., 1985, 140, 151.

25 T.Tanase, K.Kurihara, S.Yano, K.Kobayashi, T.Sakurai, and S.Yoshikawa, J. Chem. Soc., Chem. Commun., 1985, 1562.

26 S.Yoshikawa, S.Yano, K.Kurihara, T.Tanase, K.Kobayashi, and T.Sakurai, Kenkyu Hokoku-Asahi Garasu Kogyu Gijutsu Shoreikai, 1984, 44, 189 (Chem. Abstr., 1985, 102, 214 136).

27 S.Yano, S.Takizawa, H.Sugita, T.Takahashi, H.Shioi, T.Tsubomura, and S.Yoshikawa, Carbohydr. Res., 1985, 142, 179.

28 S.J.Angyal, L.Littlemore, and P.A.J.Gorin, Aust. J. Chem., 1985, 38, 411.

29 S.J.Angyal, D.Greeves, and L.Littlemore, Aust. J. Chem., 1985, 38, 1561.

30 A.G.Pierce, Jr., Inorg. Chim. Acta, 1985, 106, L9 (Chem. Abstr., 1985, 102, 167 056).

31 A.S.Cohen and E.Grushka, J. Chromatogr., 1985, 318, 221.

32 M.D.Reily, K.Wilkowski, K.Shinozuka, and L.G.Marzilli, Inorg. Chem., 1985, 24, 37 (Chem. Abstr., 1985, 102, 68 048).

33 P.Arrizabalaga, M.Geoffroy, J.M.J.Tronchet, and G.Zozino-Landolfo, Inorg. Chim. Acta, 1984, 93, L31 (Chem. Abstr., 1985, 102, 113 805).

# 18
# Alditols and Cyclitols

## 1 Alditols

6-Deoxy-altritol has been shown to be present in the cell wall polysaccharide of Nocardia asteroides, along with arabinose, galactose, and glucose, the first reported occurrence of this alditol in nature.[1] D-Sorbitol and its mono-phosphate have been detected in brown seaweeds.[2] Anhydro aminoalditols have also been characterized in natural sources. (2R,3S)-2-Hydroxymethyl-3-hydroxypyrrolidine (1) occurs in Castanospermum australe,[3] and a 4-hydroxy analogue (2) has been isolated from the african tree Angylocalyx boutiqueanus (stereochemistry not assigned).[4] A piperidine carboxylic acid derivative is mentioned in Chapter 16.

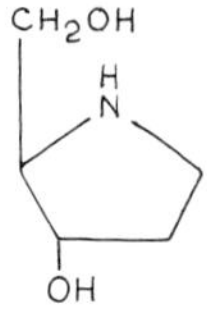


(1)

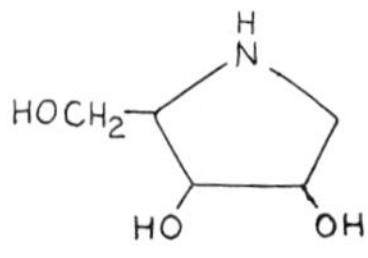


(2)

The preparation of D-mannitol from standard monosaccharides[5] and the processing of hardwoods for producing xylitol and other polyols and monosaccharides[6] have been reviewed. The hydrogenation of xylose to xylitol over a molybdenum-promoted copper catalyst[7] and of glucose over a nickel-titanium catalyst[8] has been studied. The hydrogenation of D-fructose and D-fructose/D-glucose mixtures over various metal catalysts has been investigated. The selectivity for D-mannitol from D-fructose was highest for copper/silica, and fructose is preferentially hydrogenated via furanose forms; a proposed mechanism of attack at the anomeric centre with inversion also explains enhanced mannitol yields using borate esters of fructose, and the diastereoselectivity of the hydrogenation of seven other ketoses.[9] A further study examined the combined action of an enzyme with a metal catalyst; good yields of D-mannitol resulted from glucose-fructose mixtures using D-glucose isomerase immobilized on silica in conjunction with a copper/silica catalyst.[10]

Glyceraldehyde has been used as a chiral source for several

polyol derivatives. Tin-catalysed condensation with a bromomethyl ketone led to the *threo* product (3), which was converted in standard steps to pentaacetyl 2-amino-2-deoxy-D-arabinitol (4)(Scheme 1).[11]

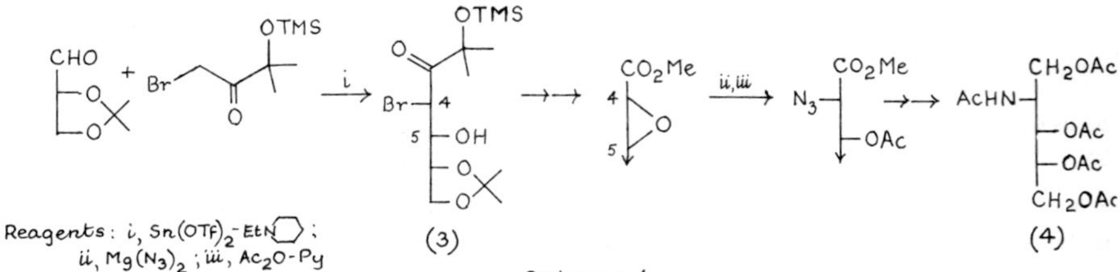


Scheme 1

A stereoselective aldol condensation with the boryl thio-ester derivative (5) gave the *syn* isomer (6) of the 2-deoxy or 2-C-branched pentitol (Scheme 2).[12] High *syn* stereoselectivity was also shown in reaction with organocopper reagents, leading to the anhydropentitols illustrated in Scheme 3.[13]

The use of (*S*)-malic acid as a source of 2-deoxy-D-*erythro*-tetrose and hence 1,2-anhydro-3-deoxypentitols *en route* to milbemycin is mentioned in Chapter 24.

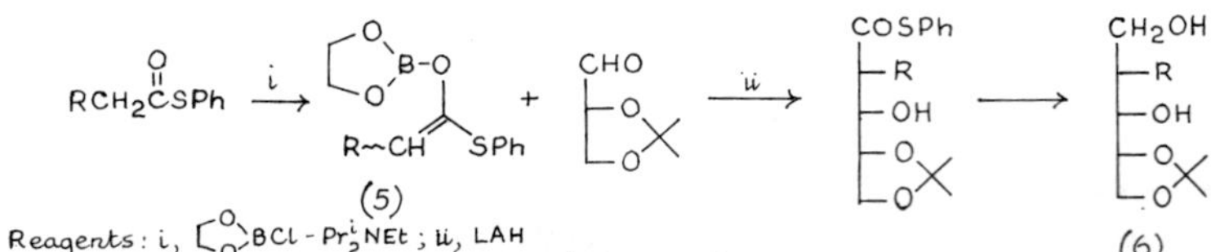


Scheme 2

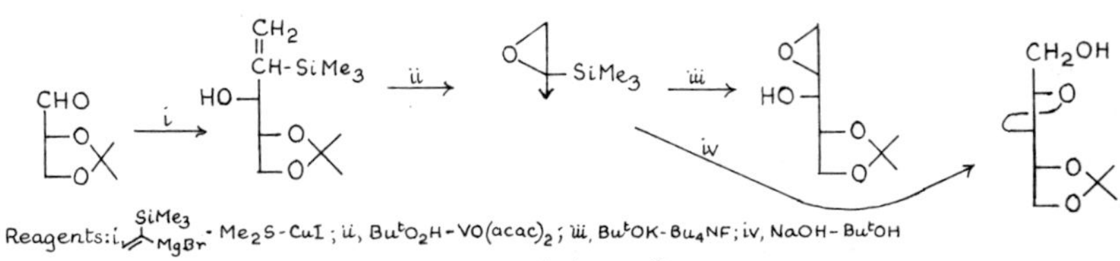


Scheme 3

The synthesis of (4*S*)-pentane-1,2,3,4-tetrol (7) from (*S*)-2-benzyloxypropanal by condensation with benzyloxyacetic ester has been reported (Scheme 4).[14]

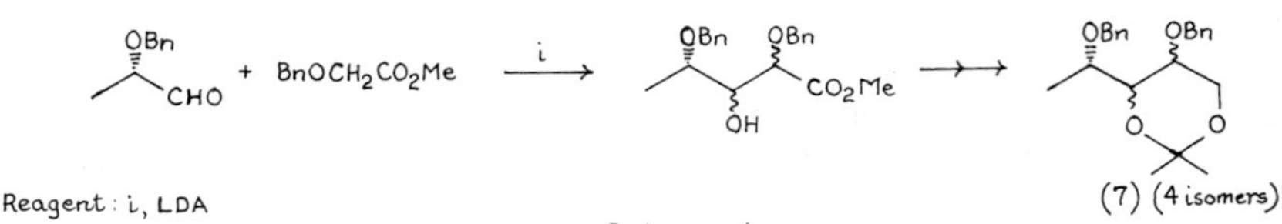


Scheme 4

The deamination of 2-amino-2-deoxy-D-galactitol has been studied; both C-3 epimers (xylo and lyxo) of the corresponding 2-deoxy-hexitol could be detected, depending on the conditions, besides other minor constituents; these results are significant for deamination studies on mucus glycoproteins containing such amino-hexitol units.[15]

Reaction of per-O-diethylboryl glycosides with diethylborane-9 BBN mesylate causes mostly endocyclic cleavage of C-O bonds leading to alditol derivatives; thus the derivative of methyl β-D-glucopyranoside gave 90% D-glucitol and only 10% exocyclic cleavage to 1,5-anhydro-D-glucitol.[16]

The furoisoxazoline (8) has been used for the total synthesis of amino-deoxy pentose and pentitol derivatives, synthesis of the latter being outlined in Scheme 5.[17]

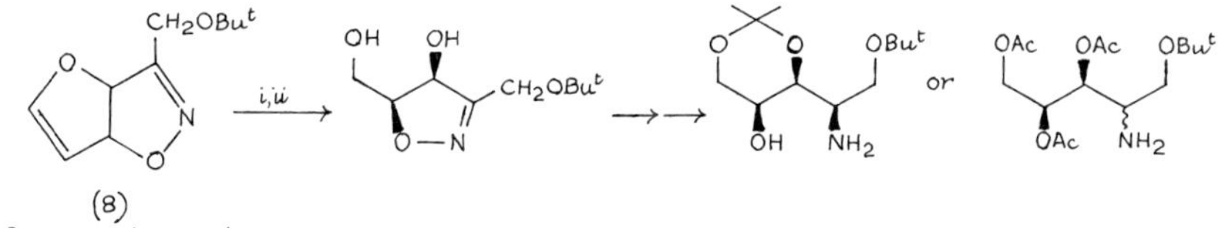


Reagents: i, $O_3$ ; ii, $NaBH_4$

Scheme 5

The racemic 2-amino-2-deoxy-erythro-tetritol derivative (9) has been prepared by iodocyclization of the acylamino-alkene (10) as indicated in Scheme 6.[18]

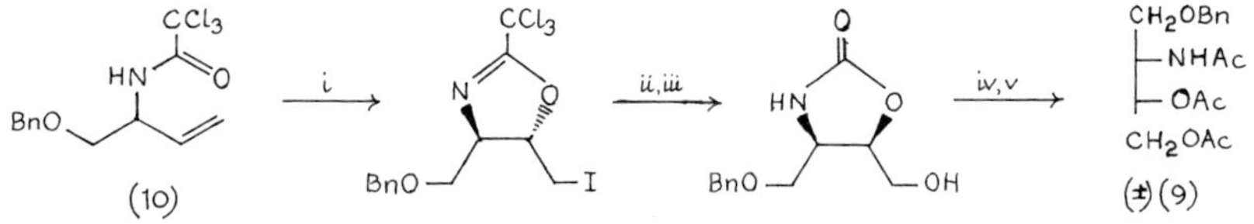


Reagents: i, NIS ; ii, $MeOH-H^+$ ; iii, Resin($CO_3^{2-}$) ; iv, KOH-MeOH ; v, $Ac_2O$-Py

Scheme 6

1-Amino-N-octyl-1-deoxy-D-glucitol has been prepared from octylamine and D-glucose and used as a chiral resolving agent for preparing Naproxen (D-2-(6-methoxy-2-naphthyl)propanoic acid).[19]

A number of 1-deoxy-1-nitro-hexitol and -heptitol derivatives have been prepared from D-glucose, and their spectral data recorded (including CD and ORD).[20] A synthesis of the higher-carbon alditol, L-galacto-D-galacto-decitol, has been described, using the dialdose prepared from galactose and Wittig reactions to extend the sugar chain two carbon units at a time, as outlined in Scheme 7; the D-galacto-D-galacto isomer was also obtained.[21]

A standard sequence of conversions has been used to convert 1,6-dibromo-1,6-dideoxy-3,4-O-isopropylidene-D-mannitol by chain elongation to the 3,6-diamino-1,4;5,8-dianhydro-L-manno- and L-ido-octitols (11).[22] Acyclic sugar nitro-alkenes undergo Michael addition with cyclohexa-1,3-dione and then cyclize to the corresponding heterocyclic compound (12)(Scheme 8).[23] The 3-deoxy-3-

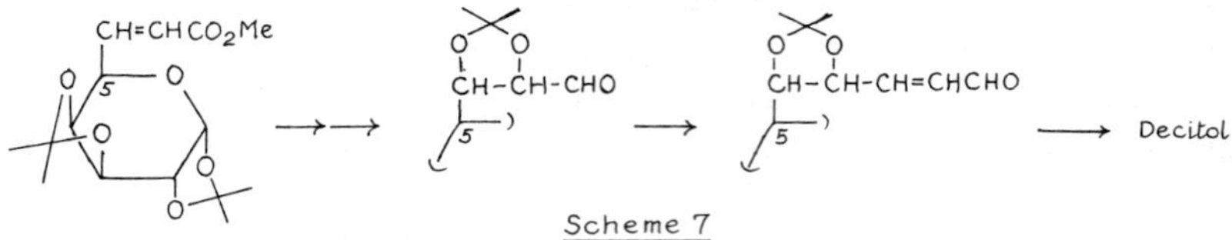

Scheme 7

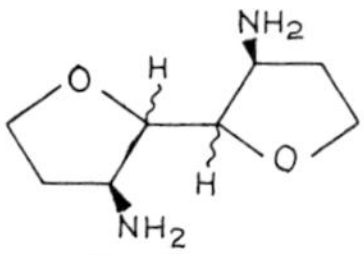

(11) (H syn)

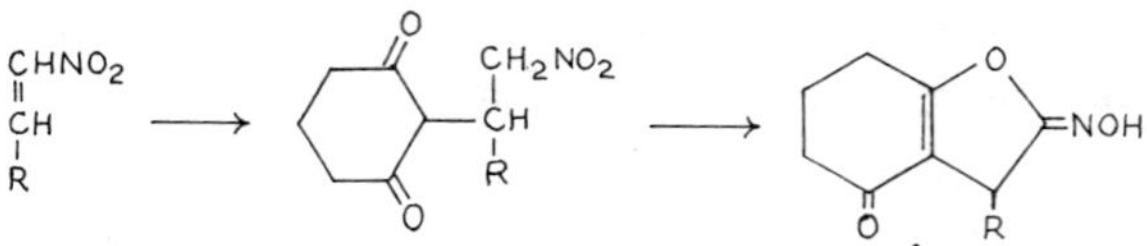

R= Peracetylated sugar chain

(12)

Scheme 8

hydrazino-D-glucose derivative (13) gives the trihydroxypropyldiazole (14) on treatment with acid, which was converted to a series of analogues of 9-(2,3-dihydroxypropyl)adenine.[24] Other

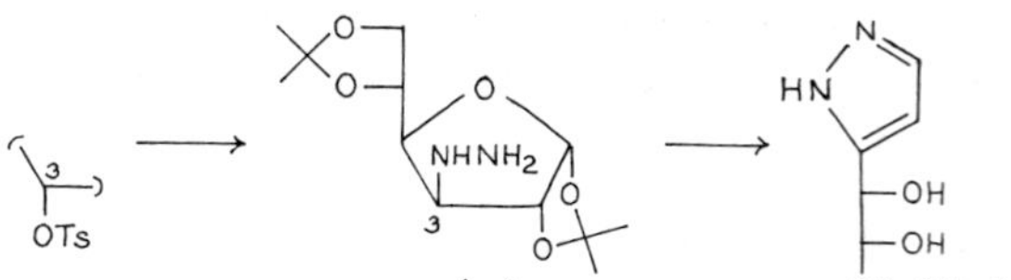

(13) (14)

polyhydroxyalkyl derivatives of nitrogen heterocycles are mentioned in Chapter 10.

Anhydro derivatives of alditols.- A four-step synthesis of 2,5;3,6-dianhydro-1-deoxy-D-glucitol (15) has been described, starting from 1,6-dibromo-1,6-dideoxy-D-mannitol; (15) was also converted to the 5-amino analogues (16) and (17) in standard steps.[25] The hexenitol (18) prepared from ribonolactone can be cyclized under electrophilic catalysis to give the 2,5-anhydro-hexitols (19).[26]

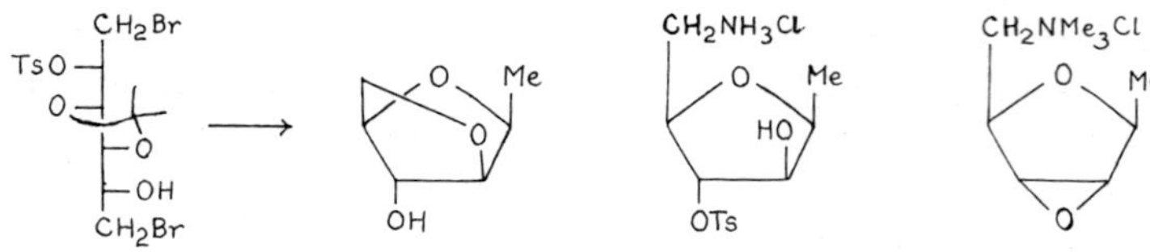

(15) (16) (17)

1,5-Anhydrocellobiitol (20) has been used as a model for investigating the oxidative degradation of cellulose promoted by iron. Oxidation of either sugar unit occurs (at C-6 or C-3' respectively) with release of the other unit as glucose or 1,5-anhydro-glucitol.[27]

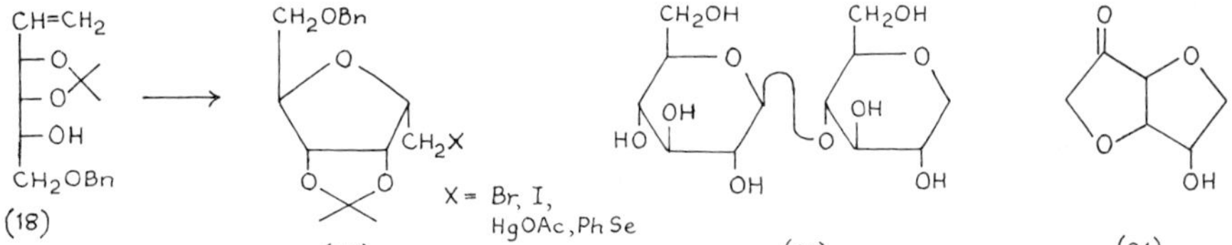


(18) (19) (20) (21)

A new method of D-glucan analysis involves reductive cleavage of permethylated compounds with triethylsilane and trimethylsilyl triflate or boron trifluoride-etherate as catalyst, followed by acetylation, which yields a series of acetylated, methylated 1,5-anhydro-glucitols. A number of isomeric substituted 1,5-anhydro-glucitols were synthesized as references.[28] β-Glycosylmethylamines of lactose, cellobiose, and maltose have been prepared by reduction of the corresponding nitromethyl C-glycosides.[29] The electrochemical oxidation of isosorbide (1,4;3,6-dianhydro-sorbitol) to the corresponding dianhydro-hexulose (21) has been reported.[30]

The reduction of glycosyl isoselenocyanates to 1,5-anhydrohexitols is mentioned in Chapter 10.

Cyclic imino derivatives of amino-alditols.- Several laboratories have reported studies on these compounds, which can show glycosidase inhibitory activity. Fleet's group has studied the plant alkaloids (22)-(24) which are analogues of glucose, fructose, and mannose respectively; analysis of the inhibitory activity of these compounds towards a variety of glycosidases has shown that the previously reported activity of compound (24) is actually due to the fructose analogue (23).[31] They have used a 2-azido-2-deoxy-

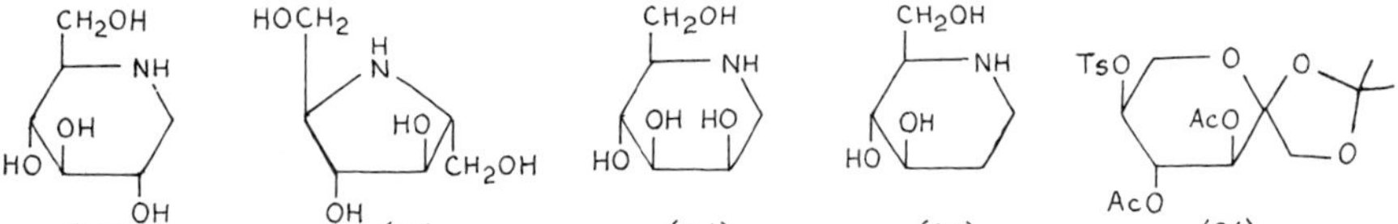

(22) (23) (24) (25) (26)

mannofuranoside derivative to synthesize the mannose analogue (24) (deoxymannojirimycin), the fructose analogue (23) and fagomine (25), the analogue of 2-deoxy-D-glucose.[32] Neither fagomine itself nor its 4-O-β-D-glucopyranosyl derivative, which is found in buckwheat seed, showed any activity.[33] Another laboratory has also reported the synthesis of the fructose analogue (23)(via azide displacement on the sorbose tosylate (26)).[34] The 4-O-α-D-glucopyranosyl derivative of 1-deoxynojirimycin (22) has been synthesized enzymatically from the amine using α-cyclodextrin as glucose donor.[35] Fleet's group have also reported the synthesis of 1,5-dideoxy-1,5-imino-L-fucitol (27) from glucose[36] and the 1,4-dideoxy-1,4-imino-D-lyxitol (28) and D- and L-arabinitols (29) and (30).[37] Another synthesis

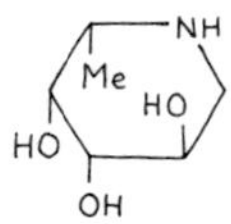
(27)

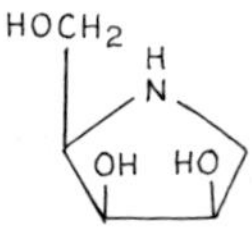
(28)

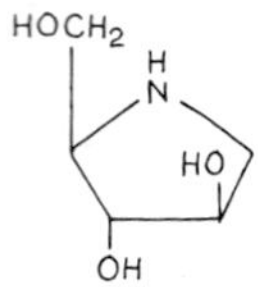
(29)

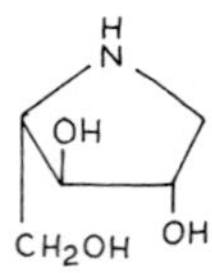
(30)

of the L-*arabino* enantiomer (30) by Williams' group[38] has shown that the alkaloid present in *Angylocalyx Boutiqueanus* is the D-*arabino* isomer (29), and that the isomer that occurs in *Arachniodes Standishii*, claimed to have the *xylo* configuration,[39] is in fact also the same D-*arabino* compound (29). Another report describes the synthesis of 1-deoxynojirimycin (22) and the *manno* isomer (24) from glucose and mannose respectively, using the sequence illustrated for the *gluco* isomer in Scheme 9.[40] D-Perosamine (31) serves as a convenient source for a synthesis of 1,4,6-trideoxy-1,4-imino-D-mannitol (32), a potent mannosidase inhibitor.[41]

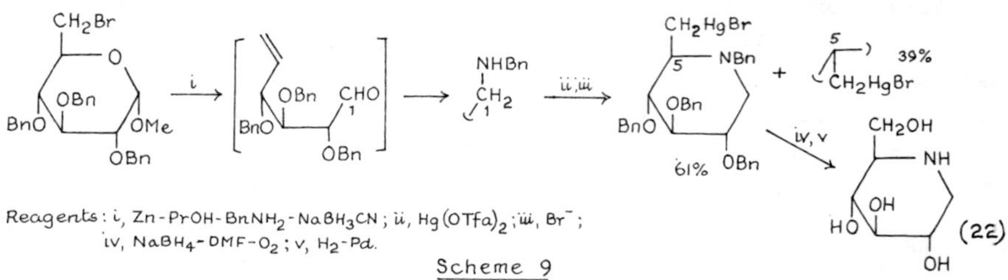

Scheme 9

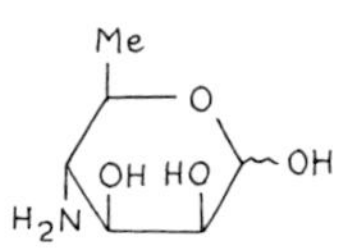
(31)

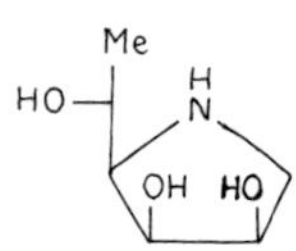
(32)

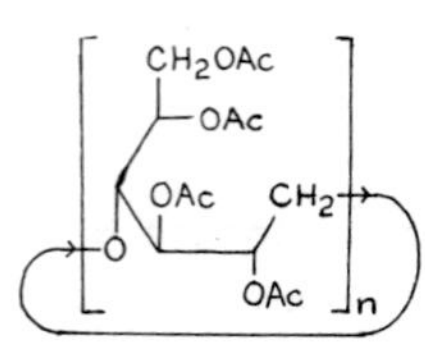
(33) n = 6 or 7

Full details of the synthesis of 4-amino-1,5-dideoxy-1,5-iminohexitols reported earlier (Vol.16,p.94) have been published.[42]

Reaction of methyl hepta-*O*-diethylboryl-β-maltoside with diethylborane and 9-BBN mesylate gave a 1-*O*-methyl-4-*O*(D-glucit-1-yl)-D-glucitol derivative as the major product isolated as its nona-acetate, together with glucitol derivatives as byproducts; similar products were obtained from methyl and benzyl β-lactosides.[43] The same reagent mixture has been used to cleave perborylated α- and β-cyclodextrins, the products then being chiral macrocyclic polyhydroxy ethers (33) together with acyclic analogues.[44]

Acetonation of D-mannitol with methyl isopropenyl ether in DMF at room temperature with acid catalysis gave principally the 1,2;5,6-di-*O*-isopropylidene diacetal with small amounts of the 1,2;4,6- and 1,2;3,4-isomers.[45]

4-*O*-(2-Acetamido-2-deoxy-D-glucopyranosyl)-D-ribitol derivatives have been prepared from D-ribitol as artificial antigens for an immunogenicity study; mono-allylation of 2,3,5-tri-*O*-benzyl-D-ribitol gave a separable mixture of the 1-*O*- and 4-*O*-allyl ethers, both of which were used in the syntheses.[46]

Free radicals produced from xylitol and standard hexitols on reaction with hydroxy radicals, or by direct γ- radiolysis, have been investigated by spin-trapping and e.s.r..[47]

The ability of electrolytes to complex with sorbitol and mannitol has been studied by polarimetry; anions were more effective than cations, and group IIA elements than IA.[48]

*N*-(1-Deoxyhexitol-1-yl)-aminoacids, as per-trimethylsilylated derivatives, have been identified by m.s.; the synthesis of reference compounds was also described.[49]

400 Mhz $^1$H N.m.r. spectra of nine monodeoxyalditols have been reported, supporting previous conformational conclusions.[50]

The relative stereochemistry of alditols can be deduced by formation of borate complexes, acetylation, partial hydrolysis to the polyol acetates, and then n.m.r. analysis. Borate complexes form preferentially in the sequence 1,2-syn > 1,3-syn > 1,2-anti, 1,3-anti, and terminal diols (in staggered conformation).[51]

The structure and energy of D-sorbitol has been calculated using an empirical force-field method; bond lengths and angles are in reasonable agreement with crystal structure *X*-ray and neutron diffraction evidence, but calculated energies do not suggest the observed conformation.[52]

The conversion of a mannitol derivative to antibiotic A32390A is mentioned in Chapter 19.

## 2 Cyclitols

Dambonitol (1,3-di-*O*-methyl-*myo*-inositol) has been identified as a constituent of glycosyl nervogenic acid esters from *Anodendron affine*.[53] Ononitol (4-*O*-methyl-*myo*-inositol) has been characterized as a major carbohydrate in pea nodules, and a minor constituent of soya bean nodules; it has previously been mistaken chromatographically for the 1-*O*-methyl-*myo*-inositol isomer.[54]

Improved procedures have been described for the synthesis of (±)-1,2;4,5-di-*O*-isopropylidene-*myo*-inositol (34), which was

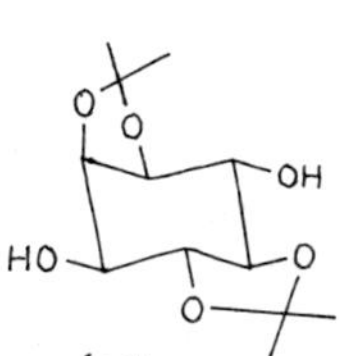


(34)

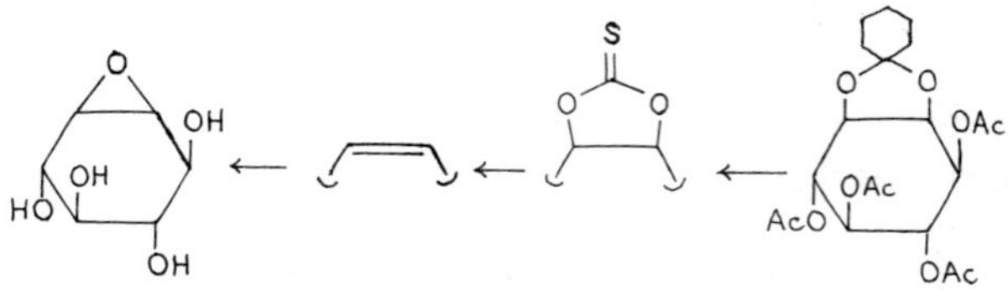


(35) (36)

purified as its dibenzoate,[55] and for conduritol β-epoxide (35), an inhibitor of glucosidases, from the *myo*-inositol acetal tetraacetate (36).[56] The derivative (34) also serves as a source for the preparation of (±)-1,2,4-tri-*O*-benzyl-*myo*-inositol by a sequence of standard reactions; stereospecific cleavage of the dibutylstannylene derivative (37) with allyl bromide gives access to the unsymmetrically substituted tri-allyl ether which could be benzylated and

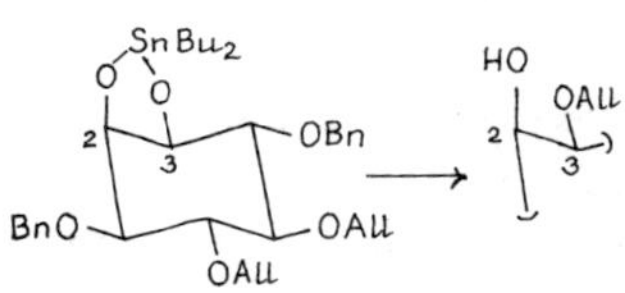


(37)

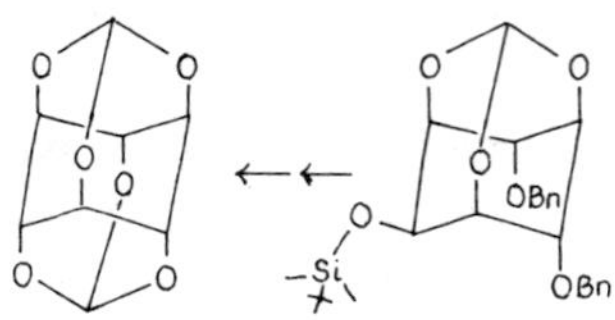


(38) (39)

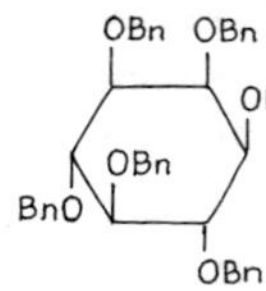


(40)

de-O-allylated to give the product.[57] The *scyllo*-inositol di-ortho-formate ester (38) has been prepared from the *myo*-inositol derivative (39), inverting the C-2 hydroxy group by a standard oxidation/reduction sequence.[58]

Penta-*O*-benzyl-*sn*-*myo*-inositol (40) has been resolved by formation of a diastereoisomeric mixture of α-D-mannopyranoside derivatives (ortho-ester route), which could be separated and hydrolysed to yield both enantiomers.[59] Syntheses have also been described for 1D- and 1L-4-*O*-benzyl-*myo*-inositol and their 4-*O*-α-L-fucopyranosyl derivatives; (±)-4-*O*-benzyl-1,6;2,3-di-*O*-cyclohexylidene-*myo* inositol was resolved using (+)-*O*-acetyl-mandelic acid and deacetalated to give the required chiral mono-benzyl ether compounds.[60]

In the Ferrier rearrangement, the orientation of the hydroxy group at C-5 in the product cyclohexanone depends on the conformation of the starting alkene, illustrated in Scheme 10.[61] A synthesis of the aminocyclohexanone (41) from 2-amino-2-deoxy-D-glucose also employs the Ferrier carbocyclization reaction.[62]

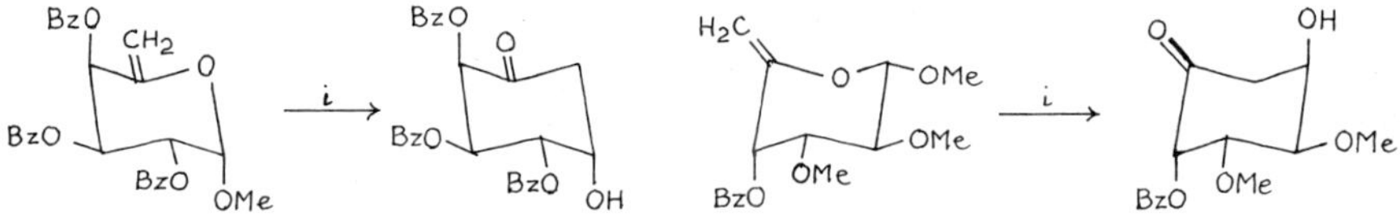


Reagents: i, $HgSO_4$-Dioxan.

Scheme 10

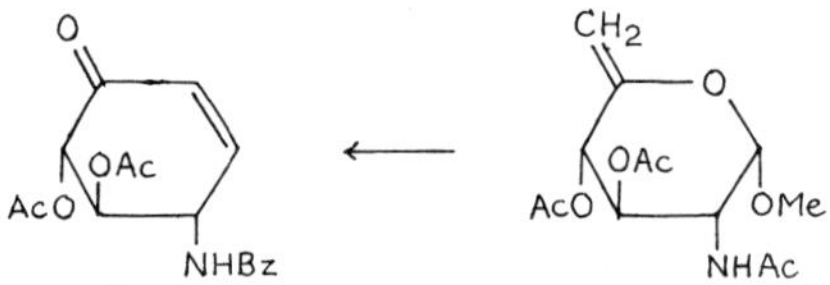


(41)

The synthesis of pseudo-hexopyranoses (cyclohexylmethanols) has been reviewed (in Japanese).[63] Suami's group has described the syntheses of four new compounds, α- and β-*allo*, α-*gulo*, and β-*ido* analogues of hexopyranoses, all as racemates,[64] and has also reported syntheses of the sweet-tasting pseudo-β-fructopyranose,[65] pseudo-α-L-altropyranose (42)(Scheme 11),[66] and derivatives of the

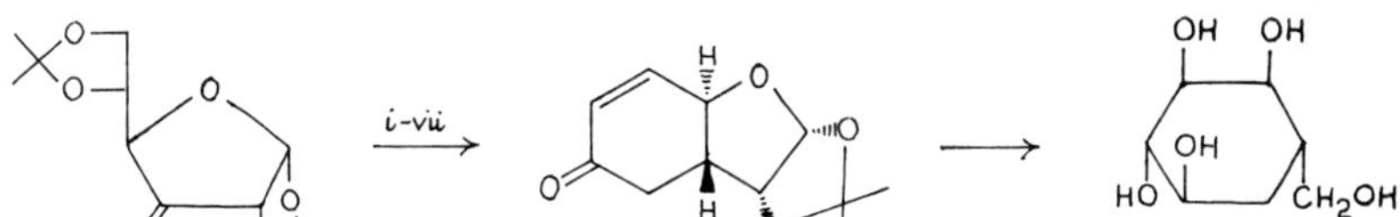


(42)

Reagents: i, $Ph_3P$=CHAc; ii, $H_2$-Ni; iii, PCC; iv, HOAc-$H_2O$; v, $NaIO_4$; vi, DBU; vii, $Ac_2O$-Py

Scheme 11

amino-pseudo-saccharides (+)-valienamine (43),[67] the glucosidase inhibitor valiolamine (44),[68] and (+)-validamine (45), the latter from the resolved oxabicycloheptene (46), which also furnished pseudo-galactopyranose.[69] Another paper describes the stereoselective conversion of valienamine and validamine to valiolamine.[70]

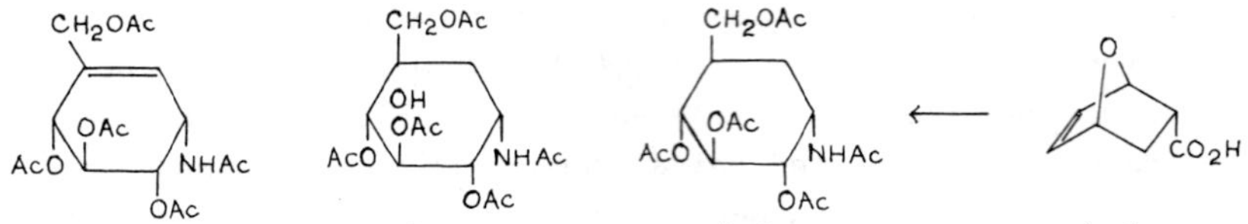

(43) (44) (45) (46)

Lukacs' group has reported the synthesis of the carbocyclic analogue (47) of daunosamine from the 2-deoxy-hexose (48), again utilizing a Ferrier carbocyclization (Scheme 12); hydroxy-group-directed homogeneous hydrogenation gave the required methyl stereoisomer exclusively, whereas traditional palladium-catalysed hydrogenation gave largely the alternative epimer.[71] The synthesis of

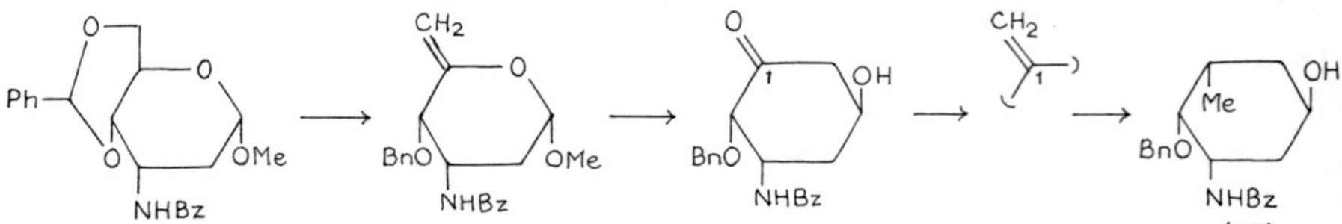

(48) (47)

Scheme 12

the carbocyclic analogue (49) of β-KDO (3-deoxy-β-D-manno-2-octulopyranosonic acid) from (-)-quinic acid (50) has been described,[72] and a study of the mono-caffeoylquinic acids (51) and some related dicaffeoyl analogues has shown that they possess antihistamine activity.[73] D-Ribonolactone has been used to synthesize the

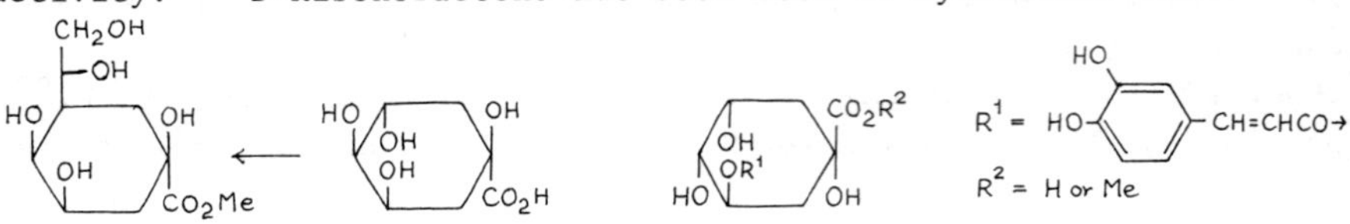

(49) (50) (51)

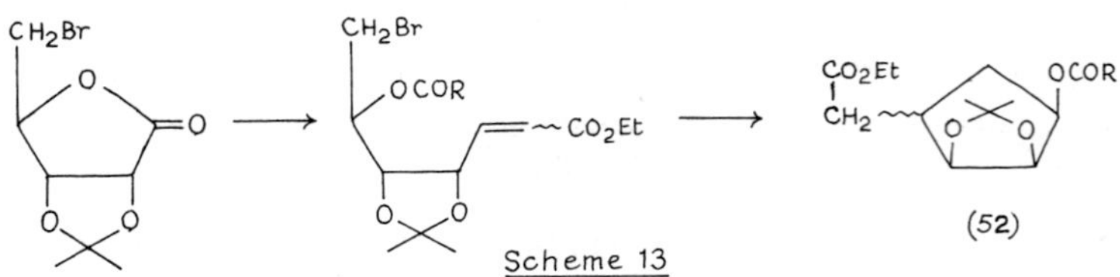

(52)

Scheme 13

pseudofuranose compounds (52)(Scheme 13).[74] Other references to pseudofuranose sugars occur in Chapters 19 and 20.

$^{13}$C N.m.r. chemical shifts for epi-inositol are altered by addition of lanthanide ions, those for ytterbium, thulium, and erbium changing in anomalous fashion.[75]

Calculations show that the conformational energies for myo-inositol 4,5-diphosphate increased from 840 to 1369 $kJmol.^{-1}$ as the charge on phosphorus increases from zero to -4.[76]

## References


1 A.Voiland and G.Michel, Carbohydr. Res., 1985, 141, 283.
2 M.Quillet, V.Imhoff, and G.de Lestang-Bremond, Phytochemistry, 1985, 24, 67.
3 R.J.Nash, E.A.Bell, G.W.J.Fleet, R.H.Jones, and J.M.Williams, J. Chem. Soc., Chem. Commun., 1985, 738.
4 R.J.Nash, E.A.Bell, and J.M.Williams, Phytochemistry, 1985, 24, 1620.
5 M.Makkee, A.P.G.Kieboom, and H.Van Bekkum, Starch/Staerke, 1985, 37, 136 (Chem. Abstr., 1985, 102, 205 679).
6 N.V.Lebedev, V.I.Sarzhe, V.E.Kononov, and V.M.Kuznetsov, Kompleksn. Ispol'z. Drev. Syr'ya, 1984, 92 (Chem. Abstr., 1985, 102, 115 324).
7 N.K.Nadirov, A.M.Ashirov, B.Sh.Kedel'baev, and V.F.Pechenkina, Gidroliz. Lesokhim. Prom-st., 1985, 8 (Chem.Abstr., 1985, 102, 133 792).
8 N.F.Shumateva, F.B.Bizhanov, and Z.Prnazarov, Izv. Akad. Nauk Kaz. SSR., Ser. Khim., 1984, 27 (Chem. Abstr., 1985, 102, 79 223).
9 M.Makkee, A.P.G.Kieboom, and H.Van Bekkum, Carbohydr. Res., 1985, 138, 225.
10 M.Makkee, A.P.G.Kieboom, and H.Van Bekkum, Carbohydr. Res., 1985, 138, 237.
11 T.Mukaiyama, T.Yura, and N.Iwasawa, Chem. Lett., 1985, 809.
12 C.Gennari, A.Bernardi, S.Cardani, and C.Scolastico, Tetrahedron, 1984, 40, 4059.
13 F.Sato, Y.Kobayashi, O.Takahashi, T.Chiba, Y.Takeda, and M.Kusakabe, J. Chem. Soc., Chem. Commun., 1985, 1636.
14 K.Takai and C.H.Heathcock, J. Org. Chem., 1985, 50, 3247.
15 S.R.Carter, J.M.Williams, and J.R.Clamp, J. Chem. Soc.,Perkin Trans. 1, 1985, 2775.
16 R.Köster, S.Penedés-Ullate, and W.V.Dahlhoff, Angew. Chem. Int. Ed. Engl., 1985, 24, 519.
17 V.Jäger, I.Müller, and E.F.Paulus, Tetrahedron Lett., 1985, 26, 2997.
18 G.Cardillo, M.Orena, S.Sandri, and C.Tomasini, Tetrahedron, 1985, 41, 163.
19 L.Zhang, S.Lao, S.Huang, and G.Liang, Yiyao Gongye, 1985, 16, 145 (Chem. Abstr., 1985, 103, 215 685).
20 K.Satsumabayashi and Y.Nomoto, Nippon Shika Daigaku Kiyo, Ippan Kyoiku-kei, 1983, 12, 131 (Chem. Abstr., 1985, 102, 132 392).
21 J.S.Brimacombe, A.K.M.S.Kabir, and I.D.Taylor, Carbohydr. Res., 1985, 140,C9.
22 J.Kuszmann, G.Medgyes, and P.Dvortsak, Carbohydr. Res., 1985, 142, 61.
23 A.Gomez-Sanchez, J.Galan, M.Rico, and J.Bellanato, J. Chem. Soc.,Perkin Trans. 1, 1985, 2695.
24 J.G.Buchanan, A.Millar, R.H.Wightman, and M.R.Harnden, J. Chem. Soc.,Perkin Trans. 1, 1985, 1425.
25 J.Kuszmann, Carbohydr. Res., 1985, 142, 71.
26 F.Freeman and K.D.Robarge, Carbohydr. Res., 1985, 137, 89.
27 R.Blattner and R.J.Ferrier, Carbohydr. Res., 1985, 138, 73.
28 D.Rolf, J.A.Bennek, and G.R.Gray, Carbohydr. Res., 1985, 137, 183.
29 L.Petrus, J.Alföldi, M.Matulova, and V.Bilik, Chem Papers, Chem. Zvesti, 1985, 39, 401.
30 F.Jacquet, A.Gaset, J.Simonet, and G.Lacoste, Electrochim. Acta, 1985, 30, 477 (Chem. Abstr., 1985, 103, 196 344).
31 S.V.Evans, L.E.Fellows, T.K.M.Shing, and G.W.J.Fleet, Phytochemistry, 1985, 24, 1953.
32 G.W.J.Fleet and P.W.Smith, Tetrahedron Lett., 1985, 26, 1469.
33 S.V.Evans, A.R.Hayman, L.E.Fellows, T.K.M.Shing, A.E.Dermone, and G.W.J.Fleet, Tetrahedron Lett. 1985, 26, 1465.

34 P.J.Card abd W.D.Hitz, J. Org. Chem., 1985, 50, 891.
35 Y.Ezure, Agric. Biol. Chem., 1985, 49, 2159.
36 G.W.J.Fleet, A.N.Shaw, S.V.Evans, and L.E.Fellows, J. Chem. Soc., Chem. Commun., 1985, 841.
37 G.W.J.Fleet, S.J.Nicholas, P.W.Smith, S.V.Evans, L.E.Fellows, and R.J.Nash, Tetrahedron Lett., 1985, 26, 3127.
38 D.W.C.Jones, R.J.Nash, E.A.Bell, and J.M.Williams, Tetrahedron Lett., 1985, 26, 3125.
39 J.Furukawa, S.Okuda, K.Saito, and S.Hatonoka, Phytochemistry, 1985, 24, 593.
40 R.C.Bernotas and B.Ganem, Tetrahedron Lett., 1985, 26, 1123.
41 M.J.Eis, C.J.Rule, B.A.Wurzburg, and B.Ganem, Tetrahedron Lett., 1985, 26, 5397.
42 G.Augelmann, J.Streith, and H.Fritz, Helv. Chim. Acta, 1985, 68, 95.
43 M.Alonso-Lopez, M.Bernabé, M.Martin-Lomas, and S.Penadés-Ullate, Carbohydr. Res., 1985, 142, 135.
44 M.Bernabe, M.Martin-Lomas, S.Penadés, R.Köster, and W.V.Dahlhoff, J. Chem. Soc., Chem. Commun., 1985, 1001.
45 I.E.Valashek, T.M.Filippova, M.K.Shakhova, N.A.Bogoslovskii, M.G.Romanyuk, and G.I.Samokhvalov, Zh. Obshch. Khim., 1984, 54, 2775 (Chem. Abstr., 1985, 102, 185 390).
46 P.Boulanger, C.André, and G.Descotes, J. Carbohydr. Chem., 1985, 4, 279.
47 M.Kuwabara, Z.Y.Zhang, O.Imanami, and G.Yoshii, Radiat. Phys. Chem., 1984, 24, 217 (Chem. Abstr., 1985, 102, 204 199).
48 J.G.Dawber and G.E.Hardy, J. Chem. Soc., Faraday Trans. 1, 1984, 80, 2467.
49 D.J.Walton, W.A.Szarek, D.B.Maclean, and R.V.Gerard, Carbohydr. Res., 1985, 137, 31.
50 D.G.Gillies and D.Lewis, J. Chem. Soc.,Perkin Trans. 2, 1985, 1155.
51 R.E.Moore, J.J.Barchi, and G.Bartolini, J. Org. Chem., 1985, 50, 374.
52 Y.J.Park, Taehan Hwahakhoe Chi, 1985, 29, 104 (Chem.Abstr., 1985,103,22867).
53 F.Abe and T.Yamauchi, Chem. Pharm. Bull., 1985, 33, 2712.
54 J.G.Streeter, Phytochemistry, 1985, 24, 174.
55 J.Gigg, R.Gigg, S.Payne, and R.Conant, Carbohydr. Res., 1985, 142, 132.
56 K.J.Lee, S.A.Boyd, and N.S.Radin, Carbohydr. Res., 1985, 144, 148.
57 J.Gigg, R.Gigg, S.Payne, and R.Conant, Carbohydr. Res., 1985, 140, Cl.
58 H.W.Lee and Y.Kishi, J. Org. Chem., 1985, 50, 4402.
59 Yu.I.Sibrikov, A.E.Stepanov, and V.I.Shvets, Zh. Org. Khim., 1984, 20, 979 (Chem. Abstr., 1985, 102, 6991).
60 P.J.Garegg, B.Lindberg, I.Kvarnström, and S.C.T.Svensson, Carbohydr. Res., 1985, 139, 209.
61 A.S.Machado, A.Olesker, and G.Lukacs, Carbohydr. Res., 1985, 135, 231.
62 M.Mádi-puskás, I.Pelyvás, and R.Bognár, J. Carbohydr. Chem., 1985, 4, 323.
63 S.Ogawa, Yuki Gosei Kagaku Kyokaishi, 1985, 43, 26 (Chem. Abstr., 1985, 102, 221 105).
64 S.Ogawa, Y.Tsukiboshi, Y.Iwasaka, and T.Suami, Carbohydr. Res., 1985,136,77.
65 T.Suami, S.Ogawa, M.Takata, K.Yasuda, A.Suga, K.Takei, and Y.Uematsu, Chem. Lett., 1985, 719.
66 T.Suami, K.Tadano, Y.Ueno, and C.Fukabori, Chem. Lett., 1985, 1557.
67 S.Ogawa, Y.Shibata, T.Nose, and T.Suami, Bull. Chem. Soc. Jpn., 1985, 58, 3387.
68 S.Ogawa and Y.Shibata, Chem. Lett., 1985, 1581.
69 S.Ogawa, Y.Iwasawa, T.Nose, T.Suami, S.Ohba, M.Ito, and Y.Saito, J. Chem. Soc. Perkin Trans. 1, 1985, 903.
70 S.Horii, H.Fukase, and Y.Kameda, Carbohydr. Res., 1985, 140, 185.
71 A.S.Machado, A.Olesker, S.Castillon, and G Lukacs, J. Chem. Soc., Chem. Commun., 1985, 330.
72 H.Molin and B.G.Pring, Tetrahedron Lett., 1985, 26, 677.
73 Y.Kimura, H.Okuda, T.Okuda, T.Hatano, I.Agata, and S.Arichi, Chem. Pharm. Bull., 1985, 33, 690.
74 C.S.Wilcox and L.M.Thomasco, J. Org. Chem., 1985, 50, 546.
75 S.J.Angyal, L.Littlemore, and P.A.J.Gorin, Aust. J. Chem., 1985, 38, 411.
76 S.M.Shevchenko, S.G.Semenov, and G.V.Kiselev, Teor Eksp. Khim., 1984, 20, 223 (Chem. Abstr., 1985, 102, 6990).

# 19
# Antibiotics

## 1 Aminoglycoside and Aminocyclitol Antibiotics

Reviews have appeared on new semi-synthetic aminoglycoside antibiotics, with particular reference to habekacin and polydeoxykanamycins,[1] and on the production of aminocyclitol antibiotics by fermentation and chemical synthesis.[2]

New compounds reported include inosamycin, a complex produced by _S.hygroscopicus_, which is related to neomycin and paromomycin, but containing 2-deoxy-_scyllo_-inosamine (1) instead of 2-deoxystrept-

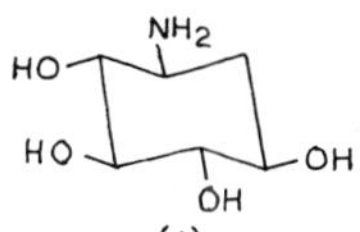


(1)

amine (_i.e._ hydroxy replaces amino at C-1).[3] The pseudotrisaccharide 1-_N_-amidino-1-_N_-demethyl-2-hydroxydestomycin has been characterized as an antibiotic elaborated by _Saccharopolyspora hirsuta_.[4] (See Vol.9,p.131, for the structure of destomycin). A total synthesis of destomycin C has been described, the formation of the crucial orthoester linkage being indicated in Scheme 1.[5]

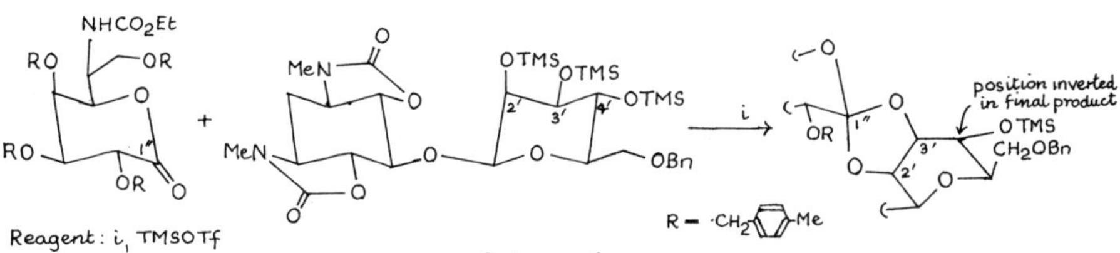


Scheme 1

A synthesis of the pseudodisaccharide 3-_O_-methylsporaricin A has been reported, coupling suitably protected derivatives of the diamino sugar 1-acetate and the diaminocyclitol using trimethylsilyl triflate.[6]

A synthesis of 6-deoxy-6-hydroxyamino-α-D-glucopyranoside from the corresponding 6-aminosugar by the sequence of oxidation with hydrogen peroxide - sodium tungstate to the oxime followed by cyanoborohydride reduction was copied in a synthesis of the pseudotrisaccharides 6'-_N_-hydroxykanamycin A and 6'-_N_-hydroxydibekacin from the corresponding parent antibiotics, but the products were only very weakly active.[7] 3'-Deoxy-3'-fluorokanamycins A and B, which show activity against resistant bacteria, have been prepared by condensation of a 6-azido-3-fluoro-D-glucosyl bromide derivative with an appropriate 3-amino-glucosyl - 2-deoxystreptamine derivative.[8] A series of 3''-*N*-trifluoroacetylkanamycin A derivatives have been prepared, possessing $C_5$- $C_{22}$ acyl residues, or related alkyloxy-,-amino-, or -thio-carbonyl groups, all attached at N-1; the optimum antiviral activity was shown for the $C_{14}$- $C_{16}$ acyl derivatives.[9,10] A number of 4''-epi-amikacin derivatives (OH,F,$N_3$) have been synthesized, which should be less sensitive to enzymatic deactivation, but all were much less active than the parent.[11]

Conventional methods have been used to convert kanamycin A to its 4''-deoxy-4'',5''-didehydro derivative, and hence to 4''-deoxy-5''-epikanamycin A.[12]

The synthesis of 5-*O*- and 6-*O*-methyl-dihydrostrepomycin has been reported, using diazomethane - stannous chloride on selectively protected derivatives.[13]

Hanessian's group has reported another synthesis of spectinomycin (2), coupling a 4,6-dideoxy-*ribo*-hexosyl chloride to a streptamine derivative, and then elaborating the internal dioxan ring; the synthesis and reactions of analogues are also described.[14] Spectinomycin has also been modified at C-3'(desoxo, monohalo, and dihalo analogues) via diazoketone intermediates,[15] which were also used to make some C-3' branched-chain analogues (Scheme 2), the most potent antibiotic being the diamine derivative (3), which was more active than the parent antibiotic.[16]

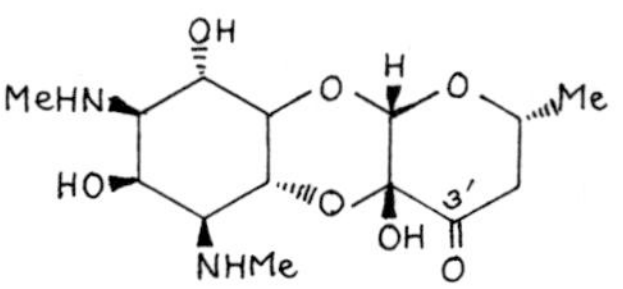

(2)

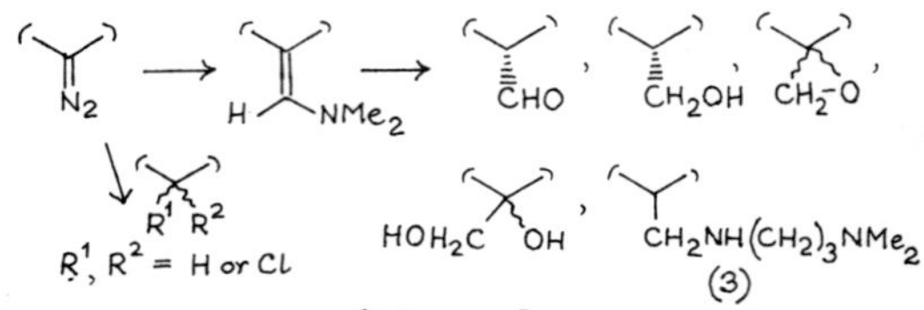

Scheme 2

The structure of *N*-acylkanosamine has been established as 4,6-dideoxy-4[(*R*)-2-methoxypropanamido]-3-*C*-methyl-2-*O*-methyl-L-mannose by a synthesis from rhamnose.[17]

Fluorescent-labelled aminoglycosides have been prepared from sulphate salts with fluorescamine, some retaining their biological activity.[18]

Validamycin has been synthesized by 4-*O*-β-D-glucosylation of validoxylamine A (using a modified Koenigs Knorr procedure), and its

7-deoxy analogue was similarly prepared.[19] The synthesis of valiolamine derivatives is mentioned in Chapter 18.

Sisamine has been modified by conventional methods to give the fortimicins 3,4-di-demethyl-, 4-demethyl-, and 3-amino-4-demethoxy-4-deoxy-fortimicin $KG_2$.[20]

A further study on the biosynthesis of 2-deoxystreptamine has been reported, showing that enzymes from S. fradiae catalyse three steps in the proposed sequence (Scheme 3).[21]

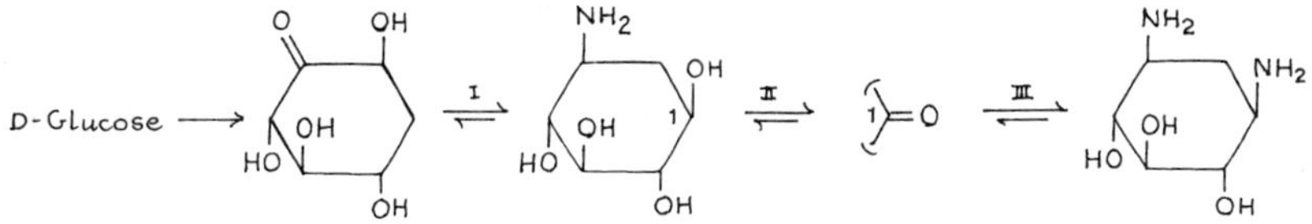


Scheme 3

The inactivation of astromicin (fortimicin A) by a gentamicin-resistant strain of Staphylococcus aureus involves 6'-N-acetylation by a 6'-N-acetyl-transferase.[22]

Perfluorinated carboxylic acid counter ions ($C_2$- $C_4$ acids) have been found to provide for easy combination of ion-pair HPLC and FD m.s. in analysis of amino-glycoside antibiotics, allowing early identication in culture broths.[23] FAB m.s. of seven nebramycin and lividomycin antibiotics have been recorded.[24]

## 2 Macrolide Antibiotics

A review on the total synthesis of macrolide antibiotics covers glycosidation.[25]

A new antifungal antibiotic, 3'-O-decarbamoyl-irumamycin, containing 2-deoxy-β-D-rhamnopyranose linked to a 20-membered macrolide, has been isolated together with crumamycin from S. subflavus subsp. crumaensis.[26] The macrolide antibiotic, X-14952B, produced by a Streptomyces species, contains the same 3-O-carbamoyl-2-deoxy-β-D-rhamnoside present in irumamycin, attached to a closely related 20-membered macrolide.[27]

Further studies have been reported on the mycinamycin complex (see Vol.14,p.157); N-oxidation (MCPBA) followed by acetylation removed the desosamine unit, and X-ray analysis was carried out on the resulting 4-O-acetylated mycinose-macrolide derivative, giving an absolute configuration for the antibiotic;[28] CI m.s. studies indicate that the neutral mycinose sugar is cleaved more readily than the desosamine on fragmentation.[29]

Spiramycin 1 (containing mycaminose, forosamine, and mycarose) has been chemically modified. Mono-de-*N*-methylation of the 3'- or 4''-dimethylamino groups was followed by *N*-acylation or *N*-benzylation,[30] and a series of 3,3''-di-*O*-acyl-4''-*O*-sulphonyl and 3,3''-di-*O*-acyl-4''-alkyl derivatives were prepared.[31]

The hydrophobicity of 3''-*O*-acyl groups attached to the tertiary hydroxy group of mycarose in leucomycin is particularly significant for antimicrobial activity.[32]

A complete structure of amphotericin A, containing mycosamine, has been elucidated,[33] together with a m.s. analysis of dodecahydroamphotericin A.[34] 2D $^{13}C$ and $^{1}H$ n.m.r. spectra of erythromycin A have been reported.[35]

## 3 Anthracycline Antibiotics

The total synthesis of anthracycline antibiotics[36] and related hydroxynaphthalenedione glycosides[37] have been reviewed.

The structures of cosmomycins A,B,C, and D, new differentiation inducers produced by *S. cosmosus*, have been elucidated.[38-40] They contain one or two trisaccharide chains attached to β- or γ-rhodomycinone, as indicated in structure (4). Closely related anthracycline antibiotics, serirubicin (5)(identical with cosmocarcin A (see Vol.18,p.179)) and its 1-hydroxy analogue, have been found in the culture filtrate from *S. cyaneus* (which also produces ditrisarubicins (see Vol.17,p.176).[41] Mother liquors from the crystallization of steffimycin B yield a further antibiotic steffimycin C, which is 10-deoxy-steffimycin B, containing 2,4-di-*O*-methyl-L-rhamnose[42] (see Vol.11,p.41). Elloramycin (6), isolated from

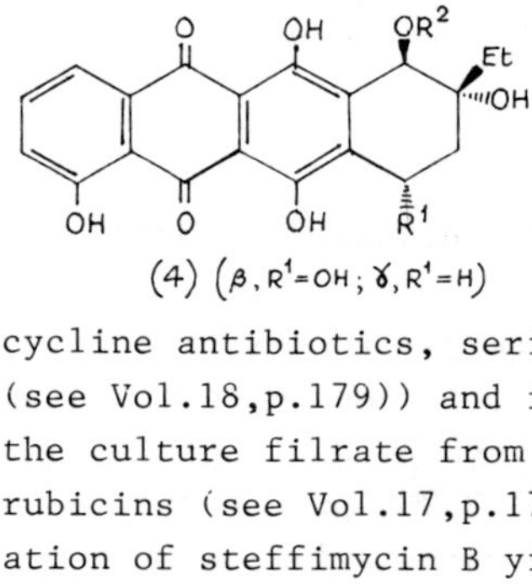

(4) (β, $R^1$=OH; γ, $R^1$=H)

A, $R^1$=H, $R^2$=$R^4$
B, $R^1$=H, $R^2$=$R^3$
C, $R^1$=$OR^3$, $R^2$=$R^4$
D, $R^1$=$OR^3$, $R^2$=$R^3$

$R^3$ = Rhn - deFuc - Rho,
$R^4$ = Rhn - Rho - Rho,
all linked α, 1→4

Rhn = Rhodosamine, Rho = Rhodinose,
deFuc = 2-deoxy-L-fucose

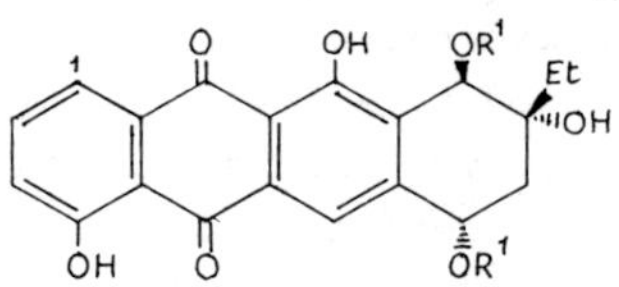

(5)

$R^1$ = Cinerulose B → 2-deoxy-L-fucose → rhodosamine

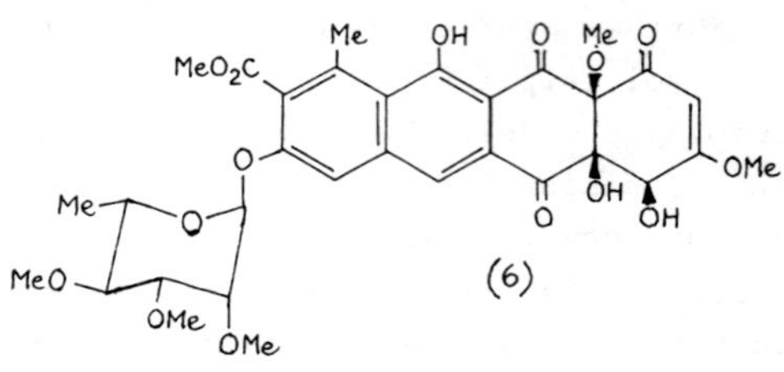

(6)

S. olivaceus, contains 2,3,4-tri-O-methyl-L-rhamnose.[43]

A new ε-rhodomycin derivative has been obtained from a Streptomyces species, containing a trisaccharide of α-1→4 linked rhodinose units attached to ε-rhodomycinone.[44] 4-O-(β-D-Glucopyranosyl)-ε-rhodomycinone has been isolated from the culture of a blocked mutant of Actinomadura roseoviolacea with ε-rhodomycinone.[45]

Analogues of daunorubicin have been prepared from modified sugars. 3'-Epi,[46] 3'-hydroxy-3',5'-diepi,[46] and 3',6'-dihydroxy-3',5'-diepi,[46] 2'-halo,[47] and 3'-C-methyl[48] analogues were prepared using appropriate glycals or glycosyl halides with daunomycinone; coupling of glycals with N-iodosuccinimide led to 2'-iodo analogues,[47] and the branched-chain sugar (7), which gives the vancosamine analogue of daunosamine, was obtained as outlined in Scheme 4.[48] The bromine or chlorine adducts of glycals (8) were

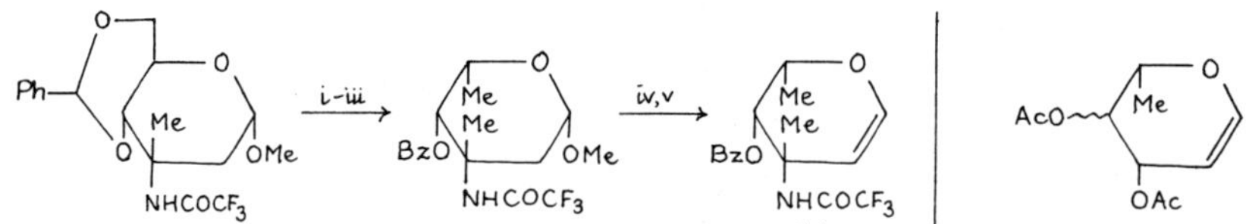


Reagents: i, NBS-$CCl_4$; ii, AgF; iii, Raney Ni; iv, $H_3O^+$; v, TsCl-Py

(7) (8)

Scheme 4

used to prepare 2'-bromo- or 2'-chloro-3'-acetoxy-3'-deamino-analogues of daunorubicin.[49] Coupling of 4-demethoxydaunomycinone with a 1-O-acyl-daunosamine derivative using trimethylsilyl triflate yielded a novel 7,9-bis-O-α-glycoside, the reagent allowing glycosidation of the tertiary C-9 hydroxy group in the aglycone, a feature which does not appear to affect the antibiotic activity of the parent daunorubicin.[50]

A 14-thio-D-glucopyranosyl analogue of daunorubicin has been synthesized from its 14-bromo derivative with a 1-thiosugar, and 3'- and 6'-azido analogues were made from corresponding azidodeoxy-sugars with daunomycinone.[51] Several other anthracycline analogues have been prepared from 2,6-dideoxy-, 2-deoxy-2-trifluoroacetamido-, and 2,3,6-trideoxy-3-trifluoroacetamido-hexopyranosyl halides and 2-deoxypentopyranosyl halides.[52] Yet other analogues have been obtained by nitrous acid deamination of daunorubicin, doxorubicin, and their configurational isomers, the range of derivatives being summarized in Scheme 5; the derivatives (9) and (10) of doxorubicin showed outstanding antileukaemic activity in mice.[53] The microbial conversion of daunorubicin to N-acetyl -13(S)-dihydrodaunorubicin has been described.[54]

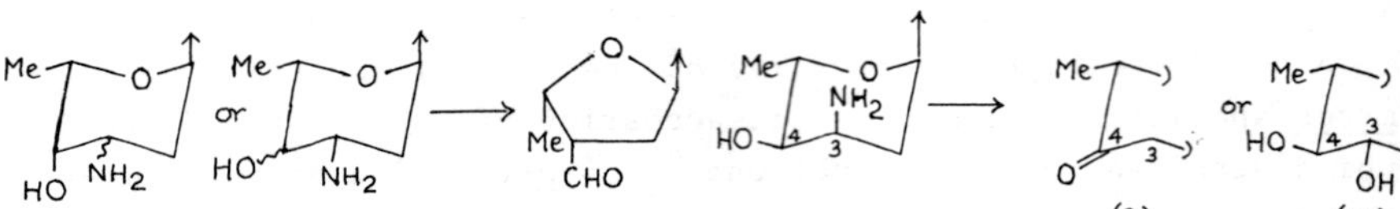

(9) (10)

Scheme 5

## 4 Nucleoside Antibiotics

Several groups have reported the isolation of 2'-chloropentostatin (11), an inhibitor of adenosine deaminase, from an Actinomadura strain.[55-57] Some new nikkomycin antibiotics (12) have been described, which are nucleoside-peptide antibiotics elaborated by a

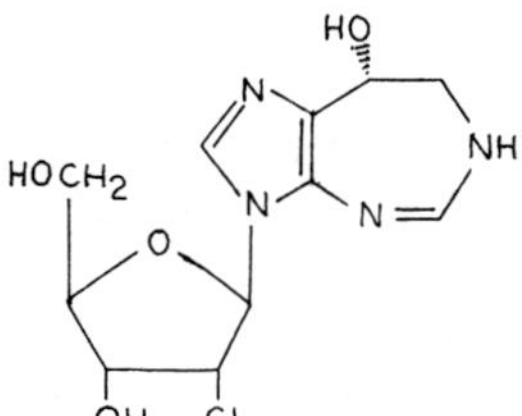

(11)

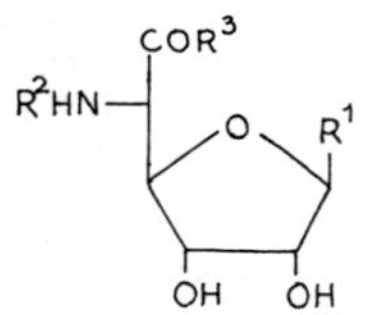

(12)

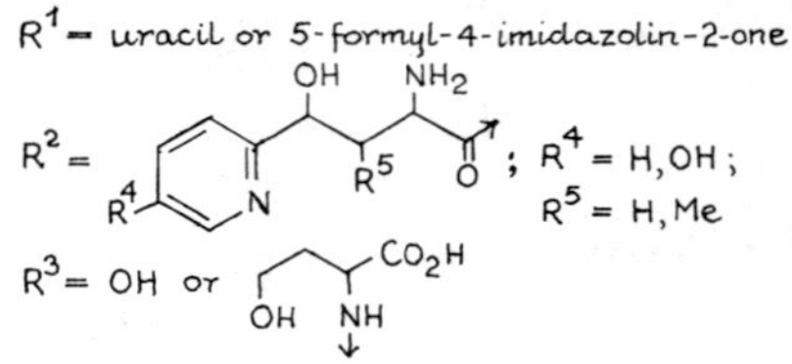


mutant strain of S. tendae[58] (see Vol.15,p.187). A novel nucleoside antibiotic, A201A (13), has been isolated from the broth of a S. capreolus strain; it contains a 3-amino-ribo-nucleoside, an enolized hexos-5-ulose, and 3,4-di-O-D-rhamnose.[59] Rebeccamycin, a new antitumour antibiotic isolated from Nocardia aerocoligenes, has been characterized as the nucleoside analogue (14) by X-ray analysis[60] and by synthesis.[61]

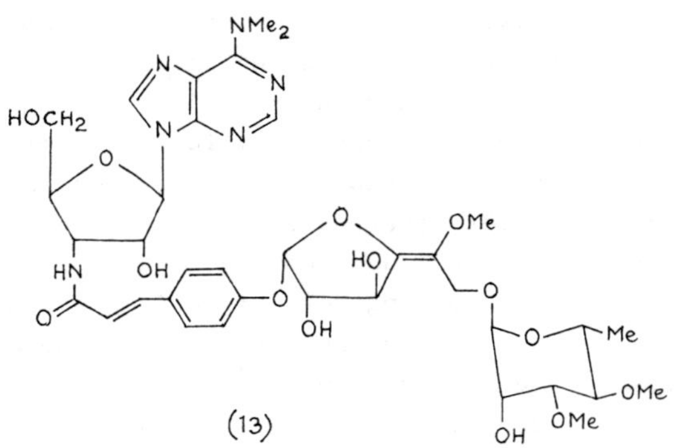

(13)

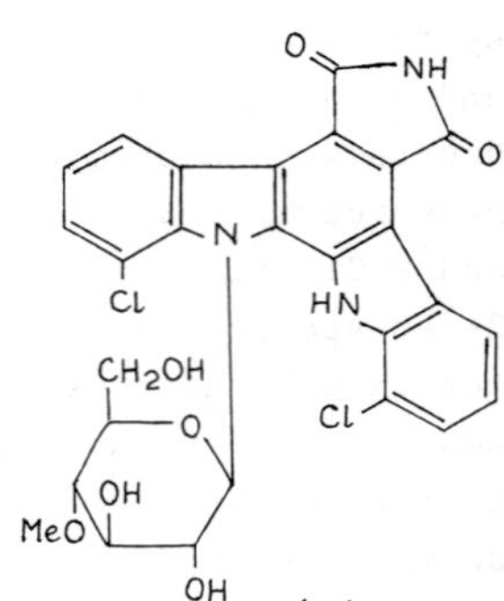

(14)

Full details of the "total" synthesis of tunicamycin V have been reported (see Vol.18,p.182).[62] The same group has also described a preparation of the nucleoside unit (15) in this synthesis from the 7-nitro-undecodialdose previously reported (see Vol.16,p.111).[63] Further evidence for the $\alpha$-glucosamine/$\beta$-tunicamine interglycosidic linkage has been adduced from COSY-2D n.m.r. data.[64] 9'-Desamino-sinefungin (16) has been synthesized, together with its 6'-epimer, using a Wittig approach previously employed for sinefungin itself[65] (see Vol.18,p.184).

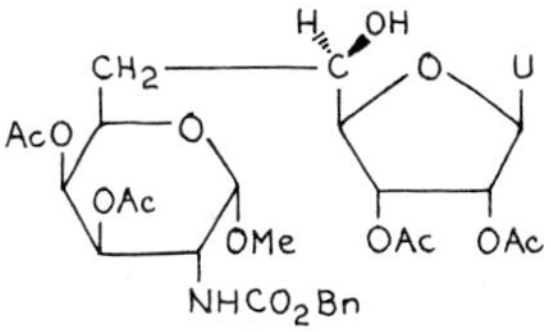


(15)

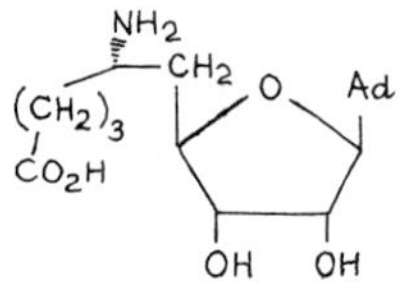


(16)

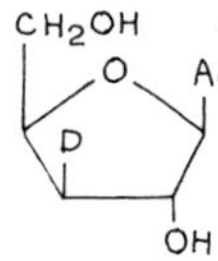


(17)

Improved syntheses have been described for cordycepin (3'-deoxy-adenosine), using 2',3'-anhydro-adenosine; in a high-yield, one-step procedure, the 3'-deoxy compound was obtained directly using lithium triethylborohydride;[66] alternatively, an $N^6,O^{5'}$-bis-monomethoxy-trityl derivative was used with LAH, with subsequent deprotection.[67] Both groups also prepared the 3-deuterio analogue (17).

Syntheses of nucleoside analogues by conventional condensations of sugar derivatives with bases have included 6-azacadeguomycin and some 3,4,6-trisubstituted pyrazolo[3,4-d]pyrimidine nucleosides,[68] and 6-mercapto- and 6-methylmercapto-purin-9-yl-$\beta$-D-glucofuranosid-uronolactones (18), which showed antitumour activity.[69] A number

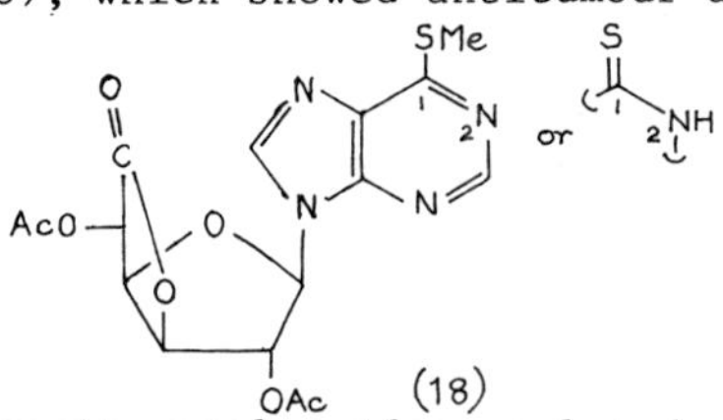


(18)

of uridine and inosine nucleosides acylated at O-5' have been prepared; 5'-O-iodoacetyl-2',3'-O-isopropylidene-uridine showed cytostatic activity against HeLa cells.[70] 3',5'-Diesters ($C_2$- $C_{12}$ carboxylic acids)[71] and 3',5'-bis-dicarboxylic acid hemiesters[72] of 5-fluoro-2'-deoxyuridine have been prepared and evaluated as pro-drugs by examination of their susceptibility to enzymic hydrolysis. Higher antitumour activity of longer-chain alkyl diesters correlates with their slower enzymic hydrolysis.

Treatment of tubercidin with thionyl chloride gave the 5'-chloro-

5'-deoxy analogue, reduced by tributylstannane to 5'-deoxytubercidin.[73] Glycosylation of the sodium salt of 6-substituted and 2,6-disubstituted-7-deazapurine derivatives with a protected 2-deoxyribosyl chloride gave a series of corresponding 2'-deoxytubercidin derivatives, but they did not show comparable antibiotic activity to ribavirin.[74]

An extensive review on the synthesis and antiviral properties of 5-vinylpyrimidine nucleosides has been published.[75] A range of 5-halovinyl-2'-deoxyuridines have been prepared by a standard glycosidation procedure; all showed significant antiviral activity, the most potent being 5-(2-chloroethyl)-2'-deoxyuridine (19).[76]

Carbocyclic compounds - Carbocyclic analogues of (*E*)-5-(2-halovinyl)-2'-deoxyuridines and corresponding cytidines have been made from 4-amino-2,3-dihydroxycyclopentyl-methanol,[77] and this cyclopentylamine has also been used for the synthesis of (±)-1-deaza-aristeromycin (20), a potent enzyme inhibitor,[78] and carbocyclic analogues of nucleosides from adenine, $N^6$-methyladenine, 8-azaadenine, and 3-deazaadenine, whose antiviral activity was also investigated.[79]

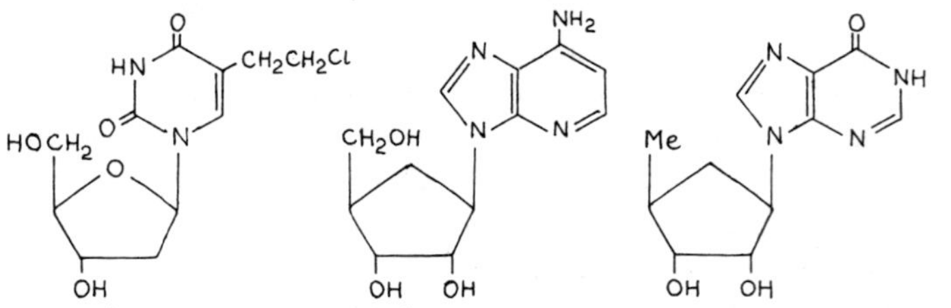


(19) (20) (21) (22)

There has been a further report on the synthesis of aristeromycin and neplanocin A, including a cytosine analogue (see Vol.17,p.178, ref.72).[80] The mother liquors left after aristeromycin crystallization derived from *S. citricolor* have yielded an analogue, aristeromycin M (21), which could be synthesized from aristeromycin by standard reactions.[81] Racemic aristeromycin has been resolved by conversion to its 3',5'-cyclic phosphate, which was then selectively hydrolysed enzymatically to give (-)-aristeromycin and the unhydrolysed (+)-cyclic phosphate, the latter then being separately hydrolysed to (+)-aristeromycin; the latter showed no antibiotic activity.[82] Feeding experiments with 6-$^{13}$C-D-glucose indicated that biosynthesis of aristeromycin proceeds by bonding C-2 to C-6, the label becoming the ring methylene in the product.[83]

The structure of neplanocin C has been determined to be (22) by *X*-ray analysis.[84] An improved synthesis of neplanocin and of related cyclopentenyl nucleosides has been described, which includes direct reaction of a cyclopentenyl allylic tosylate with the sodium salt of 6-chloropurine.[85] A possible approach to neplanocin analogous to one to aristeromycin, involving hydroxylation of the bicycloalkene (23), foundered when only endo-addition could be induced, the required exo-addition presumably hindered by the syn-bromine substituent.[86]

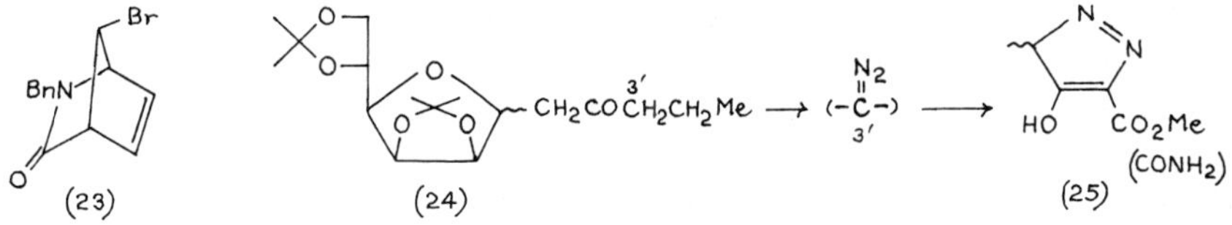

(23) (24) (25)

C-Nucleosides - A review of the work by Buchanan's group on the synthesis and biosynthesis of C-nucleosides, including antibiotics, has been published as a lecture abstract.[87]

An easy synthesis of the Wittig product (24) from 2,3;5,6-di-O-isopropylidene-D-mannofuranose gives a new approach to pyrazofurin analogues (25).[88] Ribosyl cyanide provides a straightforward synthesis of tiazofurin (26) and selenazofurin (27).[89] 3'-Deoxy-tiazofurin has been prepared from tiazofurin by tributylstannane reduction of a 3'-O-imidazole thiocarbonyl derivative.[90] 2'-Epi-tiazofurin, together with its α-anomer, has been synthesized from 2,3,5-tri-O-benzyl-1-O-(p-nitrobenzoyl)-D-arabinofuranose via 1-cyano and 1-thioamide intermediates.[91]

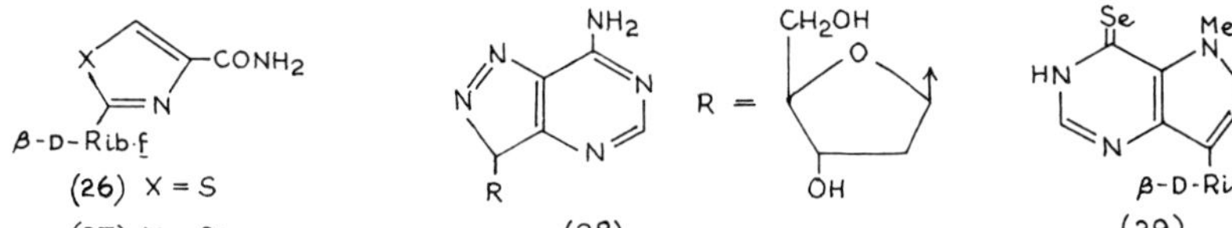

(26) X = S

(27) X = Se

(28)

(29)

An improved conversion of formycin A to its 2'-deoxy analogue (28) has been described (standard steps), and the latter was found to be a potent inhibitor of mouse lymphoma.[92] 2-Fluoro- and 2-amino-formycin have been prepared from the ribosyl base which is the analogue of guanosine in this series.[93] Similar base modification has furnished the selenium analogue (29), as well as its 5'-phosphate; both were found to be potent antileukaemic agents.[94]

Other references to nucleoside antibiotics, including microbial synthesis of 2'-amino-2'-deoxy-nucleosides and Ara-A, the synthesis of α-2'-deoxynucleosides, and the synthesis of radio-labelled

3'-halo-3'-deoxy-ara-U compounds, can be found in Chapter 20.

## 5 Miscellaneous Antibiotics

Formacidins A and B, new monocyclic β-lactam antibiotics produced by the new bacterial species Flexibacter alginoliquefaciens sp., have been structurally characterized; they contain D-glucopyranosiduronic acid glycosidically β-linked through phenol to the β-lactam component (30;part structure); they are remarkably stable to β-lactamases.[95] Chaetiacandin, a papulacandin produced by Monochaetia dimorphospora, has been identified as the pseudo-disaccharide derivative (31)[96] (for papulacandins, see Vol.14,p.163).

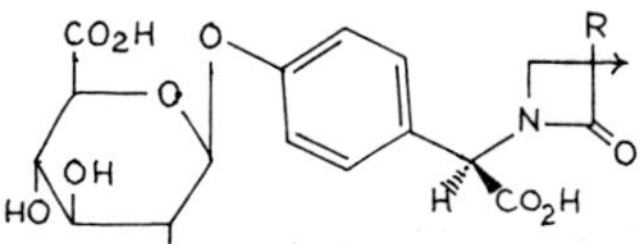

(30) R = H or NHCHO

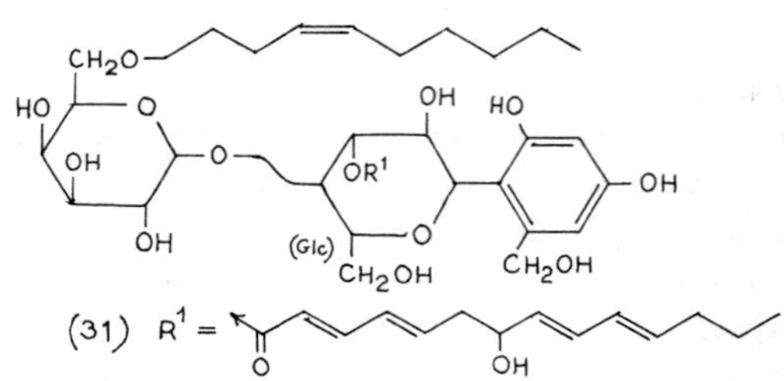

(31)

Spicamycin (32),found in the culture broth of S. alanosinicus, is a pseudo-nucleoside antibiotic, apparently the 2'-epimer of septacidin.[97]

Glycocinnaspermimicin D (33) has been characterized as a new

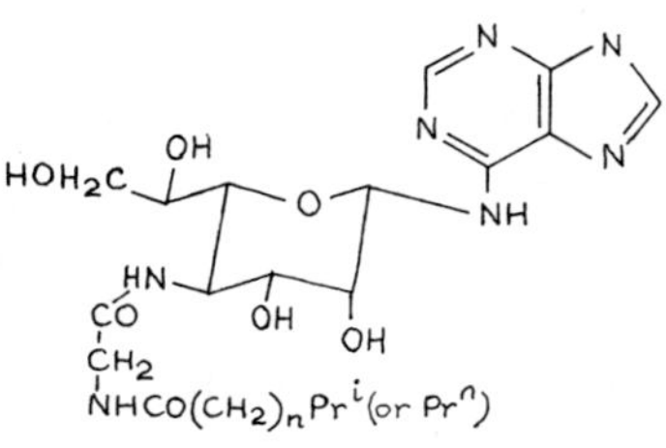

(32) $n = 8-14$

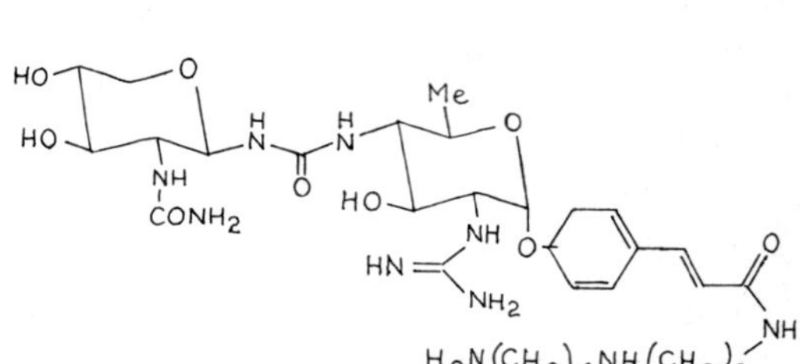

(33)

member of the glycocinnamoylspermidine group of antibiotics, produced by Nocardia sp.[98]

Several new glycosides involving polycyclic aromatic aglycones have been described this year. The anti-cancer antibiotic lactoquinomycin, produced by S. tanashiensis, is the 3-amino-2,3,6-trideoxy-β-arabino-hexose C-glycoside (34)(D or L configuration not established).[99] The isotetracenone anti-tumour antibiotic capoamycin, obtained from S. capoamus, has been shown to be a C-glycoside of β-2,6-dideoxy-D-arabino-hexose (35).[100] Kerriamycins A,B, and C are closely related isotetracenone compounds, containing a disaccharide or trisaccharide C-glycoside unit with or without a further L-rhodinose O-glycoside component, summarized in structure (36);

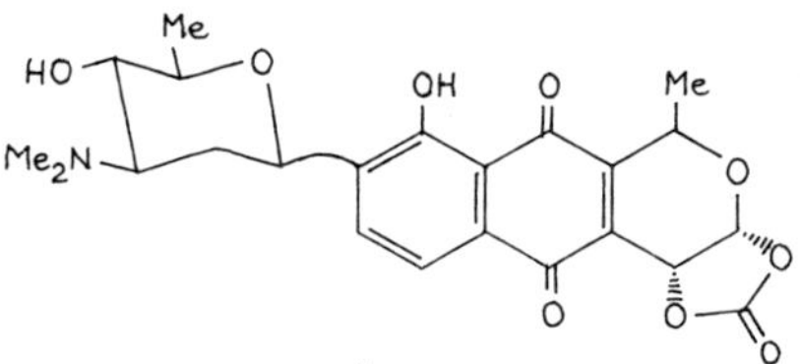

(34)

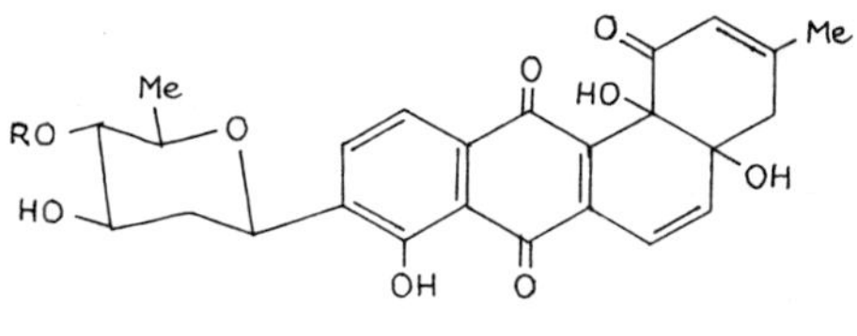

(35) R = E,E-2,4-decadienoyl

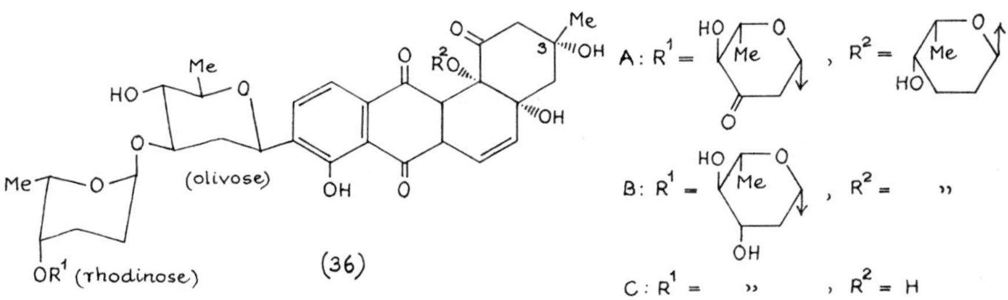


(36)

kerriamycin A contains the glycos-3-ulose kerriose (2,6-dideoxy-L-erythro-hexopyranos-3-ulose).[101] The saquayamycins, produced by S. nodosus, are close relatives in the same family; four components, A-D, have been characterized, which comprise the same C-glycoside (aquayamycin) linked with three other sugars, L-rhodinose being attached at O-3, and the glycos-4-uloses L-cinerulose or L-aculose attached at O-4 of either the D-olivose or L-rhodinose units.[102] Another new anti-tumour antibiotic, virenomycin A, has the branched-chain sugar 6-deoxy-3-C-methyl-D-gulopyranoside (37) β-linked as a C-glycoside to the same aglycone present in gilvocarcin V and ravidomycin [103] (see Vol.17,p.181,182).

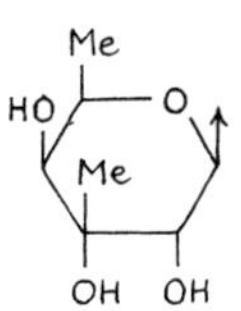

(37)

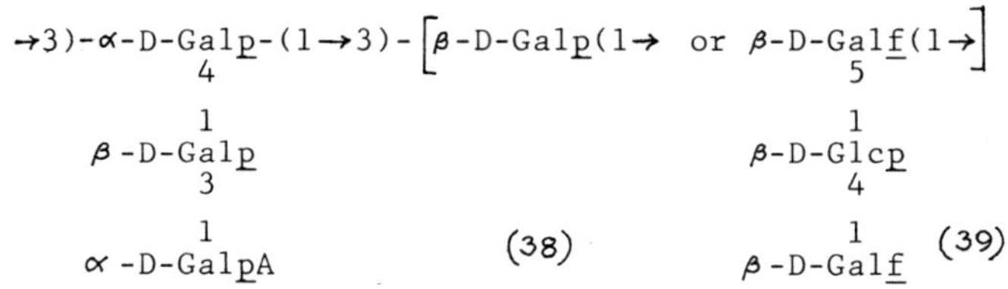


(38) (39)

A number of polysaccharide antibiotics have been reported. The 5'-nucleotidase inhibitor nucleoticidin, isolated from a Pseudo-

*monas* sp., contains D-glucose and D-mannose in ratio varying from 1.7 to 1.0. Evidence suggests D-man*p* 1→2 D-man*p* 1→3 chains are attached to a β-1→4 glucan backbone, with an average 1 in 4 glucose units substituted at O-3, with m.w. > $10^6$.[104] Melanocidin A and B, likewise 5'-nucleotidase inhibitors, obtained from *Nocardioides* sp., also m.w. > $10^6$, are acidic polysaccharides, containing glucose, galactose, and galacturonic acid, with repeating units (38) and (39) respectively.[105] The fruiting body and growing mycelium of *Ganoderma lucidum* (*Reishi*) produces 1→3 linked β-D-glucans, carrying single D-glucose units, mainly at O-6, and a few short-chain 1→4 linked D-glucose oligosaccharides attached at O-2, which show anti-tumour activity; interestingly, modification of the polysaccharides to D-glucan polyols ($NaIO_4$-$NaBH_4$) increased their antibiotic activity.[106]

New bacterial metabolites with bulge-inducing activity, bulgecins A,B,and C (40), contain D-glucosamine linked to proline derivatives.[107] A new glycosylated polyether antibiotic, moyukamycin, from *S. hygroscopicus* TM-581, contains 2,3,6-trideoxy-4-*O*-methyl-β-D-*erythro*-hexopyranosyl units, as in dianemycin.[108] The anti-tumour compounds FR-900405 and FR-900406, obtained from *Actinomadura pulveracea*, contain the 2-deoxy-fucose residue (41) as a part structure.[109] Methanolysis of the esperamicins from *Actinomadura*

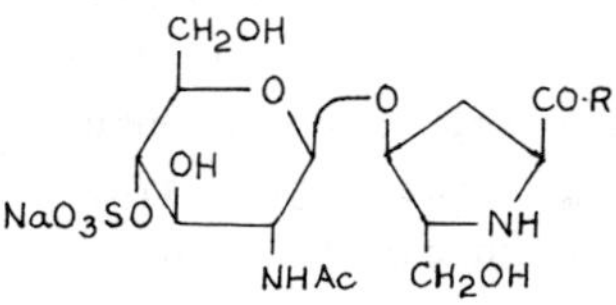

A : R = $-NH(CH_2)_2SO_3H$

B : R = $-NH(CH_2)_2CO_2H$

C : R = OH

(40)

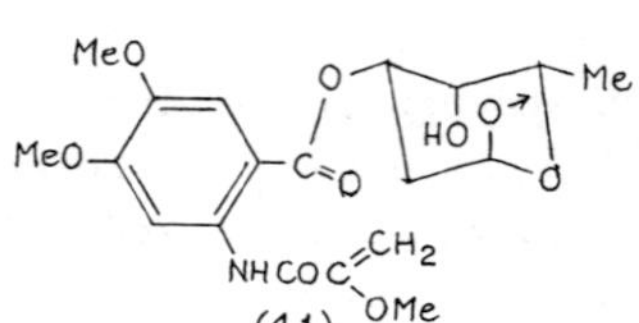

(41)

*verrucosospora*,[110] and of compounds PD114,759 and PD 115,028, from *Actinomadura* sp.,[111] yielded the methyl glycoside of the same fucoside (41).

Nojirimycin bisulphite adduct adopts a $^4C_1$(D) β-pyranose conformation in the crystal. The compound provides a suitable derivative for storing the antibiotic,which is unstable in free form, and its glycosidase-inhibiting activity was studied.[112] An improved procedure for preparing moranoline (1-deoxy-nojirimycin) from *Streptomyces* strains has been described, giving 27-33-fold increase in yields.[113] The synthesis of 4-*O*-β-D-glucopyranosyl-moranoline by enzymatic transglycosylation is referred to in Chapter 3.

The production of eight new avermectins (which contain an oleandrose disaccharide unit) from *S. avermitilis* has been reported;

sinefungin was added to a mutant strain, giving avermectins lacking methoxy groups in both the oleandrose moiety and the aglycone.[114]

An alternative synthesis has been described for amylostatin (42) using a derivative of maltose with valienamine [115] (see Vol.16,p.97). The same group has also reported an analogous synthesis of the pseudo-trisaccharide adiposin-1 (43).[116]

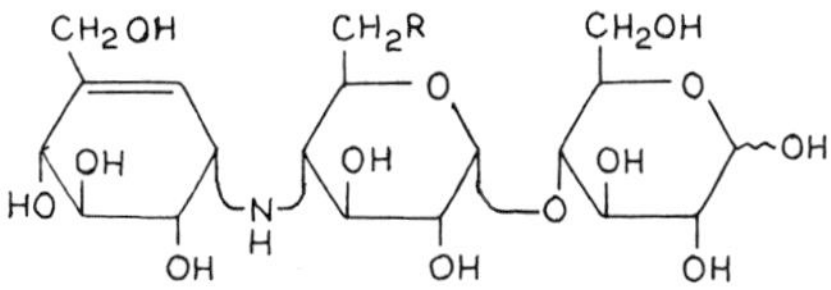

(42) R = OH
(43) R = H

The synthesis of higher-carbon carbohydrates has been reviewed (in Japanese); there is particular reference to tunicamine (an undecose) and coverage of octoses, nonoses, and decoses.[117]

Syntheses have been reported for the pseudo-disaccharides (44)[118] and (45),[119] which are derivatives of the component in oligostatin α-glucosidase inhibitors; the protected derivative (44) could not be converted to the free form without spontaneous rearrangement to the tricyclic pyrrolidine sugar isomer (46). Likewise derivatives of

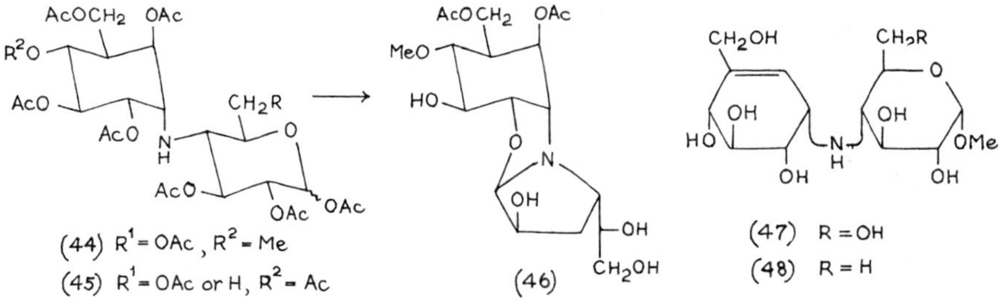

(44) $R^1$ = OAc, $R^2$ = Me
(45) $R^1$ = OAc or H, $R^2$ = Ac
(46)
(47) R = OH
(48) R = H

acarviosin (47) have been prepared, the dehydrated form of this pseudo-disaccharide being produced by methanolysis of oligostatin. The 6-deoxy analogue (48) and the C-1' and C-2' epimers were also synthesized.[120] In these syntheses, a free amine was condensed with an oxiran to generate the inter-residue linkage; use of a dianhydro sugar with amino-cyclitol[118] gave better regio-control than converse use of an aminosugar with anhydro-inositol, which was further complicated by using racemic inositol derivatives.[119,120]

Antibiotic A32390A (1,6-di-O-(2-cyano-3-methyl-but-2-enoyl)-D-mannitol) has been synthesized using standard procedures.[121]

Glycosylated cephalosporins have been prepared from cephalosporanic acid with 1-thio-glycosyl derivatives , giving acylated thio-glycosides of glucose, galactose, and glucosamine.[122]

FAB m.s. has been used to elucidate the structure of everninomycins[123] and of glycopeptide antibiotics related to vancomycin, including the newly discovered compounds aridicins A, B, and C (isolated from the new genus Kibdelosporangium aridum), which contain mannose and 2-deoxy-2-alkylamido-glucuronic acid units, the latter involving $C_9$-$C_{11}$ acyl units on the amino group.[124] The role of the sugar moieties in teichoplanin and ristocetin A and their derivatives in binding to bacterial cell walls has been probed; only D-mannose appears to be involved.[125]

The copper(II) complexes of D-glucosone bis-thiosemicarbazones have been shown to have anti-tumour activity.[126]

## References


1 H.Umezawa,S.Kondo, and I.Kitasato, Drugs Exp. Clin. Res., 1984, 10, 631 (Chem. Abstr., 1985, 102, 149 641).

2 S.D.Gero, J. Cleophax, D.Mercier, M.Philippe, C.J.Pearce, B.Quiclet-Sire, D.Semeria, and A.M.Sepulchre, Stud. Org. Chem. (Amsterdam),, 1984, 17, 79 (Chem. Abstr., 1985, 102, 22 722).

3 M.Tsunakawa, M.Hanada, H.Tsukiura, M.Konishi, and H,Kawaguchi, J. Antibiot., 1985, 38, 1313.

4 Y.Ikeda, S.Kondo, F.Kanai, T.Sawa, M.Hamada, T.Takeuchi, and H.Umezawa, J. Antibiot., 1985, 38, 436.

5 J.Yoshimura, S.Horito, J.Tamura, and H.Hashimoto, Chem. Lett., 1985, 1335.

6 N.Yasuda, K.Matsuda, H.Tsutsumi, and T.Takaya, J. Antibiot., 1985, 38, 1512.

7 T.Tsuchiya, M.Nakano, T.Torii, Y.Suzuki, and S.Umezawa, Carbohydr. Res., 1985, 136, 195.

8 T.Tsuchiya, Y.Takahashi, Y.Kobayashi, S.Umezawa, and H.Umezawa, J. Antibiot., 1985, 38, 1287.

9 K,Matsuda, N.Yasuda, H.Tsutsumi, and T.Takaya, J. Antibiot., 1985, 38, 1050.

10 K.Matsuda, N.Yasuda, H.Tsutsumi, and T.Takaya, J. Antibiot., 1985, 38, 547.

11 R.Albert, K.Dax, A.E.Stütz, and J.Hildebrandt, J. Antibiot.,1985, 38, 275.

12 R.Albert, K.Dax, N.A.Gassner, A.E.Stutz, and H.Weidmann, Liebigs Ann. Chem., 1985, 644.

13 T.Kishi, T.Tsuchiya, N.Koga, Y.Kobayashi, and S.Umezawa, Bull. Chem. Soc. Jpn., 1985, 58, 3087.

14 S.Hanessian and R.Roy, Can. J. Chem., 1985, 63, 163.

15 R.C.Thomas and E.L.Fritzen, J. Antibiot., 1985, 38, 197.

16 R.C.Thomas and E.L.Fritzen, J. Antibiot., 1985, 38, 208.

17 J.Yoshimura, K.-I.Sato, A.Aqeel, R.Bir Singh, and H.Hashimoto, Carbohydr. Res., 1985, 144, C15.

18 J.Kusnir and K.Barna, Biologia (Bratislava),, 1984, 39, 813 (Chem. Abstr., 1985, 102, 191).

19 S.Ogawa, T.Nose, T.Ogawa, T.Toyokuni, Y.Iwasawa, and T.Suami, J. Chem. Soc. Perkin Trans. 1, 1985, 2369.

20 H.Paulsen and H.Schmidt-Lewerkühne, Liebigs Ann. Chem., 1985, 959.

21 K.Suzukake, K.Tokunaga, H.Hayashi, M.Hori, Y.Uehara, D.Ikeda, and H.Umezawa, J. Antibiot., 1985, 38, 1211.

22 T.Okubo, M.Inoue, M.Nagashima, M.Okii, T.Iida, and S.Mitsuhashi, J. Antibiot., 1985, 38, 122.

23 G.Ihchauspe, C.Deshayes, and D.Samain, J. Antibiot., 1985, 38, 1526.

24 Y.Fang, Z.Lu, and G.Lin, Huaxue Xuebao, 1985, 43, 406 (Chem. Abstr., 1985, 103, 215 686).

25 S.Masamune and P.A.McCarthy, Macrolide Antibiot., 1984, 127 (Chem, Abstr., 1985, 102, 221 110).
26 A.Nakagawa, Y,Tanaka, K.Otoguro, S.Omura, T.Takiguchi, and Y.Arai, J. Antibiot., 1985, 38, 1266.
27 S.Omura,A.Nakagawa, N.Imamura, K.Kushida, C.-M.Liu, L.H.Sello, and J.W.Westley, J. Antibiot., 1985, 38, 674.
28 K.Kinoshita, S.Satoi,M.Hayashi, K.-I.Harada, M.Suzuki, and K.Nakatsu, J. Antibiot., 1985, 38, 522.
29 K.-I.Harada, N.Takeda, M.Suzuki, M.Hayashi, M.Ohno, and S.Satoi, J. Antibiot. 1985, 38, 868.
30 H.Sano, H.Tanaka, K.Yamashita, R.Okachi, and S.Omura, J. Antibiot., 1985, 38, 186.
31 H.Sano, T.Sunazuka, H.Tanaka, K.Yamashita, R.Okachi, and S.Omura, J. Antibiot., 1985, 38, 1350.
32 H.Tanaka, I.Moriguchi, S.Hirono, and S.Omura, Chem. Pharm. Bull., 1985, 33, 2803.
33 P.Sowinski, J.K.Pawlak, E.Borowski, and T.Iwashita, J. Antibiot., 1985, 38, 175.
34 P.Sowinski, P.Kolodziejczyk, and E.Borowski, J. Antibiot., 1985, 38, 169.
35 J.R.Everett and J.W.Tyler, J. Chem. Soc., Perkin trans. 1, 1985, 2599.
36 A.V.R.Rao, Proc.-Indian Acad. Sci., Chem. Sci., 1984, 93, 1059 (Chem.Abstr., 1985, 102, 24926).
37 F.Arcamone and G.Franceschi, Stud. Org. Chem. (Amsterdam), 1984, 17, 287 (Chem. Abstr., 1985, 102, 113 793).
38 H.Morioka, Y.Etoh, I.Horino, M.Takezawa, T.Ando, K.Hirayama, H.Kano, and H.Shibai, Agric. Biol. Chem., 1985, 49, 1951.
39 T.Ando, K.Hirayama, R.Takahashi, I.Horino, Y.Etoh, H.Morioka, H.Shibai, and A.Murai, Agric. Biol. Chem., 1985, 49, 1207.
40 T.Ando, K.Hirayama, R.Takahashi, I.Horino, Y.Etoh, H.Morioka, H.Shibai, and A.Murai, Agric. Biol. Chem., 1985, 49, 259.
41 T.Uchida, M.Imoto, T.Masuda, H.Yoshimoto, K.Imamura, T.Sawa, H.Naganawa, T.Takeuchi, and H.Umezawa, J. Antibiot., 1985, 38, 795.
42 T.F.Brodasky, S.Mizsak, and J.R.Hoffstetter, J. Antibiot., 1985, 38, 849.
43 H.Drautz, P.Reuschenbach, and H.Zähner, J. Antibiot., 1985, 38, 1291.
44 G.C.S.Reddy, R.Sahai, H.W.Fehlhaber, and B.N.Ganguli, J. Antibiot., 1985, 38, 1423.
45 M,Nakagawa, H.Kawai, Y.Hayakawa, H.Seto, and N.Otake, J. Antibiot., 1985, 38, 1623.
46 T.-M.Cheung, D.Horton, W.Priebe, W.R.Turner, and W.Weckerle, J. Antibiot., 1985, 38, 683.
47 D.Horton and W.Priebe, Carbohydr. Res., 1985, 136, 391.
48 T.T.Thang, J.-L.Imbach, C.Fizames, F.Lavelle, G.Ponsinet, A.Olesker, and G.Lukacs, Carbohydr. Res., 1985, 135, 241.
49 D.Horton, W.Priebe, and O.Varela, Carbohydr. Res., 1985, 144, 305.
50 Y.Kimura, T.Matsumoto, M.Suzuki, and S.Terashima, J. Antibiot., 1985, 38, 1277.
51 F.Sztaricskai, M.Menyhart, J.Borda, and R.Bognar, Stud. Biophys., 1984, 104, 103 (Chem. Abstr., 1985, 103, 105 247).
52 J.W.Lown and S.M.Sondhi, J. Org. Chem., 1985, 50, 1413.
53 G.Cassinelli, M.Ballabio, F.Arcamone, A.M.Casazza, and A.Podesta, J. Antibiot., 1985, 38, 856.
54 K.Dornberger, R.Hübener, W.Ihn, H.Thrum, and L.Radics, J. Antibiot., 1985, 38, 1219.
55 J.P.Schaumberg, G.C.Hokanson, J.C.French, E.Smal, and D.C.Baker, J. Org. Chem., 1985, 50, 1651.
56 S.Omura, N.Imamura, H.Kuga, H.Ishikawa, Y.Yamazaki, K.Okano, K.Kimura, Y.Takahashi, and H.Tanaka, J. Antibiot., 1985, 38, 1008.
57 J.B.Tunac and M.Underhill, J. Antibiot., 1985, 38, 1344.
58 C.Bormann, W.Huhn, H.Zähner, R.Rathmann, H.Hahn, and W.A.König, J. Antibiot., 1985, 38, 9.
59 H.A.Kirst, D.E.Dorman, J.L.Occolowitz, N.D.Jones, J.W.Paschal, R.L.Hamill, and E.F.Szymanski, J. Antibiot., 1985, 38, 575.

60 D.E.Nettleton, T.W.Doyle, B.Krishnan, G.K.Matsumoto, and J.Clardy, Tetrahedron Lett., 1985, 26, 4011.
61 T.Kaneko, H.Wong, K.T.Okamoto, and J.Clardy, Tetrahedron Lett., 1985, 26, 4015.
62 T.Suami, H.Sasai, K.Matsuno, and N.Suzuki, Carbohydr. Res., 1985, 143, 85.
63 H.Sasai, K.Matsuno, and T.Suami, J. Carbohydr. Chem., 1985, 4, 99.
64 T.Ito, S.Zushi, K.Kawamura, A.Takatsuki, and G.Tamura, Agric. Biol. Chem., 1985, 49, 2257.
65 N.Dodic, M.Gèze, P.Blanchard, J.L.Fourrey, and M.Robert-Gero, Nucleosides Nucleotides, 1985, 4, 275.
66 F.Hansske and M.J.Robins, Tetrahedron Lett., 1985, 26, 4295.
67 H.Bazin and J.Chattopadhyaya, Synthesis, 1985, 1108.
68 C.R.Petrie III, H.B.Cottam, P.A.McKernan, R.K.Robins, and G.R.Revankar, J. Med. Chem., 1985, 28, 1010.
69 J.Maurins, R.Paegle, A.Zidermane, M.Lidaks, E.I.Kvasyuk, and I.A.Mikhailopulo, Nucleosides Nucleotides, 1984, 3, 147 (Chem. Abstr., 1985,102,167099).
70 M.T.Garcia-Lopez, P.Fernandez-Resa, F.G.De las Heras, and P.P.Mendez-Castrillon, An. Quim., Ser. C, 1984, 80, 168 (Chem.Abstr., 1985, 102,149700).
71 T.Kawaguchi, Y.Suzuki, Y.Nakahara, N.Nambu, and T.Nagai, Chem. Pharm. Bull., 1985, 33, 301.
72 T.Kawaguchi, Y.Suzuki, N.Nambu, and T.Nagai, Chem. Pharm. Bull., 1985, 33, 2956.
73 A.F.Cook and M.J.Holman, Nucleosides Nucleotides, 1984, 3, 401 (Chem. Abstr., 1985, 102, 185 415).
74 H.B.Cottam, Z.Kazimierczuk, S.Geary, P.A.McKernan, G.R.Revankar, and R.K.Robins, J. Med. Chem., 1985, 28, 1461.
75 E.De Clercq and R.T.Walker, Pharmacol. Ther., 1984, 26, 1 (Chem. Abstr., 1985, 102, 132 351).
76 H.Griengl, M.Bodenteich, W.Hayden, E.Wanek, W.Streicher, P.Stütz, H.Bachmayer, I.Ghazzouli, and B.Rosenwirth, J. Med. Chem., 1985, 28, 1679.
77 P.Herewijn, E.De Clercq, J.Balzarini, and H.Vanderhaeghe, J. Med. Chem., 1985, 28, 550.
78 H.Ohno, T.Itoh, A.Nomura, and Y.Mizuno, Nucleosides Nucleotides, 1984, 3, 345 (Chem. Abstr., 1985, 102, 185 412).
79 D.M.Houston, E.K.Dolence, B.T.Keller, U.Patel-Thombre, and R.T.Borchardt, J. Med. Chem., 1985, 28, 471.
80 M.Ohno, Nucleosides Nucleotides, 1985, 4, 21.
81 O.Miyashita, F.Kasahara, T.Kusaka, and R.Marumoto, J. Antibiot., 1985, 38, 981.
82 P.Herdewijn, J.Balzarini, E.De Clercq, and H.Vanderhaeghe, J. Med. Chem., 1985, 28, 1385.
83 R.J.Parry and V.Bornemann, J. Am. Chem. Soc., 1985, 107, 6402.
84 K.Kinoshita, S.Yaginuma, M.Hayashi, and K.Nakatsu, Nucleosides Nucleotides, 1985, 4, 661.
85 C.K.H.Tseng and V.E.Marquez, Tetrahedron Lett., 1985, 26, 3669.
86 W.C.Faith, C.A.Booth, B.M.Foxman, and B.S.Snider, J. Org. Chem., 1985, 50, 1983.
87 J.G.Buchanan, Nucleosides Nucleotides, 1985, 4, 13.
88 F.J.Lopez Herrera and G.Uraga Baelo, Carbohydr. Res., 1985, 139, 95.
89 W.J.Hennen, B.C.Hinshaw, T.A.Riley, S.G.Wood, and R.K.Robins, J. Org. Chem., 1985, 50, 1741.
90 R.H.Baur and D.C.Baker, Nucleosides Nucleotides, 1984, 3, 77.
91 C.Jiang, R.H.Baur, J.J.Dechter, and D.C.Baker, Nucleosides Nucleotides, 1984, 3, 123.
92 A.Rosowsky, V.C.Solan, and L.J.Gudas, J. Med. Chem., 1985, 28, 1096.
93 J.A.Secrist III, A.T.Shortnacy, and J.A.Montgomery, J. Med. Chem., 1985, 28, 1740.
94 B.G.Ugarkar, R.K.Robins, and G.R.Revankar, Nucleosides Nucleotides, 1984, 3, 233.
95 T.Hida, S.Tsubotani, N.Katayama, H.Okazaki, and S.Harada, J. Antibiot., 1985, 38, 1128.
96 T.Komori and Y.Itoh, J. Antibiot., 1985, 38, 544.

97 Y.Hayakawa, M.Nakagawa, H.Kawai, K.Tanabe, H.Nakayama, A.Shimazu, H.Seto, and N.Otake, Agric. Biol. Chem., 1985, 49, 2685.
98 K.Dobashi, K.Nagaoka, Y.Watanabe, M.Nishida, M.Hamada, H.Naganawa, T.Takita, T.Takeuchi, and H.Umezawa, J. Antibiot., 1985, 38, 1166.
99 T.Okabe, K.Nomoto, H.Funabashi, S.Okuda, H.Suzuki, and N.Tanaka, J. Antibiot., 1985, 38, 1333.
100 Y.Hayakawa, T.Iwakiri, K.Imamura, H.Seto, and N.Otake, J. Antibiot., 1985, 38, 959.
101 Y.Hayakawa, T.Iwakiri, K.Imamura, H.Seto, and N.Otake, J.Antibiot., 1985, 38, 960.
102 T.Uchida, M.Imoto, Y.Watanabe, K.Miura, T.Dobashi, N.Matsuda, T.Sawa, H.Naganawa, M.Hamada, T.Takeuchi, and H.Umezawa, J. Antibiot., 1985,38,1171.
103 M.G.Brazhnikova, M.K.Kudinova, V.V.Kulyaeva, N.P.Potapova, L.M.Rubasheva, B.V.Rozynov, and G.Horvath, Antibiotiki (Moscow), 1984, 29, 884 (Chem. Abstr., 1985, 102, 185 366).
104 K.Uchino, H.Ogawara, T.Akiyama, A.Fukuchi, S.Shibata, K.Takahashi, and T.Narui, J. Antibiot., 1985, 38, 157.
105 K.Uchino, H.Ogawara, T.Akiyama, A.Fukuchi, S.Shibata, K.Takahashi, and T.Narui, J. Antibiot., 1985, 38, 592.
106 Y.Sone, R.Okuda, N.Wada, E.Kishida, and A.Misaki, Agric. Biol. Chem., 1985, 49, 2641.
107 S.Shinagawa, M.Maki, K.Kintaka, A.Imada, and M.Ashi, J. Antibiot., 1985, 38, 17.
108 H.Nakayama, H.Seto, N.Otake, M.Yamagishi, A.Kawashima, T.Mizutani, and S.Omura, J. Antibiot., 1985, 38, 1433.
109 S.Kiyoto, M.Nishikawa, H.Terano, M.Kohsaka, H.Aoki, H.Imanaka, Y.Kawai, I.Uchida, and M.Hashimoto, J. Antibiot., 1985, 38, 840.
110 M.Konishi, H.Ohkuma, K.-I.Saitoh, H.Kawaguchi, J.Golik, G.Dubay, G.Groenewold, B.Krishnan, and T.W.Doyle, J. Antibiot., 1985, 38, 1605.
111 J.H.Wilton, G.C.Hokanson, and J.C.French, J. Chem. Soc.,Chem. Commun.,1985,919.
112 Y.Kodama, T.Tsuruoka, T.Niwa, and S.Inouye, J. Antibiot., 1985, 38, 116.
113 Y.Ezure, S.Maruo, K.Miyazaki, and M.Kawamata, Agric. Biol. Chem,1985,49,1119.
114 M.D.Schulman, D.Valentino, D.D.Hensens, D.Zink, M.Nallin, L.Kaplan, and D.A.Ostlind, J. Antibiot., 1985, 38 , 1494.
115 S.Ogawa, H.Sugizaki, Y.Iwasawa, and T.Suami, Carbohydr. Res., 1985, 140, 325.
116 S.Ogawa, Y.Iwasawa, T.Toyokuni, and T.Suami, Carbohydr. Res., 1985, 141, 29.
117 Y.Fukuda and T.Suami, Yuki Gosei Kagaku Kyokaishi, 1985, 43, 836 (Chem. Abstr., 1985, 103, 196 329).
118 H.Paulsen and W.Röben, Liebigs Ann. Chem., 1985, 974.
119 S.Ogawa, Y.Iwasawa, T.Toyokuni, and T.Suami, Carbohydr. Res., 1985, 144, 155.
120 S.Ogawa, K.Yasuda, T.Takagaki, Y.Iwasawa, and T.Suami, Carbohydr. Res., 1985, 141, 329.
121 H.-H. Wust, J.Bardenhagen, and U.Schollkopf, Liebigs Ann. Chem., 1985, 1825.
122 I.Miskolczi, F.Sztaricskai, P.Herczegh, R.Bognar, and I.Koczka, J. Antibiot., 1985, 38, 1273.
123 A.K.Ganguly, B.N.Pramanik, V.M.Girijavallabhan, O.Sarre, and P.L.Bartner, J. Antibiot., 1985, 38, 808.
124 G.D.Roberts, S.A.Carr, S.Rottschaefer, and P.W.Jeffs, J. Antibiot., 1985, 38, 713.
125 J.C.J.Barna, D.H.Williams, and M.P.Williamson, J. Chem. Soc., Chem. Commun., 1985, 254.
126 E.R.Dilanyan, T.R.Ovsepyan, F.G.Arsenyan, G.M.Stepanyan, E.Mironov, and M.E.Vol'pin, Khim.-Farm. Zh., 1984, 18, 927 (Chem. Abstr., 1985, 102,132 366).

# 20
# Nucleosides

## 1 General

The new cytokinin (1) has been isolated from the culture filtrates of Pseudomonas syringae.[1] Adenosine has been isolated as the platelet aggregation inhibitor present in aqueous extracts of the fungus Ganoderma lucidum.[2]

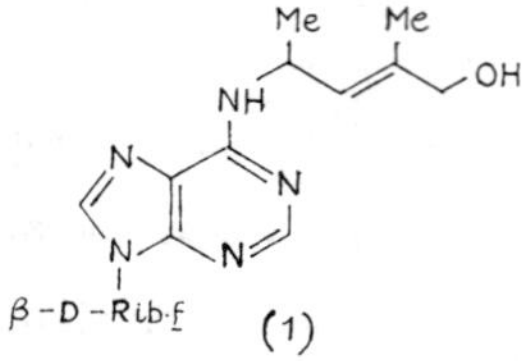


(1)

The synthesis and antiviral properties of 5-vinylpyrimidine nucleosides have been reviewed, with detailed attention being paid to the potential of (E)-5-bromovinyl-deoxyuridine (BVDU) in human medicine.[3] Although acyclic nucleoside analogues are generally considered to be outside the scope of this volume, we note the publication of a bibliography on such compounds, other than acyclovir.[4]

## 2 Synthesis

Standard condensation procedures have been used to prepare β-D-ribofuranosyl derivatives of the pyridone (2), a metabolite of NAD(P), and its analogous 5-carboxylic acid,[5] various 5-substituted barbituric acids,[6] 5-substituted-6-azauracils,[7,8] a number of 2-pyrimidinones(other pentafuranosyl derivatives were also reported),[9] 1,2,4-triazol-3(2H)-one (3) and its 5-methyl derivative,[10] the imidazoles (4)[11] and (5),[12] 3-arylisoxazol-5-ones,[13] the triazole (6) and the corresponding amide and nitrile,[14] 3-phenyl-3-methylsuccinimide,[15] thieno[2,3-d]pyrimidin-4-one (7) and related pyrrolopyrazolo- and thiazolo- systems,[16] and lin-naphthimidazole.[17] 2-Chloro- and 4-desamino-2,4-dichloro-1-deazaadenosine have been synthesized,[18] as have the β-D-ribonucleosides of 5,6-tetramethyl-

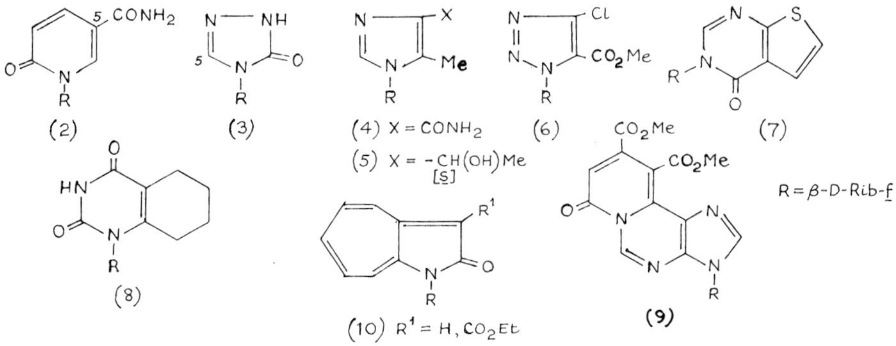


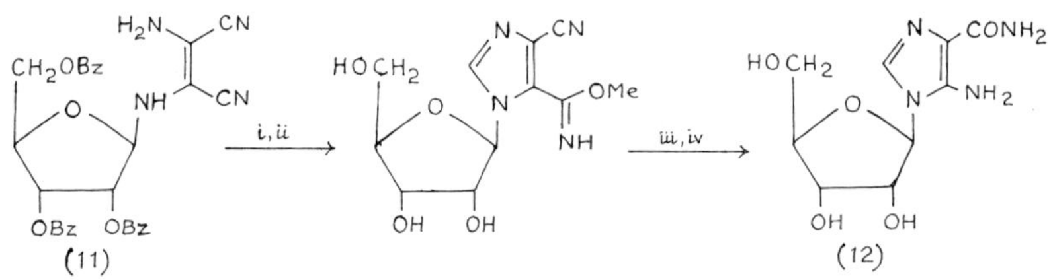

Reagents: i, $HC(OEt)_3$ ; ii, $^{-}OMe$ ; iii, $^{-}OCl$-$^{-}OH$ ; iv, $^{-}OH$ (Δ)

Scheme 1

eneuracil (8),[19] the pyrido[2,1-i]purine (9),[20] and the 1,2-dihydrocyclohepta[b]pyrrol-2-ones (10).[21] The synthesis of 6-azacytidine has been monitored by h.p.l.c.[22] The synthesis of 6-azacadeguomycin and related systems is discussed in Chapter 19.

The β-D-ribofuranosylderivative of diaminomaleonitrile (11) was produced by condensation and then elaborated into the aminoimidazole nucleoside (12) (Scheme 1). The analogous β-D-ribopyranosyl compound was also reported; the regioselectivity in imidate formation was ascribed to intramolecular catalysis by the 2'-oxyanion.[23] 2',3',5'-Tri-O-benzoyl-β-D-ribofuranosylisocyanate was converted into a series of $N^4$-substituted 5-azacytidines,[24] and 1,4-dihydropyridine nucleosides have been prepared by an ingenious photochemical cycloaddition[25] (Scheme 2).

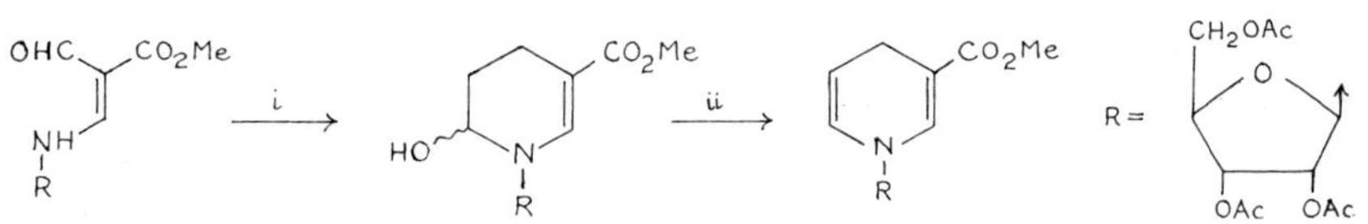

Reagents: i, $CH_2{=}CH_2$, hν, -20°C ; ii, Δ, toluene, mol. sieves

Scheme 2

The t-RNA constituent wyosine (13) has been prepared from an imidazole nucleoside, and its lability under acidic or alkaline con-itions was demonstrated.[26]

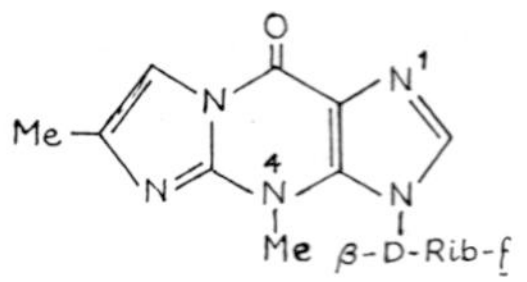


(13)

α-Nucleosides of pyrimidine bases are accessible with good stereoselectivity by the process illustrated by the example in Scheme 3.[27,28] Some α-ribonucleotides have been prepared by inter-

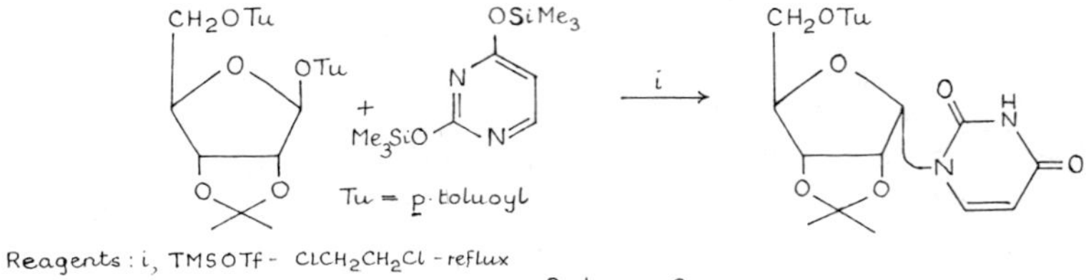


Scheme 3

action of a tin (II) derivative of the base with a 1,5-cyclic phosphate of 2,3-O-isopropylidene ribofuranose,[29] and the α-anomer of $N^6$-benzyladenosine, which possesses weak cytokinin activity, has been prepared in low yield.[30]

A full account has been given of a microbiological procedure (see Vol. 14, p. 170) for the synthesis of 9-β-D-arabinofuranosyladenine (Ara-A), by transglycosylation between adenine and arabinofuranosyl-uracil. Various bacterial strains can achieve this, and, using a strain of Enterobacter aerogenes, ara-A was produced in 87% yield.[31] The same method also works with various other purine bases, both natural and unnatural.[32] The mechanism of the transarabinosylation involves the intermediate formation of arabinose-1-phosphate, with uridine phosphorylase and purine nucleoside phosphorylase acting in concert.[33] The β-D-arabinofuranosyl analogue of doridosine has been prepared; it is resistant to adenosine deaminase.[34] A number of 5-alkyl arabinofuranosyluracils have been synthesized, including both α- and β-anomers, by coupling silylated bases with tribenzyl-arabinofuranosyl chloride; deprotection was effected with boron trifluoride etherate and ethanethiol.[35] The imidazole nucleoside (14) has been prepared using the sodium salt of the aglycone.[36]

Stereospecific syntheses of the β-D-xylofuranosyl derivatives of adenine and guanine have been reported, although in the case of the guanosine analogue the regiochemistry favours the $N^7$-isomer.[37]

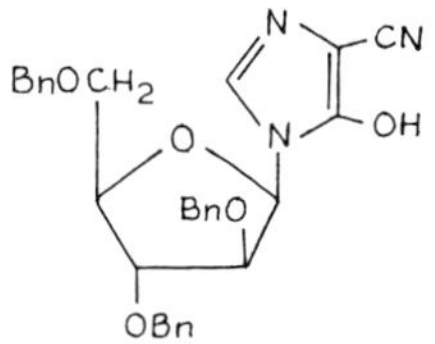

(14)

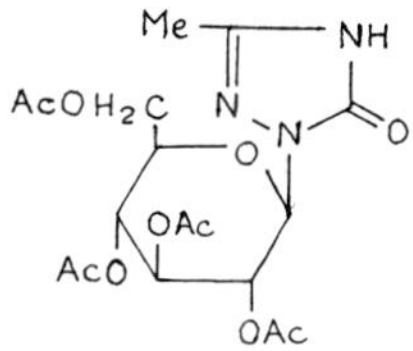

(15)

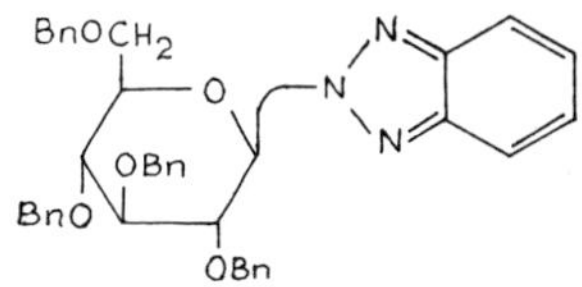

(16)

The protected glycoside (15) was prepared conventionally along with the $N^4$-regioisomer,[38] and β-D-glucopyranosides of various benzimidazoles have been reported,[39] as has the $N^1$, $N^3$-bisglucoside of uracil.[40] $N^7$-β-D-glucopyranosyl-1,3,8-trimethylxanthine was prepared by cyclization of a glucosylamine derivative.[41] When tetra-O-benzyl-α-D-glucopyranosyltrichloroacetimidate reacted with benzotriazole under boron trifluoride-etherate catalysis, the glucoside (16) and its $N^1$-isomer were produced in a 2:1 ratio; similar chemistry gave 1-β-D-glucosides of uracil and thymine.[42] β-D-Allofuranosylthymine derivatives have been prepared from compounds of gluco configuration by inversion at C-3'.[43]

## 3 Anhydro- and Cyclonucleosides

A full account has been given of the synthesis of 2',5'-anhydro-1-β-D-ribofuranosyluracil and its 5-substituted analogues (see Vol 18, p. 194-5).[44] Similar chemistry has also been used to prepare the analogous isocytidine system (17).[45]

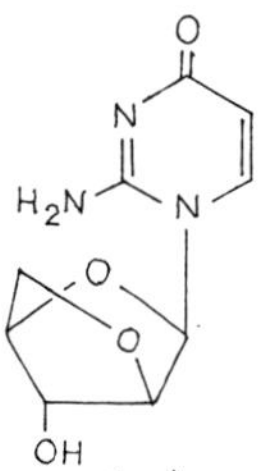

(17)

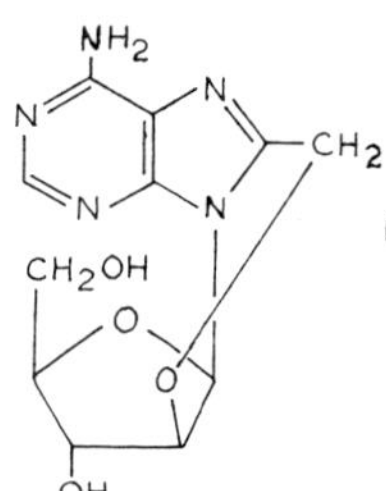

(18)

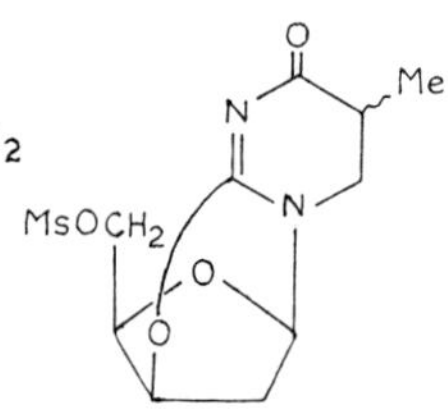

(19)

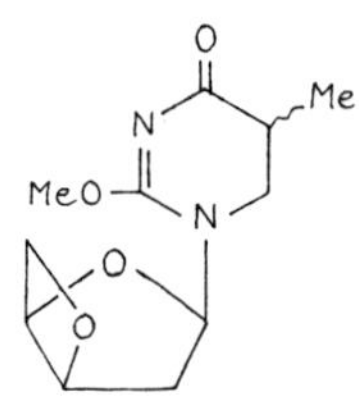

(20)

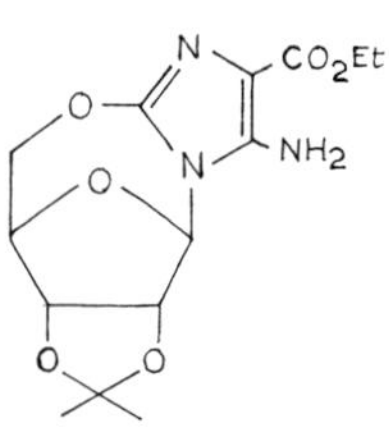

(21)

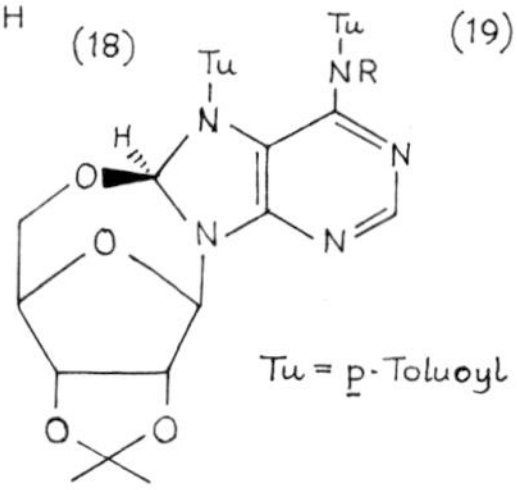

(22)

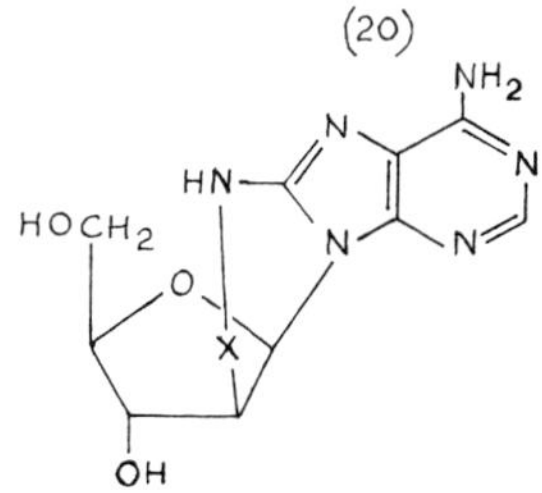

(23) X = NH or O

8,2'-Anhydro-8-hydroxymethyl-9(β-D-arabinofuranosyl)adenine (18) was prepared by monotosylation of the 2',3'-dibutylstannylene derivative of the ribonucleoside, followed by intramolecular sulphonate displacement.[46] Cyclization of the 3',5'-dimesylate of dihydrothymidine gave the cyclonucleoside (19), which with methoxide gave the 3',5'-anhydro system (20).[47] The 2,5'-cyclonucleoside (21) was formed on oxidation of the imidazole nucleoside with alkaline hypobromite or N-chlorosuccinimide.[48] Formation of 8,5'-O-cyclonucleosides can accompany the acylation of isopropylideneadenosine, as in the production of (22; R=H or p-Tol) in a combined yield of 45% by use of p-toluoyl chloride and triethylamine in dichloromethane.[49]

A number of purine 8,5'-imino and substituted imino cyclonucleosides have been prepared,[50] as have a variety of 8,2'-hydrazo and oximido- bridged purines such as (23).[51] The 2,3'-iminohexosyluracil nucleoside (24) was produced as shown in Scheme 4; some similar cyclonucleosides were prepared and a mechanistic picture emerged.[52] The iminocyclonucleosides (25) were prepared by reactio

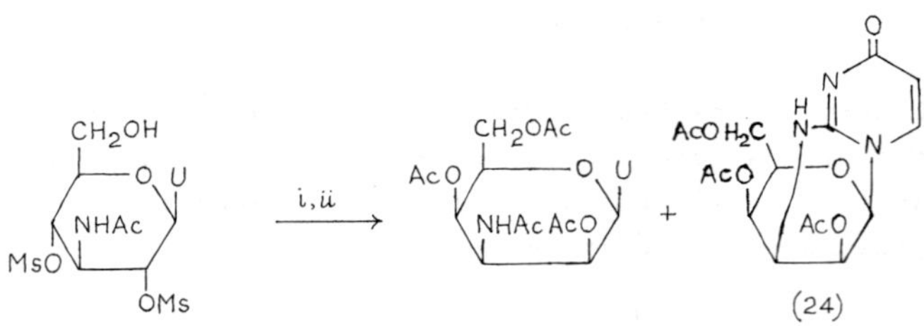


(24)

Reagents : i, NaOAc - methylcellosolve - $H_2O$ ; ii, $Ac_2O$-Py

Scheme 4

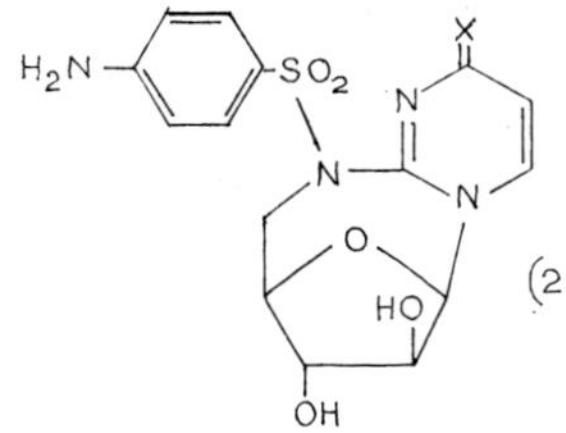


(25) X = O or NH

of 2,2'-anhydro-5'-chloro-β-D-arabinofuranosyluracil or -cytosine with p-aminobenzenesulphonamide.[53]

In an abstract of a plenary lecture, Ueda has reviewed his work on carbon-bridged cyclonucleosides.[54] His group have reported the synthesis of 6,2'-methanecyclouridine (26) as outlined in Scheme 5,[55] whilst Scheme 6 indicates a route to the 2'-deoxy derivative

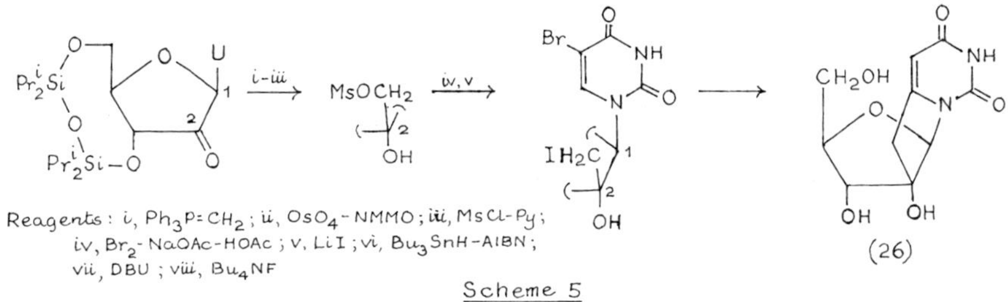

Scheme 5

(27);[56] the starting material in Scheme 6 was produced from the 2'-ketonucleoside of Scheme 5. Similar chemistry has been employed for the synthesis of 2'-deoxy-6,2'-ethanocyclouridine.[57] A free-radical cyclization was employed to prepare 6'-deoxy-6,6'-methanocyclouridine.[58]

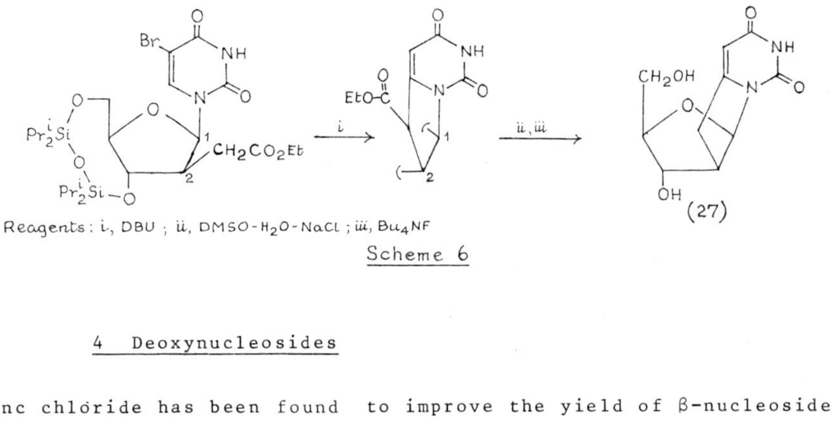

Scheme 6

## 4 Deoxynucleosides

Zinc chloride has been found to improve the yield of β-nucleoside produced from α-2'-deoxyglycosyl chlorides with less reactive bases, which otherwise undergo competing anomerization rather than $S_N2$ reaction with the base.[59]

Standard procedures were used to obtain α- and β-2-deoxy-D-erythro-pentofuranosyl nucleosides of 5-methylimidazole-4-carboxamide,[60] 5-halovinyl-6-azauracils,[61] 5-trimethylsilyluracil and -cytosine,[62] 5-halo-2(1H)-pyrimidinones (both anomers showing antiherpetic activity),[63] and 5-alkynyl-2(1H)-pyrimidinones, these last being better prepared from the 5-iodo compound and an alkyne in the presence of $Cu^{I}$ and $Pd^{II}$ complex.[64] Some 5-substituted 2'-deoxyuridines (28; X=H, R=Br, OBu, SMe, $NMe_2$) have been prepared by silyl-Hilbert-Johnson procedures, but attempts to make analogous methyl orotidines (28; X=$CO_2$Me) gave only the $N^3$-riboside or $N^1,N^3$-bis-ribosides.[65]

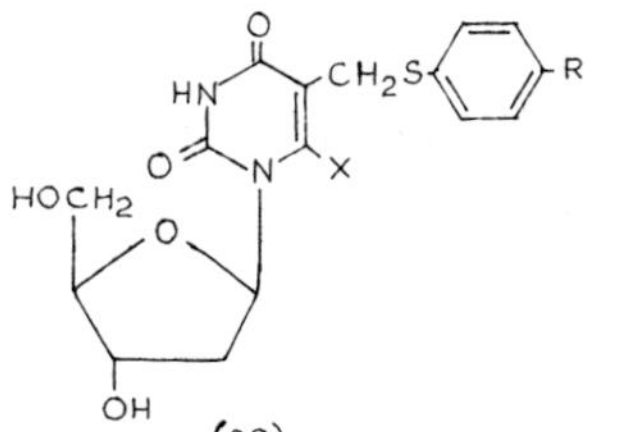
(28)

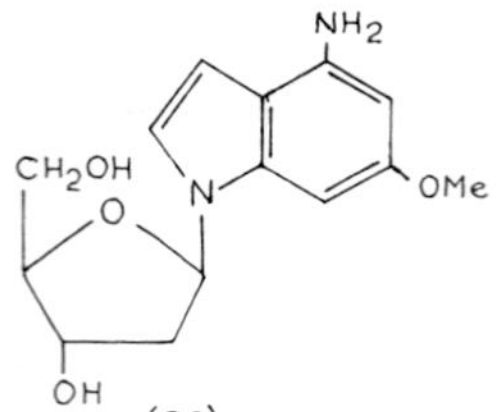
(29)

Phase-transfer glycosylation was employed in the synthesis of 2'-deoxyribofuranosides of allopurinol[66] and various other 4-substituted-1H-pyrazolo[3,4-d]pyrimidines,[66,67] and for the preparation of 7-deaza-2'-deoxyinosine,[68] whilst glycosylation of a sodium salt was used in the synthesis of 7-deaza-2'-deoxyxanthosine and 7-deaza-2'-deoxyspongosine (29),[69] 2'-deoxy-1-deazawyosine (see 13), which was shown to be much more stable than wyosine itself,[70] and 2'-deoxyribofuranosides of some benzimidazoles.[71] 1-(2'-Deoxy-β-D-ribofuranosyl)benzimidazole was also prepared by Raney nickel desulphurization of a 2,2'-cyclic sulphide.[72]

The synthesis of 2',3'-dideoxynucleosides can be achieved by Barton-type deoxygenation of 2'-deoxynucleosides protected at O-5'.[73]

## 5 Halonucleosides

2'-Chloro-2'-deoxy- and 2'-deoxy-2'-fluorouridine labelled with radiohalogen have been prepared from 2,2'-anhydro-1-β-D-arabinofuranosyluracil.[74] 1-(3'-Halo-3'-deoxy-β-D-arabinofuranosyl)uracils labelled with $^{36}$Cl, $^{82}$Br or $^{123}$I were synthesized from the 2',3'-anhydronucleoside.[75]

Derivatives of (5-chloro-5-deoxy-β-D-arabinofuranosyl)isocytosine have been prepared as shown in Scheme 7.[45] A derivative of 3-chloro-3-deoxyallose has been converted into $N^1$-(3-chloro-3-deoxy-2,4,6-tri-O-acetyl-β-D-allopyranosyl)-5-fluorouracil.[76]

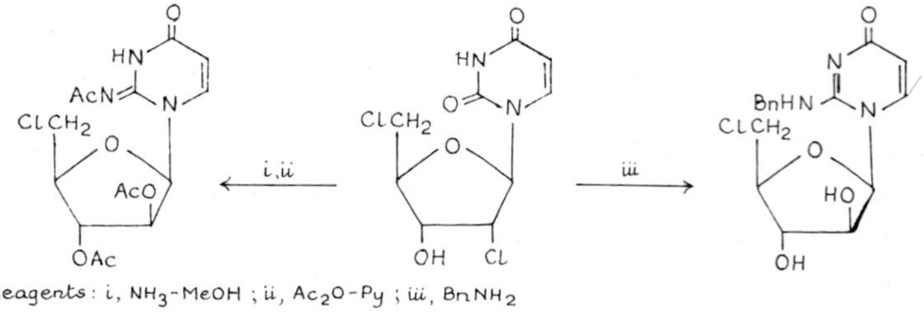

Reagents: i, $NH_3$-MeOH ; ii, $Ac_2O$-Py ; iii, $BnNH_2$

Scheme 7

References to the isolation of 2'-chloropentostatin from natural sources are given in Chapter 19.

## 6 Nucleosides with Nitrogen-substituted Sugars

A number of purine 2'-amino-2'-deoxy-β-D-ribonucleosides have been prepared by microbial transglycosylation between the purine base and 2'-amino-2'-deoxyuridine.[77] The 3'-azido-3'-deoxy-β- and 3'-azido,2',3'-dideoxy-α- and β-D-ribofuranosides of benzimidazole were synthesized conventionally from 3-azido sugar derivatives.[78] A range of 3'-O-acyl derivatives of 5'-amino-5'-deoxythymidine have been reported, and shown to have anti-herpes activity,[79] whilst displacement of a 5'-tosylate was employed in the synthesis of 5'-O-aminothymidine and 5'-deoxy-5'-hydrazinothymidine.[80]

Reaction of 3,4,6-tri-O-acetyl-2-deoxy-2-nitroso-α-D-hexopyranosyl chloride with pyrazole gave as major product (30), together with epimers at C-1 and C-3, the latter epimerization presumed to occur via the 2-nitroso-2-ene. Treatment of (30) with sodium azide in ethanol (Scheme 8) gave (31) as major product, presumably also by elimination-addition, and hydrogenation-acetylation then gave the aminooxime (32).[81]

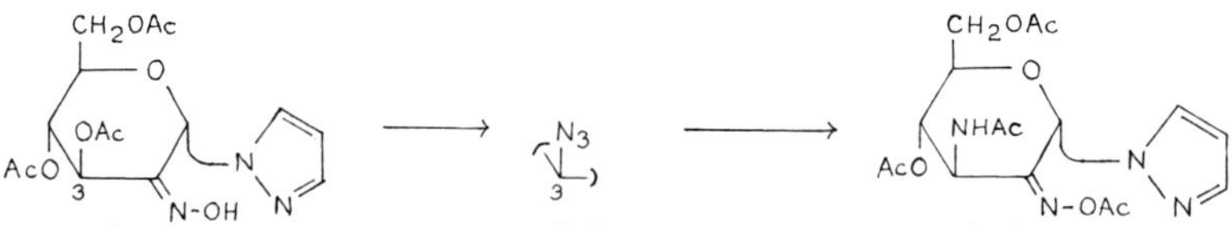

(30) (31) (32)

Scheme 8

Further references to 3'-amino-3'-deoxynucleosides are in Chapter 19.

## 7 Thionucleosides

A convenient preparation has been reported of S-adenosylhomocysteine (SAH, 33) from adenosine and a homocysteine derivative by a Mitsunobu-type reaction (Scheme 9). The 3- and 7-deazaanalogues

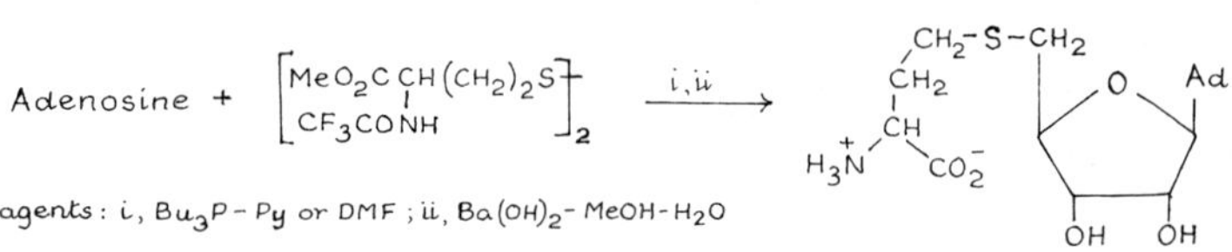

Reagents: i, $Bu_3P$-Py or DMF; ii, $Ba(OH)_2$-MeOH-$H_2O$

(33)

Scheme 9

were also prepared.[82] Another improved procedure for the synthesis of SAH has also been reported, involving more conventional reaction of the disodium salt of homocysteine (analogues can also be used) and 5'-chloro-5'-deoxy-2',3'-O-isopropylideneadenosine.[83] Use of the salt of 2-methylhomocysteine in a similar procedure gives the appropriate SAH analogue;[84] the same approach was used in a much improved route to S-3-aminopropyl-5'-deoxy-5'-thioadenosine (decarboxy-SAH),[85] and for the synthesis of the $N^6,N^6$-dimethyl analogues of SAH and of S-adenosylmethionine(SAM).[86] The natural form of SAM in various tissues has been shown to have S-configuration at sulphur by an h.p.l.c. method that separates the diastereoisomers.[87]

A number of nucleoside analogues of type (34) containing the 1-oxa-4-thiacyclohexane ring have been reported,[88] and their n.m.r. spectra studied.[89]

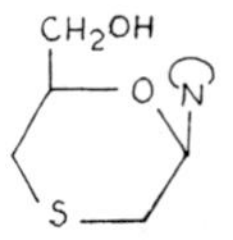


(34)

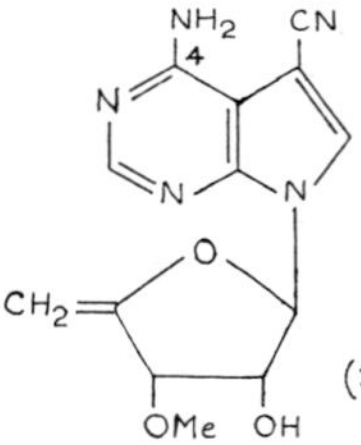


(35)

## 8 Nucleosides of Unsaturated Sugars, Ketosugars and Uronic Acids

Anomeric 2,3-dideoxy-D-erythro-hex-2-enopyranosyl nucleosides have been synthesized by Lewis-acid-catalysed condensation of triacetylglucal with silylated heterocycles;the C.D. spectra of the products were investigated.[90] Mycalisine A, isolated from a marine sponge, has the structure (35), related to known antibiotics such as tubercidin. Mycalisine B, from the same source, is the corresponding 4-oxo compound; neither material has antimicrobial activity, but they inhibit cell division of fertilized starfish eggs.[91]

The 2'-keto-3'-deoxy nucleoside (36) has been prepared by an abnormal Grignard reaction[92] and by a Perkow reaction,[93] as outlined in Scheme 10; the latter approach was unsuccessful when attempted for the synthesis of 3'-keto systems, due to β-elimination. The unsaturated theophylline ketonucleoside (37), with a spacer-arm suitable for linkage to macromolecules, has been prepared by a multistep sequence.[94]

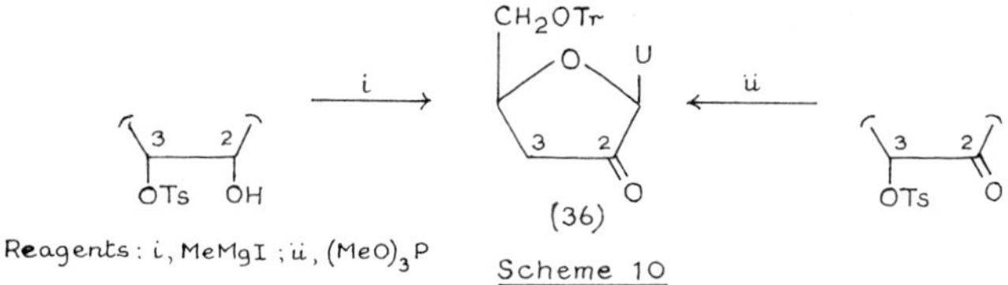


(36)

Scheme 10

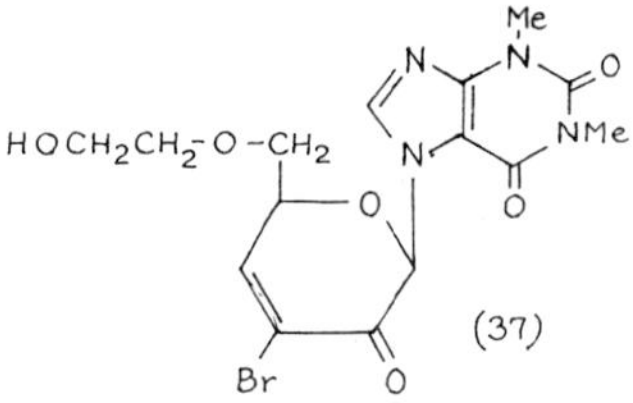


(37)

The uronic acid analogue (38) of the antiviral agent IDU and the corresponding 2'-deoxy system were prepared via the oxidation procedure shown in Scheme 11, which should be of wide utility.[95]

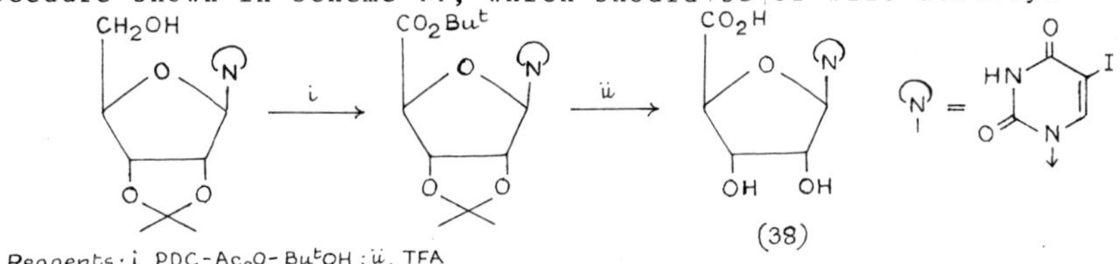


(38)

Scheme 11

## 9 C-Nucleosides

The reaction of pseudouridine with α-acetoxyisobutyryl chloride has been reinvestigated, and some of the findings are outlined in Scheme 12. The major product had previously been assigned a ribo

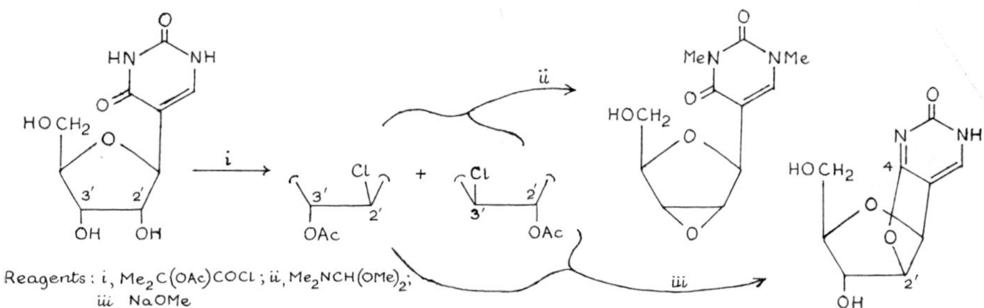


Scheme 12

configuration and the intermediacy of the 4,2'-anhydro system had been presumed; this system has now been shown to be stable to HCl in acetonitrile.[96]

The alkene (39), produced by Wittig reaction, gave exclusively β-anomers (40) on cyclization and hydrolysis. The α-chloroacid was converted (Scheme 13) via the chloroketene to pyrido[1,2-a]pyrimidine C-nucleosides,but in low yields.[97] The spiro 1,3-oxazine (41),

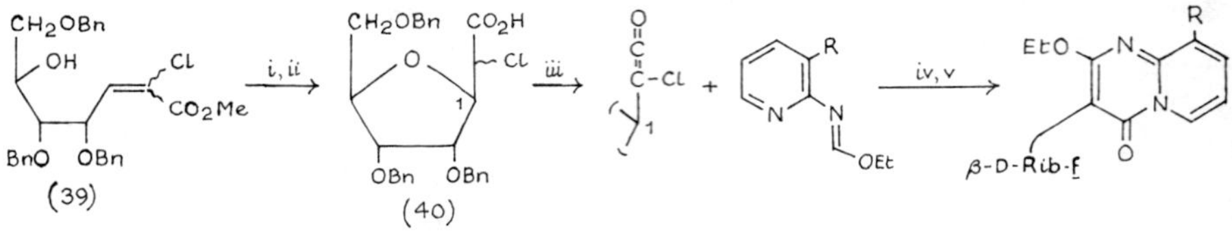

Reagents: i, $Et_3N$ ; ii, $NaOH-H_2O$-Dioxan ; iii, $SOCl_2-Et_3N$ or $Tf_2O-Et_3N$ ; iv, formimidate ; v, $BCl_3$

Scheme 13

available from the 2,5-anhydroallononitrile via a cyclic imidoester, was converted into pyrimidine and 1,2,4-triazole C-nucleosides,[98] and imidazo[1,2-a]pyrazine systems such as (42) have been

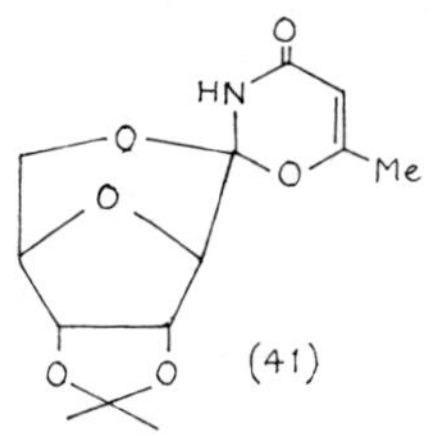

(41)

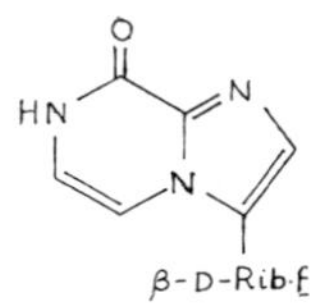

(42)

elaborated from the nitroaldol product of nitromethane and tri-O-benzoyl-2,5-anhydro-D-allose.[99] Some 2-(α-and β-D-ribofuranosyl) pyrroles have been synthesized by dehydration of pentitols under kinetic control; under conditions of thermodynamic control, pyranosyl systems predominated.[100]

5-(β-D-Ribofuranosyl)-1,2,3,4-tetrahydrophthalazine-1,4-diones have been prepared by means of cycloaddition to derivatives of 2-(β-D-ribofuranosyl)furan,[101] and related cycloadducts with maleimide have been described.[102] The pyrazolo[3,4-e][1,3]-oxazine (43) and protected forms of the 5-thione and 5-aminocompound have been obtained by cyclization of the antibiotic pyrazofurin,[103] and 3-deazaoxanosine (44) and its 1-methyl derivative have been reported.[104]

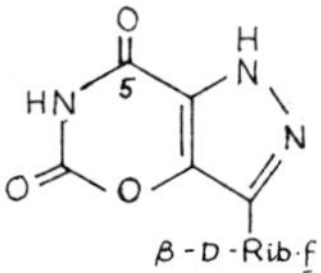

(43)

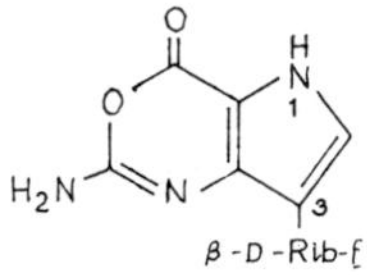

(44)

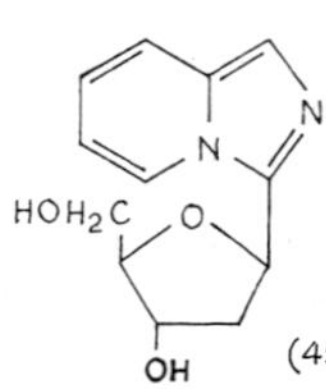

(45)

A series of 2'-deoxyribofuranosyl C-nucleosides of imidazo-fused bridgehead nitrogen systems, typified by (45), have been prepared starting from tri-O-benzoyl-2,5-anhydro-D-allonic acid.[105]

A route has been developed for the synthesis of a number of 2'-substituted-2'-deoxy-β-D-arabinofuranosyl derivatives of 1-methyl-ψ-uridine (46). As outlined in Scheme 14, participation of the carbonyl oxygen of the aglycone in reactions at C-2' was prevented by O-2/C-5' cyclization.[106]

4-(D-Erythrofuranosyl)imidazolethiones have been prepared by cyclization of the corresponding tetritols,[107] and α-D-erythrofuranosyl C-nucleosides of pyrimidines have been reported.[108] 3-(β-D-Erythrofuranosyl)pyridazines can be prepared from glycosylfurans.[109] The synthesis of 4-acetyl-2-(3-deoxy-3-nitro-β-D-xylopyranosyl)-6-methylpyrrole has been accomplished by periodate oxidation of the β-D-erythrosylpyrrole followed by condensation with nitromethane.[110]

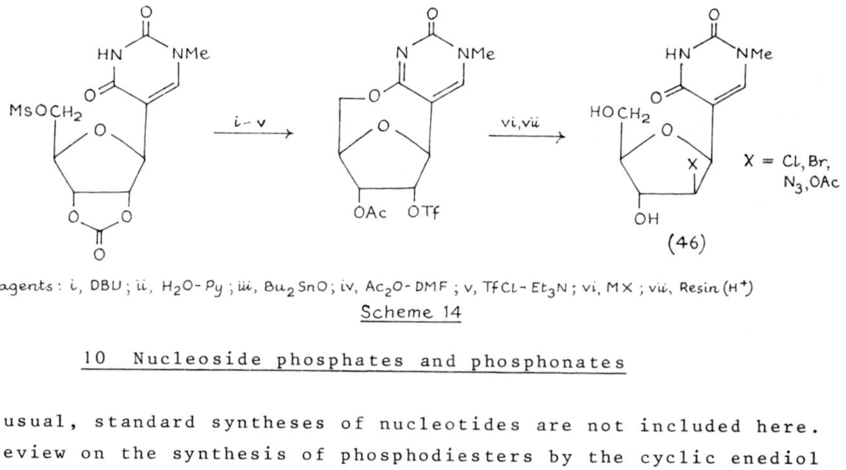


(46)

Reagents: i, DBU; ii, $H_2O$-Py; iii, $Bu_2SnO$; iv, $Ac_2O$-DMF; v, TfCl-$Et_3N$; vi, MX; vii, Resin ($H^+$)

Scheme 14

## 10 Nucleoside phosphates and phosphonates

As usual, standard syntheses of nucleotides are not included here. A review on the synthesis of phosphodiesters by the cyclic enediol phosphate method includes syntheses of phosphatidyl sugars and nucleosides.[111] In reports of plenary lectures, Reese has discussed the current position regarding the protection of the 2'- and 5'-hydroxy functions in the synthesis of oligoribonucleotides,[112] whilst Pfleiderer and co-workers have surveyed the development of the p-nitrophenylethyl group for the protection of phosphate, phosphite and amide groups, as well as the p-nitrophenylethyloxycarbonyl group for amino and hydroxy protection, in nucleoside, nucleotide and oligonucleotide chemistry.[113]

The model dinucleotide triester (47) has been prepared. On treatment with acid, the phosphotriester was unstable, giving rise to, *inter alia*, both 3' → 5'- and 2' → 5'-phosphodiesters.[114]

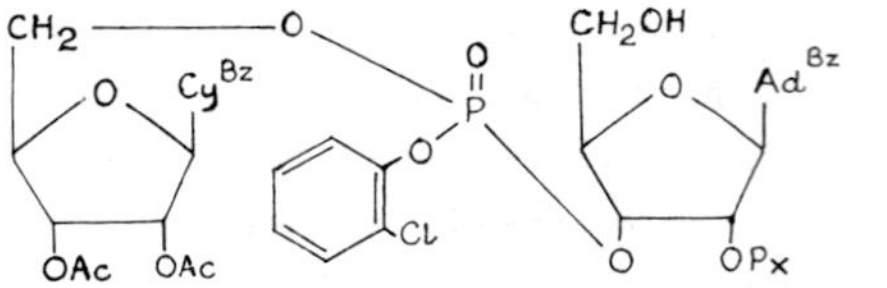

(47) (Px = 9-phenylxanthyl)

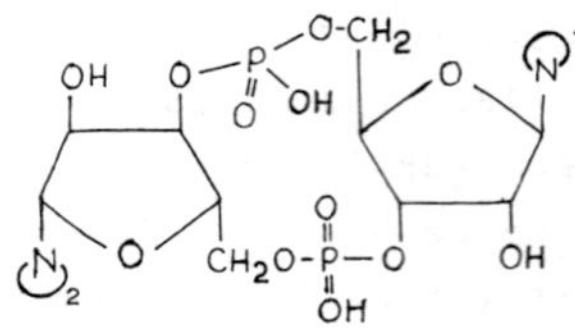

(48)

Bis-3' → 5'-cyclic dinucleotides of type (48) have been synthesized; they are competitive inhibitors of DNA-dependent RNA polymerase of E. coli.[115]

Analogues of cyclic AMP derived from the carbocyclic nucleosides aristeromycin and neplanocin have been prepared by intramolecular phosphodiester bond formation catalysed by divalent lead ions,[116] and studies have been reported on the synthesis and biological activity of nucleoside 3',5'-cyclic phosphotriesters and phosphoramidates.[117]

Enzymic phosphorylation has been used to prepare 5'-phosphates of 5-aminoimidazole nucleosides of D-ribo, D-xylo, and D-arabino configurations.[118]

The two signals in the $^{31}P$ n.m.r. spectrum of the mixed diastereomers of 2'-deoxyadenosine 5'-S-methyl-O-methylphosphorothioate have been assigned; this information, together with previous knowledge of $^{18}O$ isotope shifts and their dependence on bond order, gives rise to a simple way of determining the chirality at phosphorus of a deoxynucleoside 5'-[$^{16}O$, $^{18}O$, S]phosphorothioate.[119]

The phosphonate (49) has been synthesized from diisopropylidene glucose, and then elaborated into an isosteric bisphosphonate analogue of the trinucleotide UpUpU.[120] 2',3'-Cyclic phosphonates of type (50) and their regioisomers with the phosphonate carbon linked to O-3' have been prepared.[121] A review has appeared concerning the synthesis of 5'-O-phosphonomethyl ethers of the common nucleosides as analogues for the substrates of enzymes involved in nucleic acid metabolism.[122]

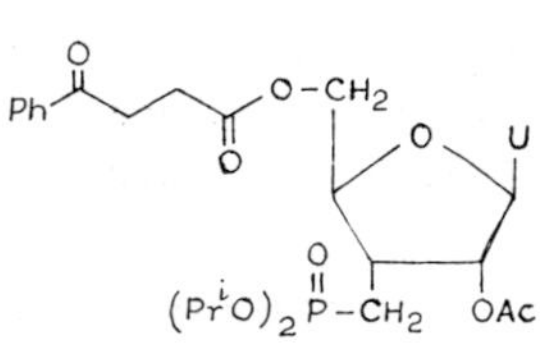

(49)

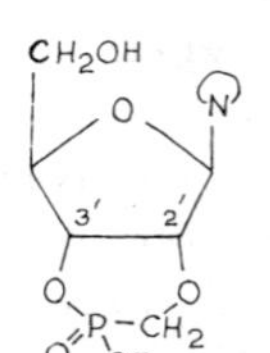

(50)

## 11 Ether, acetal and ester derivatives

A review on the use of protecting groups derived from organosilicon reagents includes a discussion of the protection of 1,2- and 1,3-diols; many examples from carbohydrate and nucleoside chemistry are included.[123]

Routes have been described for the synthesis of N-acyl-2'-O-p-methoxybenzyl ethers of ribonucleosides, these being of potential use in oligonucleotide synthesis,[124] and economical, large-scale syntheses of N-acyl-5'-O-dimethoxytrityl-2'-deoxynucleosides have been described.[125]

In connection with the synthesis of 2' → 5'-linked pyrimidine oligonucleotides, the 3'-O-(4-methoxytetrahydropyran-4-yl) derivatives of uridine and 4-N-benzoylcytidine have been prepared.[126] A similar achiral, acid-labile protecting group has been employed in oligonucleotide synthesis; this is the 3-methoxy-1,5-dicarbomethoxypentanyl (MDMP) group (51), which when present at the 2'-position was completely stable during the removal of a 5'-pixyl group.[127]

Interaction of a ribonucleoside with dibromomethane under phase-transfer conditions gave 2',3'-O-methylene acetals. In the case of the uridine derivative, glycosyl cleavage occurred in acid before complete deprotection, and, since purine nucleosides are more labile, methylene acetals are unlikely to be of any practical use as blocking groups in nucleoside synthesis.[128]

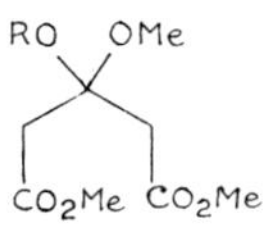


(51)

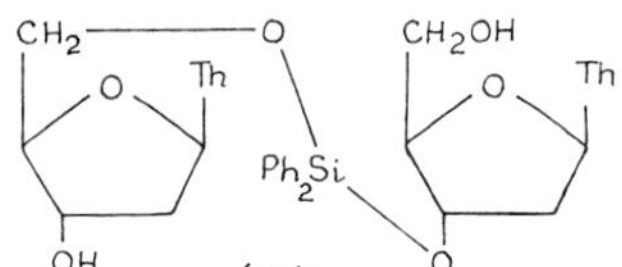


(52)

Marciewicz has reviewed the use of the 1,1,3,3,-tetraisopropyldisiloxanylidene (TIPDSi) protecting group which he has pioneered.[129] The migration of this group from the 3',5'- to the 2',3'- position in uridine has been demonstrated,[130] and its partial cleavage to 3'- and 5'-O-silyl compounds has been studied.[131] Treatment of deoxynucleosides with di-tert-butyldichlorosilane and imidazole in DMF gives sila analogues of 3',5'-cyclic nucleotides; the thymidine derivative is more stable in acid than the corresponding TIPDSi derivative.[132] The first non-ionic dinucleotide analogue with silicon in place of phosphorus (52) has been synthesized.[133]

Improved, high-yielding procedures have been given for the 2',3'-O-methoxymethylenation of N-protected ribonucleosides.[134]

2',3'-Dialkylstannylene derivatives of ribonucleosides have been prepared conventionally, and characterized by i.r. and $^{119}$Sn Mössbauer spectroscopy.[135]

A number of 3',5'-diesters[136] and 3',5'-bis(dicarboxylic acid) hemiesters[137] of 5-fluoro-2'-deoxyuridine have been synthesized and evaluated as prodrugs by examination of their susceptibility to esterase hydrolysis. The higher antitumour activity of longer-chain diesters correlated with a slower enzymic release of fluorodeoxyuridine.[136]

A high-yielding preparation of 2',3',5'-tri-O-acetylinosine has been reported,[138] and a number of partially acylated derivatives of uridine and $N^6$-benzoyladenosine were prepared for use in nucleotide synthesis.[139] Various derivatives of uridine and inosine with 5'-O-haloacetyl or -carbamoyl substituents have been prepared; 2',3'-O-isopropylidene-5'-O-iodoacetyluridine showed significant cytostatic activity.[140]

Condensation of 5'-O-trityladenosine with an aminoacid tosylate in the presence of carbonyldiimidazole followed by acidic detritylation gave a mixture of 2'- and 3'-O-(aminoacyl)adenosines, in which the 3'-isomer predominated;[141] similar chemistry was used to prepare 5'-phosphates of 2'- and 3'-aminoacylnucleosides.[142]

## 12 Miscellaneous nucleoside analogues

An abnormal Grignard reaction was used in the synthesis of 3'-C-methyl-2'-deoxyuridine, as outlined in Scheme 15.[143] The presumed

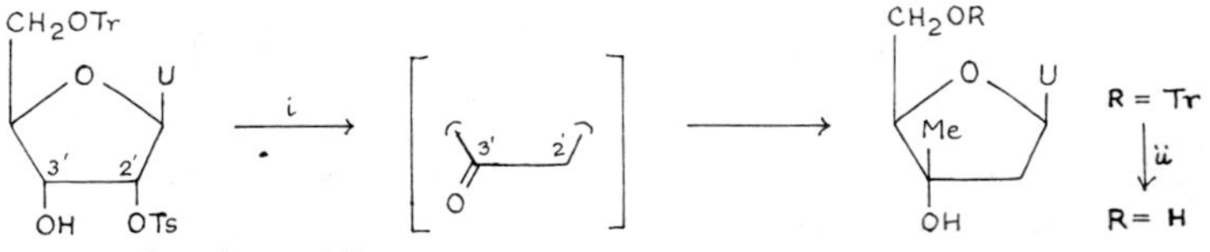


Reagents: i, MeMgI (15 eq.)-$Et_2O$ (0°) ; ii, Anthranilic acid - $ZnBr_2$ - $MeNO_2$

Scheme 15

3'-ketonucleoside intermediate reacts further with the Grignard reagent, in contrast to the 2'-keto system referred to in Scheme 10 above.[92] The conditions for detritylation are noteworthy.[143] Direct ring opening of the epoxide in 1-(5'-O-trityl-2',3'-anhydro-β-D-lyxofuranosyl)uracil with organometallics gave rise to C-alkyl, alkenyl and alkynyl-3'-deoxy-*ara*-uridine analogues such as (53).[144]

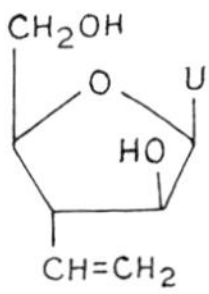
(53)

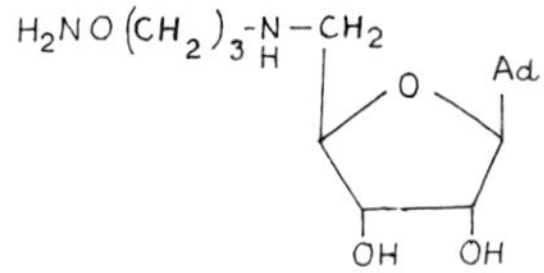
(54)

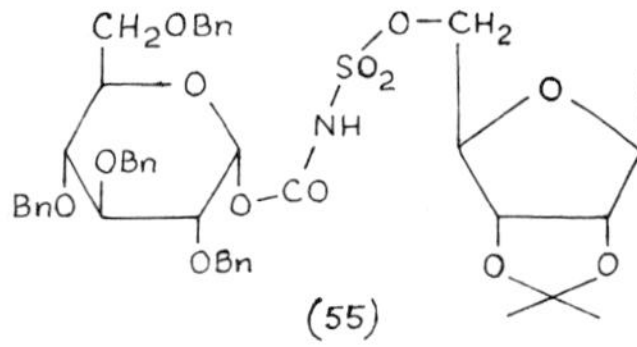
(55)

The hydroxylamine derivative (54) has been prepared as an analogue of SAH,[145] and (55), an analogue of UDPG, was synthesized from tetrabenzyl glucose,isopropylidene uridine and chlorosulphonyl isocyanate.[146]

## 13 Reactions

Convenient and high-yielding procedures have been described for the periodate oxidation-borohydride reduction of N-acyl-5'-O-monoethoxytrityl ribonucleosides.[147]

The kinetics and mechanism of the rapid acidic hydrolysis of wyosine (13) has been studied; mechanistically the process proceeds as for simple nucleosides.[148] Full details have appeared of an earlier report (Vol.12, p. 166) on the similarly rapid acid hydrolysis of 3-methylinosine.[149] The hydrolysis of 9-β-D-ribofuranosylpurine in alkali has been studied; the reaction proceeds via initial opening of the imidazole ring.[150]

2'-Deoxyguanosine was selectively labelled with deuterium at C-4', via base-catalysed exchange of the 5'-aldehyde.[151]

Further work has been reported on the interaction of the anti-tumour agent $N^2$-methyl-9-hydroxyellipticinium acetate (56) and related compounds with nucleosides (see Vol. 18). With adenosine under oxidative conditions (horseradish peroxidase-$H_2O_2$) the spiro-

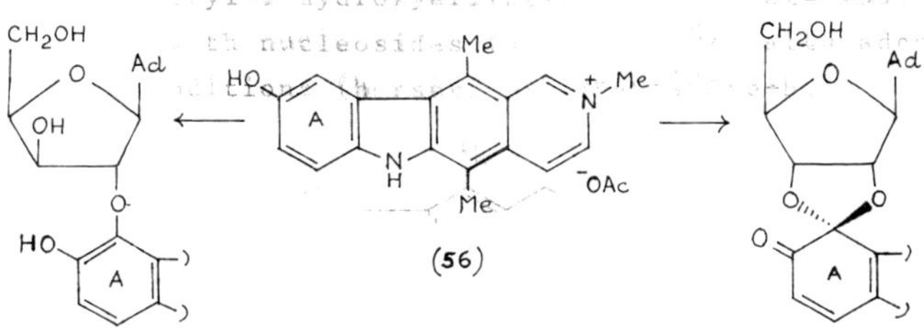
(58) (56) (57)

adduct (57) is formed, presumably via quinone imine intermediates.[152,153] When the nucleoside does not have a 2',3'-cis-diol arrangement, then the products isolated are of the type (58), or, under appropriate conditions, the corresponding, rather unstable, quinone imines.[152-154]

Reactions of arenesulphenyl chlorides with hydroxyl functions of nucleosides have been investigated; with $N^6$-phthalimido-2',3'-*O*-isopropylideneadenosine, *o*-nitrophenyl sulphenyl chloride gave the 5'-O-sulphenyl ester, whilst other sulphenyl chlorides without an *o*-nitro group gave mixtures of sulphinate esters and 5'-deoxy-5'-chloro compound. Mechanisms were suggested.[155]

## 14 Spectroscopic and other Physical Measurements

A review, in Russian, on the molecular and crystal structure of nucleosides and their derivatives includes much data on conformations and intramolecular interactions.[156]

FAB mass spectrometry of nucleosides, nucleotides and oligonucleotides has been reviewed;[157] the spectra of the major nucleosides of RNA and DNA have been reported in both positive-and negative-ion modes,and the advantages and disadvantages of the FAB method relative to other techniques was discussed.[158]

By use of $^1J$(C-H) coupling constants and other data the $^{13}$C-n.m.r. spectrum of the ribose unit of α-D-nucleosides was assigned.[159] The $^{13}$C-n.m.r. spectra of a large number of methylated ribonucleosides and deoxyribonucleosides have been measured and assigned.[160] $^{31}$P-N.m.r. was used in a study of the mutual exchange between ordered forms of GMP disodium salt in aqueous solution.[161] Three-bond coupling constants (C-6,H-1' and C-2,H-1') in the $^{13}$C-n.m.r. spectra of pyrimidine nucleosides were used to obtain information on the glycosidic bond conformation; in conjunction with *X*-ray results, a four-state conformational model was proposed consisting of an equilibrium between two *syn* and two *anti* conformations.[162] Other n.m.r.studies of conformation are discussed in the next chapter.

The stacking interaction between tryptophan and uracil was studied in the model compound (59); intramolecular stacking in

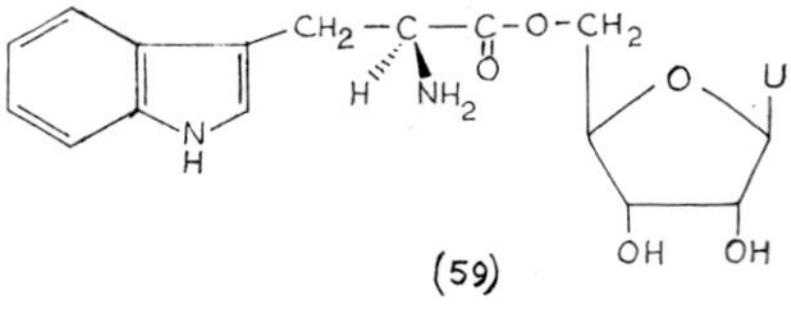

(59)

solution was demonstrated by hypochromism and fluorescence emission, but in the solid state the compound has an extended structure,the indole being stacked with a uracil in another molecule.[163] Neutron scattering has been used to study nucleoside association,and the results were consistent with the hypothesis that base stacking of nucleosides, unlike that of bare purine bases, does not lead to near-perfect parallel separations.[164] Measurements of partial molar volumes and isentropic partial molar compressibilities of some nucleosides and nucleoside bases have also been used to study stacking effects.[165]

Raman spectra of nine crystals containing guanosine moieties have been reported; useful correlations were obtained with the conformations previously established by X-ray studies.[166] The temperature-dependent CD spectra of α- and β-5-substituted-2'-deoxyuridines indicate an increase in the proportion of syn isomer in changing from aqueous to ethanolic solution.[167,168] The preferred stereostructures of two nucleoside analogues were determined by comparison of physical data with that obtained by a quantum-mechanical method of conformational analysis.[169]

As part of a study of the relationships between torsional angles and ring-puckering coordinates, it was shown that the puckering of a N-membered ring may be specified by a set of (N-3) ring-puckering coordinates computed from either the endocyclic torsion angles or the atomic Cartesians; endocyclic torsion angles were parametrized for furanose rings using 178 crystal structures of nucleic acid constituents.[170]

Theoretical calculations have been applied to adenosine and its mono-, di-, and triphosphates to calculate its vibrational and conformational behaviour and thermodynamic properties.[171]

The temperature dependence of the solubility of common nucleosides in various alcohols has been studied and used to determine the thermodynamic parameters for the dissolution process.[172]

X-ray crystal structures of nucleosides are mentioned in Chapter 22.

## References


1 G. Surico, A. Evidente, N.S. Iacobellis, and G. Randazzo, Phytochemistry, 1985, 24, 1499.

2 A. Shimizu, T. Yano, Y. Saito, and Y. Inada, Chem. Pharm. Bull., 1985, 33, 3012.

3 E. De Clercq and R.T. Walker, Pharmacol. Ther., 1984, 26, 1 (Chem. Abstr., 1985, 102, 132 351)

4 R.J. Remy and J.A. Secrist III, Nucleosides Nucleotides, 1985, 4, 411.

5 H. Frister, K. Kemper, K.-S. Boos, and E. Schlimme, Liebigs Ann. Chem., 1985, 510.

6 M.R. Harnden and R.L. Jarvest, Nucleosides Nucleotides, 1985, 4, 465.

7 C. Cristescu, Rev. Roum. Chim., 1985, 30, 117 (Chem. Abstr., 1985, 103, 142 301)

8 V.I. Kobylinskaya, T.A. Dashevskaya, A.S.Shalamai, and V.P. Chernetskii, Dopov. Akad. Nauk. Ukr. RSR, Ser. B: Geol.Khim. Biol. Nauki, 1984, 37 (Chem. Abstr., 1985, 102, 113849)

9 A. Holy, A. Ludzisa, I. Votruba, K.Sediva, and H. Pischel, Collect. Czech. Chem. Commun., 1985, 50, 393.
10 H. Hrebabecky and J. Beranek, Collect. Czech. Chem. Commun., 1985, 50, 779.
11 S.G. Wood, K.G. Upadhya, N.K. Dalley, P.A. McKernan, P.C. Canonico, R.K. Robins, and G.R. Revankar, J. Med. Chem., 1985, 28, 1198.
12 J.A. Montgomery, H.J. Thomas, A.L. Zell, H.M. Einspar, and C.E. Bugg, J. Med. Chem., 1985, 28, 1751.
13 Y. Ohba, T. Nishiwaki, H. Akagi, and M. Mori, J. Chem. Res. Synop., 1984, 276 (Chem. Abstr., 1985, 102, 79271)
14 I.D. Shingarova, I.V. Yartseva, M.P. Nemeryuk, A.L. Sedov, T.S. Safonova, G.A. Osipov, Yu. Yu. Volodin, and M.N. Preobrazhenskaya, Khim. Geterotsikl. Soedin., 1984, 1556 (Chem. Abstr., 1985, 102, 113857)
15 H. Frister and E. Schlimme, Liebigs Ann. Chem., 1985, 1704.
16 V.D. Patil, D.S. Wise, L.L. Wortring, L.C. Bloomer, and L.B. Townsend, J. Med. Chem., 1985, 28, 423.
17 A. Hijazi and W. Pfleiderer, Nucleosides Nucleotides, 1984, 3, 549.
18 G. Cristalli, M. Grifantini, S. Vittori, W. Balduini, and F. Cattabeni, Nucleosides Nucleotides, 1985, 4, 625.
19 M. Draminski, E. Frass, J. Greger, and K. Fabianowska-Majewska, Collect. Czech. Chem. Commun., 1985, 50, 280.
20 W.M. Odijk and G.J. Koomen, Tetrahedron, 1985, 41, 1893.
21 T. Nishiwaki and N. Abe, J. Chem. Res., Synop., 1984, 264 (Chem. Abstr., 1985, 102, 79270 )
22 S.V. Galushko, I.P. Shishkina, I.V. Alekseeva, and Y . I. Usatenko, Zh. Anal. Khim., 1985, 40, 926 (Chem. Abstr., 1985, 103, 160793).
23 J.P. Ferris, B. Devadas, C.-H. Huang, and W.-Y. Ren, J. Org. Chem., 1985, 50, 747.
24 A. Piskala, A. Cihar, N.B. Hanna, and J. Skutchan, Stud. Org. Chem. (Amsterdam), 1984, 18 (Bio-Org. Heterocycl.), 407 (Chem. Abstr., 1985, 103, 105251).
25 L.F. Tietze and A. Bergmann, Angew. Chem. Int. Ed. Engl., 1985, 24, 127.
26 T. Itaya, H. Matsumoto, T. Watanabe, and T. Harada, Chem. Pharm. Bull., 1985, 33, 2339.
27 S.N. Mikhailov and W. Pfleiderer, Synthesis, 1985, 397.
28 S.N. Mikhailov, Bioorg. Khim., 1984, 10, 940 (Chem. Abstr., 1985, 102, 7016).
29 T. Mukaiyama, Y. Hayashi, and Y. Hashimoto, Chem. Lett, 1985, 1087.
30 T. Hashizume, M. Hosoi, and T. Sugiyama, Agric. Biol. Chem., 1985, 49, 225.
31 T. Utagawa, H. Morisawa, F. Yoshinaga, A. Yamazaki, K. Mitsugi, and Y. Hirose, Agric. Biol. Chem., 1985, 49, 1053.
32 T. Utagawa, H. Morisawa, S. Yamanaka, A. Yamazaki, F. Yoshinaga, and Y. Hirose, Agric. Biol. Chem., 1985, 49, 2167.
33 T. Utagawa, H. Morisawa, S. Yamanaka, A. Yamazaki, F. Yoshinaga, and Y. Hirose, Agric. Biol. Chem., 1985, 49, 2425.
34 R. J. Nachman, R.Y. Wong, W.F. Haddon, and R.E.Lundin, J. Chem. Soc., Perkin Trans. 1, 1985, 1315.
35 T. Kulikowski, Z. Zawadzki, E. De Clercq, and D. Shogar, Acta Biochim. Pol., 1984, 31, 341 (Chem. Abstr., 1985, 103, 6656).
36 F. Ibrahim-Lodgemann, G. Mackenzie, and G. Shaw, Nucleosides Nucleotides, 1985, 4, 199.
37 N.E. Poopieko, E.I. Kvasyuk, I.A. Mikhailopulo, and M.J.Lidaks, Synthesis, 1985, 605.
38 M.J. Camarasa and F.G. De Las Heras, An. Quim., Ser.C, 1983, 79, 407 (Chem. Abstr., 1985, 102, 79275).
39 A.T. Ayupova, Ch.Sh. Kadyrov, L.V. Molchanov, S.S. Khalikov, Khim. Geterotsikl. Soedin., 1984, 812 (Chem. Abstr., 1985, 102, 6968).
40 Ya. V. Voznyi, L.N. Koikov, and A.A. Galoyan, Bioorg. Khim., 1985, 11, 534 (Chem. Abstr., 1985, 103, 22860).
41 R. Rico Gomez and M.L. Ruiz Mora, An. Quim., Ser. C, 1984, 80, 208 (Chem. Abstr., 1985, 102, 132407).

42 R.R. Schmidt and J. Michel, J. Carbohydr. Chem., 1985, 4, 141.
43 L.N. Kulinkovich and V.A. Timoshchuk. Zh. Obshch. Khim., 1985, 55, 205 (Chem. Abstr., 1985, 102, 167090).
44 K. Hirota, Y. Kitade, T. Tomishi, and Y. Maki, Chem. Pharm. Bull., 1985, 33, 4212.
45 K. Hirota, Y. Kitade, T. Tomishi, and Y. Maki, Nucleosides Nucleotides, 1985, 4, 681.
46 T. Ueda, Y. Nomoto, and A. Matsuda, Chem. Pharm. Bull., 1985, 33, 3263.
47 V. Skaric, M. Pavela-Vrancic, and J. Matulic-Adamic, Croat. Chim. Acta, 1984, 57, 415 (Chem. Abstr., 1985, 102, 62552).
48 G. Mackenzie, G. Shaw, and H.A. Wilson, Nucleosides Nucleotides, 1984, 3, 339.
49 K. Anzai and J. Uzawa, J. Org. Chem., 1984, 49, 5076.
50 K. Minamoto, Y. Fujiki, N. Shiomi, Y. Uda, and T. Sasaki, J. Chem. Soc., Perkin Trans. I, 1985, 2337.
51 K. Minamoto, H. Nakade, T. Tanaka, Y. Fujiki, and T. Sasaki, J. Chem. Soc., Perkin Trans. I, 1985, 2347.
52 A. Matsuda and K.A. Watanabe, J. Org. Chem., 1985, 50, 2660.
53 L. Novotny, H. Hrebebecky, and J. Beranek, Collect. Czech. Chem. Commun., 1985, 50, 383.
54 T. Ueda, Nucleosides Nucleotides, 1985, 4, 67.
55 T. Sano, S. Shuto, H. Inoue, and T. Ueda, Chem. Pharm. Bull., 1985, 33, 3617.
56 T. Sano, H. Inoue, and T. Ueda, Chem. Pharm. Bull., 1985, 33, 3595.
57 T. Ueda, S. Shuto, M. Satoh, and H. Inoue, Nucleosides Nucleotides, 1985, 4, 401.
58 T. Sano, H. Inoue, and T. Ueda, Chem. Pharm. Bull., 1985, 33, 1856.
59 A.J. Hubbard, A.S. Jones, and R.T. Walker, Nucleic Acids Res., 1984, 12, 6827 (Chem. Abstr., 1985, 102, 62549).
60 R. Alonso, J.I. Andres, M.T. Garcia-Lopez, F.G. De Las Heras, R. Herranz, B. Alarcon, and L. Carrasco, J. Med. Chem., 1985, 28, 834.
61 W.L. Mitchell, P. Ravenscroft, M.L. Hill, L.J.S.Knutsen, R.F. Newton, and D.I.C. Scopes, Nucleosides Nucleotides, 1985, 4, 173.
62 S.Ya. Mel'nik, A.A. Bakhmedova, T.D. Miniker, I.V. Yartseva, M.N. Preobrazhenskaya, O.A. Zagulyaeva, V.P. Mamaev, E.V. Chekunova, and S.S. Marennikova, Bioorg. Khim., 1984, 10, 1645 (Chem. Abstr., 1985, 102, 221 133).
63 S.M.N. Efange, E.M. Alessi, H.C. Shih, Y.-C. Cheng, and T.J.Bardos, J. Med. Chem., 1985, 28, 904.
64 S.M.N. Efange, Y.-C. Cheng, and T.J. Bardos, Nucleosides Nucleotides, 1985, 4, 545.
65 J. Okada, K. Nakano, and H. Miyake, Chem. Pharm. Bull., 1985, 33, 856.
66 F. Seela and H. Steker, Helv. Chim. Acta, 1985, 68, 563.
67 F. Seela and H. Steker, J. Chem. Soc. Perkin Trans. I, 1985, 2573.
68 F. Seela and S. Menkhoff, Liebigs Ann. Chem., 1985, 1360.
69 F. Seela H.Driller, and U. Liman, Liebigs Ann. Chem., 1985, 312.
70 F. Seela and W. Bussmann, Tetrahedron, 1985, 41, 935.
71 Z. Kazimierczuk, R. Stolarski, and D. Shugar, Z. Naturforsch, C: Biosci., 1985, 40C, 715 (Chem. Abstr., 1985, 103, 210393).
72 M. Ikehara and T. Inaoka, Nucleosides Nucleotides, 1985, 4, 515.
73 E.J. Prisbe and J.C. Martin, Synth. Commun., 1985, 15, 401 (Chem. Abstr., 1985, 103, 123846).
74 D.N. Abrams, J.R. Mercer, E.E. Knaus, and L.I. Wiebe, Int. J. Appl. Radiat. Isot., 1985, 36, 233 (Chem. Abstr, 1985, 103, 215693).
75 Y.W. Lee, J.R. Mercer, E.E. Knaus, and L.I. Wiebe, Int. J. Appl. Radiat. Isot., 1984, 35, 1053 (Chem. Abstr., 1985, 102, 167078).
76 V.A. Timoshchuk, Zh. Obshch. Khim., 1984, 54, 2646 (Chem.Abstr., 1985, 102, 113861).
77 T. Utagawa, H. Morisawa, S. Yamanaka, A. Yamazaki, and Y. Hirose, Agric. Biol. Chem., 1985, 49, 2711.
78 N.A. Dyatkina, A.A. Krayevsky, A.V. Azhayev, and I.V. Jartseva, Synthesis, 1985, 410.

79 T.S. Lin, J. Pharm. Sci., 1984, 73, 1568 (Chem. Abstr., 1985, 102, 113856).
80 L.C. Davies, Nucleosides Nucleotides, 1985, 4, 395.
81 Z. Smiatacz, R. Szweda, and J. Drewniak, Carbohydr. Res., 1985, 143, 151.
82 P. Serafinowski, Synthesis, 1985, 926.
83 A. Holy and I. Rosenberg, Collect. Czech. Chem. Commun., 1985, 50, 1514.
84 K. Ramalingam and R.W. Woodard, Tetrahedron Lett., 1985, 26, 1135.
85 P. Dewevre, C. Tapiero, and J.L. Imbach, Nucleosides Nucleotides, 1985, 4, 301.
86 K. Ramalingam and R.W. Woodard, Carbohydr.Res., 1985, 142, 123.
87 G.L. Creason, J.T. Madison and J.F. Thompson, Phytochemistry, 1985, 24, 1511.
88 W.A. Szarek, B.M. Pinto, and M. Iwakawa, Can. J. Chem., 1985, 63, 2149.
89 W.A. Szarek, B.M. Pinto, and M. Iwakawa, Can. J. Chem., 1985, 63, 2162.
90 T. Ueda and S. Watanabe, Chem. Pharm. Bull., 1985, 33, 3689.
91 Y. Kato, N. Fusetani, S. Matsunaga, and K. Hashimoto, Tetrahedron Lett., 1985, 26, 3483.
92 S. Juntunen and J. Chattopadhyaya, Acta Chem.Scand. Ser. B, 1985, 39, 149.
93 J. Theim and D. Rasch, Nucleosides Nucleotides, 1985, 4, 487.
94 T. Halmos, D. Komiotis, and K. Antonakis, Carbohydr.Res., 1985, 145, 163.
95 B. Classon and B. Samuelsson, Acta Chem. Scand. Ser. B, 1983, 39, 501.
96 K.W. Pankiewicz and K.A. Watanabe, Nucleosides Nucleotides, 1985, 4, 613.
97 N. Katagiri, T. Haneda, R. Niwa, and T. Kato, Chem. Pharm. Bull., 1985, 33, 2671.
98 N. Katagiri, N. Tabei, S. Atsuumi, T. Haneda, and T. Kato, Chem. Pharm. Bull., 1985, 33, 102.
99 P. Verschave and G. Hoornaert, Nucleosides Nucleotides, 1985, 4, 231.
100 J.A. Galbis Perez, R. Babiano Caballero, and A. Cert Ventula, Carbohydr. Res., 1985, 143, 129.
101 I. Maeba, T. Ishikawa, and H.Furukawa, Carbohydr. Res., 1985, 140, 332.
102 I. Maeba, F. Usami, T. Ishikawa, H. Furukawa, T. Ishida, and M. Inoue, Carbohydr. Res., 1985, 141, 1.
103 L.B. Townsend, V.G.Beylin, and L.L. Wotring, Nucleosides Nucleotides, 1985, 4, 29.
104 S. Niitsuma, K. Kato, T. Takita, and H. Umezawa, Tetrahedron Lett., 1985, 26, 5785.
105 L.J.S. Knutsen, B.D. Judkins, R.F. Newton, D.I.C. Scopes, and G. Klinkert, J. Chem. Soc., Perkin Trans. I, 1985, 621.
106 K.W.Pankiewicz, J.-H. Kim, and K.A. Watanabe, J. Org. Chem., 1985, 50, 3319.
107 J. Fernandez-Bolanos, J. Fuentes Mota, and J. Fernandez-Bolanos Guzman, An. Quim., Ser. C, 1983, 79, 345 (Chem. Abstr., 1985, 102, 62530).
108 F.J. Lopez Herrera, C. Gomez Perez, and M. Valpuesta Fernandez, An. Quim., Ser. C, 1984, 80, 218 (Chem. Abstr., 1985, 103, 123843).
109 F.J. Lopez Aparicio, M.T. Plaza Lopez-Espinosa, and R. Robles Diaz, An. Quim., Ser. C, 1984, 80, 215 (Chem. Abstr, 1985, 103, 71592).
110 F.J. Lopez Aparicio, M. Melgarejo Sampedro, and R.Asendro Asendro, An. Quim., Ser. C, 1983, 79, 218 (Chem. Abstr., 1985, 102, 62542).
111 F. Ramirez and J.F. Maracek, Synthesis, 1985, 449.
112 C.B. Reese, Nucleosides Nucleotides, 1985, 4, 117.
113 W. Pfleiderer, F. Himmelsbach, R. Charubala, H. Schirmeister, A. Beiter, B. Schulz, and T. Trichtinger, Nucleosides Nucleotides, 1985, 4, 81.
114 T. Pathak and J. Chattopadhyaya, Acta Chem. Scand., Ser. B, 1985, 39, 799.
115 C.-Y. J. Hsu, D. Dennis, and R.A. Jones, Nucleosides Nucleotides, 1985, 4, 377.
116 H. Sawai and M. Ohno, Chem. Pharm. Bull, 1985, 33, 432.
117 L. Zhang, L. Ma, C. Xu, L. Dong, C. Li, L. Shan, X.Wang, Y. Ge, S.Zhang, and S. He, Kexue Tonbao (Foreign Lang.Ed.), 1985, 30, 263.
118 R.W. Humble, G. Mackenzie, and G. Shaw, Nucleosides Nucleotides, 1985, 4, 281.
119 J.H. Cummins and B.V.L. Potter, J.Chem. Soc, Chem. Commun., 1985, 851.
120 A. Mazur, B.E. Tropp, and R. Engel, Tetrahedron, 1984, 40, 3949.
121 I. Rosenberg and A. Holy, Collect. Czech.Chem. Commun., 1985, 50, 1507.

122 K. Sebesta, A. Holy, K. Horska, I. Vesely, and I. Rosenberg, Fiz.-Khim. Probl. Ferment. Katal., 1984, 90 (Chem. Abstr, 1985, 102, 167029).
123 M. Lalonde and T.H. Chan, Synthesis, 1985, 817.
124 L.W. McLaughlin, N. Piel, and T. Hellmann, Synthesis, 1985, 322.
125 R. Kierzek, Nucleosides Nucleotides, 1985, 4, 641.
126 D.G. Norman and C.B. Reese, Synthesis, 1985, 874.
127 H. Sandstrom, M. Kwiatowski, and J. Chattopadhyaya, Acta Chem. Scand. Ser. B, 1985, 39, 273.
128 D.G. Norman, C.B. Reese, and H.T. Serafinowska, Synthesis, 1985, 751.
129 W.T. Markiewicz, Stud. Org. Chem. (Amsterdam), 1985, 20 (Nat. Prod. Chem), 275 (Chem. Abstr., 1985, 103, 105230).
130 W.T. Markiewicz and A. Bartoszuk, Bull. Pol. Acad. Sci., Chem., 1984, 32, 453, (Chem. Abstr., 1985, 103, 123841).
131 W.T. Markiewicz, Bull. Pol. Acad. Sci. Chem., 1984, 32, 463 (Chem. Abstr., 1985, 103, 71609).
132 K. Furusawa and T. Katsura, Tetrahedron Lett., 1985, 26, 887.
133 K.K. Ogilvie and J.F. Cormier, Tetrahedron Lett, 1985, 26, 4159.
134 A. Kume, H. Tanimura, S. Nishiyama, M. Sekine, and T. Hata, Synthesis, 1985, 408.
135 G. Ruisi, M.T. Loiudice, and L. Pellerito, Inorg. Chim. Acta., 1984, 93, 161 (Chem. Abstr., 1985, 102, 185404).
136 T. Kawaguchi, Y. Suzuki, Y. Nakahara, N. Nambu, and T. Nagai, Chem. Pharm. Bull., 1985, 33, 301.
137 T. Kawaguchi, Y. Suzuki, N. Nambu, and T. Nagai, Chem. Pharm. Bull., 1985, 33, 2956.
138 L.R. Davidenkov, E.V. Ratsino, L.D. Antonenko, and E.V. Komarov, Zh. Prikl. Khim., (Leningrad), 1984, 57, 2146 (Chem. Abstr., 1985, 102, 79273).
139 J. Lekschas, D. Cech, and A. Rosenthal, Z. Chem., 1984, 24, 329 (Chem. Abstr., 1985, 102, 167070).
140 M.T. Garcia-Lopez, P. Fernandez-Resa, F.G. De las Heras, and P.P. Mendez-Castrillon, An. Quim., Ser. C, 1984, 80, 168 (Chem. Abstr., 1985, 102, 149700).
141 P.P. Purygin, Zh. Org. Khim., 1985, 21, 918 (Chem. Abstr., 1985, 103, 54426).
142 P.P. Purygin, Zh. Org. Khim., 1984, 20, 1879 (Chem. Abstr., 1985, 102, 95983).
143 A. Grouiller, H. Essadiq, H. Pacheco, S. Juntunen, and J. Chattopadhyaya, Angew. Chem. Int. Ed. Engl., 1985, 24, 52.
144 A. Mete, J.B. Hobbs, D.I.C. Scopes, and R.F. Newton, Tetrahedron Lett., 1985, 26, 97.
145 M. Kolb and J. Barth, Liebigs Ann. Chem., 1985, 1036.
146 M.J. Camarasa, P. Fernandez-Resa, M.T. Garcia-Lopez, F.G. de las Heras, P.P. Mendez-Castrillon, B. Alarcon, and L. Carrasco, Nucleosides Nucleotides, 1985, 4, 149.
147 S.N. Mikhailov, V.L. Florentiev, and W. Pfleiderer, Synthesis, 1985, 399.
148 B. Golankiewicz, E. Zielonacka-lis, and W. Folkman, Nucleic Acids Res., 1985, 13, 2443 (Chem. Abstr., 1985, 103, 160786).
149 T. Itaya and H. Matsumoto, Chem. Pharm. Bull., 1985, 33, 2213.
150 H. Lönnberg and P. Lehikoinen, J. Org. Chem., 1984, 49, 4964.
151 M. Berger and J. Cadet, Nucleosides Nucleotides, 1985, 4, 287.
152 G. Pratviel, J. Bernadou, and B. Meunier, J. Chem. Soc., Chem. Commun., 1985, 60.
153 V.K. Kansal, S. Funakoshi, P. Mangeney, B. Gillet, E. Guittet, J.-Y. Lallemand, and P. Potier, Tetrahedron, 1985, 41, 5107.
154 G. Pratviel, J. Bernardou, B. Meunier, and C. Paoletti, Nucleosides Nucleotides, 1985, 4, 143.
155 H. Bazin, J. Heikkila, and J. Chattopadhyaya, Acta Chem. Scand., Ser. B, 1985, 39, 391.
156 A.F. Mishnev and J. Bleidezis, Zh. Strukt. Khim., 1984, 25, 122 (Chem. Abstr., 1985, 102, 113795).
157 D.L. Slowikowski and K.H. Schram, Nucleosides Nucleotides, 1985, 4, 309.
158 D.L. Slowikowski and K.H. Schram, Nucleosides Nucleotides, 1985, 4, 347.
159 F. Seela and W. Bussmann, Nucleosides Nucleotides, 1985, 4, 391.

160 C. Chang, D.J. Ashworth, L. Chern, J. DaSilva Gomes, C. Lee, P.W. Mou, and R. Narayan, Org. Magn. Reson., 1984, 22, 671 (Chem. Abstr., 1985, 102, 113855).
161 J.J. Led and H. Gesmar, J. Phys. Chem., 1985, 89, 583 (Chem. Abstr., 1985, 102, 74330).
162 D.B. Davies, P. Rajani, and H. Sadikot, J. Chem. Soc., Perkin Trans, II, 1985, 279.
163 I. Saito, H. Sugiyama, T. Matsuura, and K. Fukuyama, Tetrahedron Lett., 1985, 26, 4467.
164 P. Martel, J. Phys. Chem., 1985, 89, 230 (Chem. Abstr., 1985, 102, 46218)
165 H. Hoeiland, A. Skauge, and I. Stokkeland, J. Phys. Chem., 1984, 88, 6350 (Chem. Abstr., 1985, 102, 46214).
166 Y. Nishimura, M. Tsuboi, T. Sato, Nucleic Acids Res., 1984, 12, 6901 (Chem. Abstr., 1985, 102, 24976).
167 Yu.Yu. Volodin, B.S. Kikot, V.M. Lobachev, and M.N. Preobrazhenskaya, Bioorg. Khim., 1984, 10, 1359 (Chem. Abstr., 1985, 102, 113851).
168 Yu.Yu. Volodin, B.S. Kikot, V.M. Lobachev, and Yu. S. Mordashev, Biofizika, 1984, 29, 725 (Chem. Abstr., 1985, 102, 132404).
169 H. Broch and P.M. Kaiser, Stud. Biophys., 1984, 99, 21 (Chem. Abstr., 1985, 102, 62550).
170 E. Diez, A.L. Esteban, F.J. Bermejo, C.A. Hona, and F.A.A. M.De Leeuw, J. Mol. Struct., 1984, 125, 49 (Chem. Abstr., 1985, 102, 131282).
171 A.W. Espinosa-Muller and A.R. Bravo Nunez, THEOCHEM, 1984, 19, 341 and 349 (Chem. Abstr., 1985, 102, 132409 and 132410); 1985, 20, 241 and 251 (Chem. Abstr., 1985, 102, 149705 and 149706).
172 J. Lavrs, A. Veveris, and V. Mikstais, Latv. PSR Zinat. Akad. Vestis, Khim. Ser., 1984, 582 (Chem. Abstr., 1985, 102, 95992).

# 21 N.M.R. Spectroscopy and Conformational Features

## 1 Theoretical and General Considerations

*Ab initio* SCFMO calculations at the STO-3G level have been carried out on $HOCH_2OPO_3H^{1-}$ and $HOCH_2OPO_3^{2-}$, model systems for the C-1 phosphate moiety of sugars, to determine the anomeric effect of the phosphate group. A complete potential energy map was constructed for $CH_3OPO_3H^{1-}$ as a model for sugar C6-phosphate.[1] Calculations on co-enzyme 1 (NAD) using CNDO/2 and SCF-PPP to obtain the π-electron dipole moment and the atomic electric charge distribution have been based on nicotinamide, adenine, and 1,*N*$^6$-ethylene adenine as sections of the enzyme. The results were used to obtain quantum chemical interpretation of the n.m.r. spectra of NAD.[2] A rapid and direct method has been described for generating furanose co-ordinates from the pseudorotation angle and has thus overcome the greatest difficulty in modelling nucleic acids. The method gives co-ordinates for any point in the pseudorotational pathway from the phase angle of pseudorotation (P) and differs from previous schemes in that endocyclic bond lengths in the sugar ring are allowed to vary a small amount depending on P.[3] The relationships between torsional angles and ring-puckering co-ordinates have been studied using endocyclic torsion angles for furanose rings that were parameterized using 178 crystal structures of nucleic acid constituents, and it was shown that the puckering in an n-membered ring could be specified by n-3 ring-puckering co-ordinates computed from the endocyclic torsion angles or the Cartesian atomic co-ordinates.[4,5]

Conformational analysis calculations for determinant tri- and tetra-saccharides with Le$^a$, Le$^b$, and H group specificity showed that these oligosaccharides are structurally rigid and in aqueous solution are present in only one conformation.[6]

Detailed analysis of $^1$H n.m.r. spectra of methyl-epoxytetrahydropyrans has shown the steric requirement of a methyl group: the molecules represent simple models of anhydropyranoses. Analysis of J data for *trans*-2-methyl-3,4-epoxytetrahydropyran (1) shows that it exists entirely in the *O*-opposed conformation A whereas for the *cis*-

2-methyl compound (2) the equilibrium consists of 40% A and 60% B. The results confirm the previously assumed repulsive effect between epoxide and tetrahydropyran ring oxygen making (1) A and (2) A unexpectedly more stable.[7] Partial deuteration of exchangeable

protons manifested by a series of isotopomers was observed by n.m.r. spectroscopy under conditions of slow exchange; the method was demonstrated by $^{1}$H and $^{13}$C n.m.r. of carbohydrates such as lactose. Most signals, seen as characteristic multiplets (doublets to octets), could be assigned unequivocally in one experiment by analysis of 2- and 3-bond isotope effects. Most usefully, the method allowed study of intramolecular OH---OH hydrogen bonding, illustrated for partially acetylated sucrose.[8] Isotope multiplets in the $^{13}$C n.m.r. spectrum of partially deuterated D-idose in DMSO-propanone have also been used to assign the spectrum. Successive dilution with deuterium oxide and extrapolation gave the equilibrium composition in pure deuterium oxide. The α-pyranose: β-pyranose: α-furanose: β-furanose ratio was found to be 26.5:56:10.5:7 in the non-aqueous solvent and 38:38:10:14 in water. The α-pyranose is a mixture of conformers in water but in the non-aqueous solvents intramolecular hydrogen bonding dominated the conformation, which was largely $^{4}\underline{C}_{1}$ (D) in these media.[9]

Anomeric effects in 2-chloro-, 2-methoxy-, 2-hydroxy-, and 2-methylamino-tetrahydropyrans of $\Delta\Delta H^{o}$ = +2.12, +0.75, +0.61, and -0.33 kcal $mol^{-1}$ respectively have been calculated on the basis of variable-temperature $^{1}$H and $^{13}$C n.m.r. determination of equilibrium constants and comparison with the cyclohexane analogues. The competition between endo- and exo-anomeric effects was noted, the negative value of which indicates a preference for the equatorial orientation.[10]

A useful review of the $^{13}$C n.m.r chemical shifts of 60 naturally occurring steroidal saponins includes a tabulated compilation of data for methyl glyco-pyranoside and -furanoside pairs and a discussion of $^{13}$C n.m.r. methodology.[11] A review of the n.m.r. spectroscopy of natural macromolecules including nucleic acids and polysaccharides has appeared.[12]

Determination of $^{1}J_{C,H}$ values from the $^{13}$C n.m.r. satellites in high-field $^1$H spectra has been demonstrated for D-glucose, D-mannose, 1,2,3,4-tetra-O-acetyl-β-D-ribopyranose, and methyl 2,3,4,6-tetra-O-acetyl-α-D-glucopyranoside. The satellites occur at 80-85Hz from the central proton signal and may be observed for chemical shift separated signals within 15 min using 3-5 mg samples.[13]

## 2 Acylic Systems

The $^1$H n.m.r. spectra at 400 MHz of 9 monodeoxy alditols support previous conclusions on conformation in solution.[14] The conformations of some penta-O-benzylhexononitriles and 5-(penta-O-benzoylpentitol-1-yl)tetrazoles have been deduced from $^1$H n.m.r. spectroscopic data.[15]

A series of derivatives of N-acetylneuramino-1,4-lactone diethyl dithioacetal (3) has been investigated by n.m.r. spectroscopy. Complete assignments of $^1$H and $^{13}$C resonances were achieved by measurements of relaxation times, selective decoupling experiments, and 2D shift correlation spectroscopy. The conformational analysis of the fully acetylated derivative was based on $J_{H,H}$ values and n.O.e. using the assignments of all the acetate groups. The influence of different substituents on the conformational behaviour was discussed.[16]

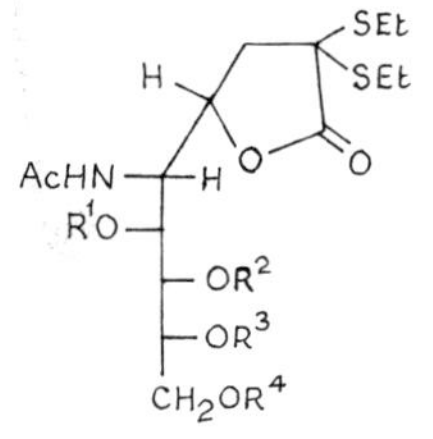


(3) $R^1=R^2=R^3=R^4=H$
$R^1=R^2=R^3=H$, $R^4=SiBu^tMe_2$
$R^1=R^2=R^3=R^4=Ac$
$R^1=R^2=R^3=Ac$, $R^4=SiBu^tMe_2$
$R^1=R^2=Ac$, $R^3=R^4=SiBu^tMe_2$
$R^1,R^2=CO$, $R^3=Ac$, $R^4=SiBu^tMe_2$

## 3 Furanose Systems

The conformations, determined by $^1$H n.m.r. spectra, of α-L-arabinofuranosyl pyridinium salt (4) and its tri-O-benzoyl derivative are

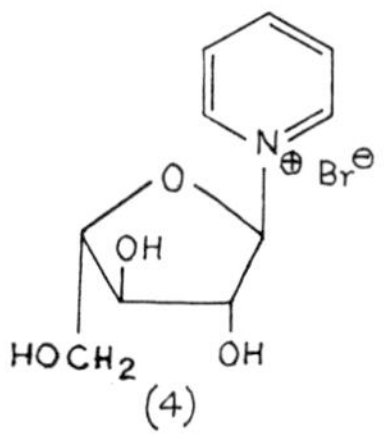


(4)

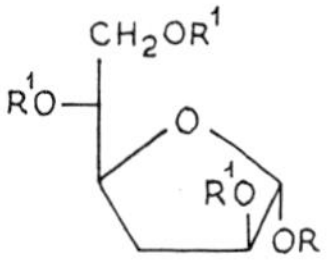


(5) R = Ac, $R^1$ = Bz
(6) R = Me, $R^1$ = Bz
(7) R = $R^1$ = H

$E_2$ and $^{O}T_1$ respectively.[17] Conformational studies by DAERM analysis of J values on the three 3-deoxy-D-arabino-hexose derivatives (5) - (7) have been carried out.[18]

The $^1H$ and $^{13}C$ n.m.r. data for the α- and β-pyranose and furanose forms of methyl L-rhamnoside have been collected and compared with those of the corresponding mannosides.[19] All signals in the $^{13}C$ n.m.r. spectra of 1,2-O-isopropylidene- and 1,2-O-exo-benzylidene-α-D-gluco- and β-L-ido-furanurono-6,3-lactones have been unambiguously assigned by means of two 2D experiments.[20] The Karplus equation applied to the conformational analysis of the sugar in sitosteryl 2,3-di-O-acetyl-α-D-riburonofuranoside methyl ester has shown that it exists preferentially in the $^{O}E$ form.[21] Reference to the conformation of the dimer of 1,2-O-isopropylidene-α-D-pentodialdo-1,4-furanose will be found in Chapter 6.

$^{13}C$-N.m.r. chemical shifts of 22 methylated ribonucleosides and deoxyribonucleosides have been measured and assigned.[22] A Karplus relationship between dihedral angle and C-2,H-1' and C-6,H-1' vicinal coupling in uridine derivatives has been found from a study of uracil cyclonucleosides of known glycosidic conformation.[23] A simple method for estimating the relative contributions of the syn- and anti-conformers for purine nucleosides and nucleotides in solution utilized three $^1H$ n.m.r. parameters: (ΔH2'-ΔH5'), $J_{1,2'}$, and $(J_{4',5'a} + J_{4',5'b})$.[24]

## 4 Pyranose Systems

The cross polarization-magic angle spinning technique used for $^{13}C$ n.m.r. of solid methyl α- and β-D-xylopyranosides has been shown to be sensitive to the crystal packings.[25]

The substitution pattern in the aromatic nucleus of 2-methoxy-4-hydroxy-phenyl β-D-glucopyranoside (8), isolated from the liverwort Isotachis japonica, has been determined with the assistance of an n.O.e. between the anomeric proton and the ortho proton on the aromatic nucleus.[26] $^{13}C$ n.m.r. data for N-(m- and p-substituted-

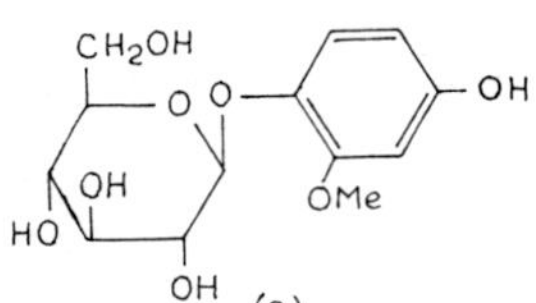


(8)

phenyl)-β-D-glucopyranosylamines, their peracetates, and the Amadori rearrangement products, 1-*N*-(*p*-methyl-, -ethyl-, and -methoxy-phenyl)amino-1-deoxy-D-fructose, have been gathered.[27] Unambiguous $^{13}C$ n.m.r. assignments have been obtained, using 2D techniques, for several glycosyl azides, 6-deoxy-, and 2-acylamino-glycosyl azides, and some 2- and 3-azido monosaccharide derivatives. Generalizations included the finding that $^{1}J_{C,He}$ values are 4-9 Hz larger than $^{1}J_{C,Ha}$ values. A substituent such as hydroxy, acetoxy, alkoxy, and azide in a 1,3-diaxial relationship with the axial proton significantly increases the value of $^{1}J_{C,Ha}$, as do the bond angle distortions in some isopropylidene fused-ring systems. Electronic and steric effects of substituents at non-anomeric carbon atoms may alter the $^{1}J_{C,H}$ values for anomeric carbon atoms to such an extent that they are no longer useful for the diagnosis of anomeric configuration. Bond angle deformations also influence the $^{13}C$ chemical shift differences in α- and β-anomers at C-5 and, to a lesser extent, C-3. A large amount of chemical shift data obtained in deuterium oxide and deuterochloroform was tabulated.[28] The effects of trifluoroacetylation on the $^{13}C$ n.m.r. spectra of 2-acetamido-2-deoxy-D-glucose, -D-galactose, and-D-mannose have been studied. In contrast to $^{1}H$ chemical shifts there was no regular trend on trifluoroacetylation.[29] The $^{13}C$ n.m.r. spectra of chlorinated methyl α-D-aldo-hexopyranoside derivatives and the parent sugars have been measured.[30]

An n.m.r. study of ethylidene acetals of model diols and various carbohydrates has shown a relationship of a number of n.m.r. parameters to the ring size of the acetal. The most useful diagnostic is the magnitude of the $^{2}J_{C,H}$ value of the ethylidene methyl carbon and the acetal proton.[31] Investigation of the conformation of dioxolan acetals of methyl α-L-rhamno- and β-L-arabino-pyranosides has been undertaken using n.m.r. spectroscopy; methylene, isopropylidene, benzylidene, and 1-phenylethylidene acetals of the two systems were studied. In each series the pyranose ring adopts a distorted chair conformation, the greatest distortion occurring for the *endo*-phenyl benzylidene acetals. The conformations were of interest in the preferred orientation of reductive ring-opening reactions.[32,33]

The conformational equilibria of peracetylated glycals have been calculated by a molecular mechanics programme, and the data compared with those determined by *X*-ray crystallography and n.m.r. spectroscopy. Calculation and experimental results were at variance for the two pentose derivatives, but matched well for the four unsaturated hexoses. The publication included a discussion of the application of molecular mechanics to designing more potent glycosidase inhibitors.[34]

The $^{13}$C n.m.r. spectra of methyl α- and β-L-diginopyranosides (9) have been assigned and the conformation of the α-anomer and its 4-O-acetyl derivative elucidated from the data.[35] Methyl 3,6-α-D-arabino-hexopyranoside (10) has been shown to exist in the $^4C_1$(D) conformation in the crystal and in solution.[36]

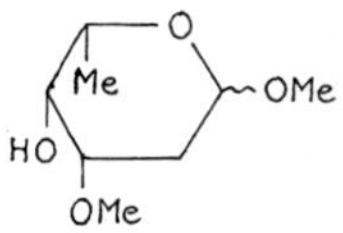
(9)

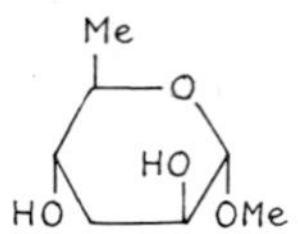
(10)

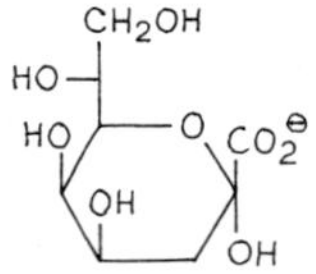
(11)

In proton-coupled $^{13}$C n.m.r. spectra of derivatives of pyranose forms of KDO (11), $^1J_{C,H}$ of C-5 is significantly larger than other $^1J$ values as expected for carbon atoms carrying axial hydroxy groups, and, in agreement with other observations, a downfield shift occurs if the 5-hydroxy group is glycosylated. Thus this coupling constant and chemical shift combined can be used to tell the position of linkage in KDO oligosaccharides.[37] The chemical shifts and coupling constants for all the protons of the galactosaminyl serine derivative (12), a model close to the glycoprotein linkages of glycoconjugates, have been determined using 2D-SECSY n.m.r. spectroscopy.[38] The same technique and 2D COSY have

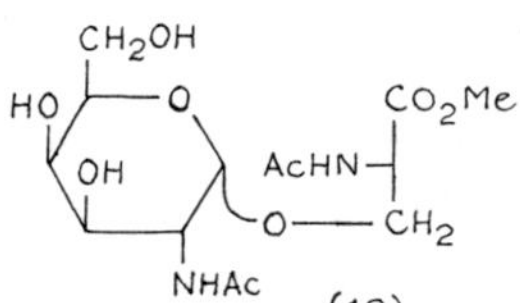
(12)

been used to determine the position of acetate and sulphate groups in glycosphingolipids.[39]

## 5 Oligosaccharides

A review on recent advances in the n.m.r. spectroscopy of natural macromolecules, including oligonucleotides, oligo- and polysaccharides with emphasis on studies related to structure, conformation, dynamics, and molecular interactions, has been published.[40] An n.m.r. pulse sequence for obtaining information on macromolecules which is essentially equivalent to that obtained with heteronuclear detection has been reported which uses a heteronuclear relay transfer technique with proton detection.[41] The type and number of isotopic multiplets for the non-anomeric secondary carbon atoms produced in $^{13}$C n.m.r. spectra of sugars with partially deuterated hydroxy groups form the basis of a new method for oligosaccharide analysis which allows determination of the sequence of glycosidic linkages.[42] A referenced computer library of high-resolution $^{1}$H n.m.r. data for oligosaccharides, determined at 500 MHz in deuterium oxide solution, has been compiled and used to analyze oligosaccharide structures and sequences found in mammalian glyco-proteins.[43]

The $^{1}$H, $^{13}$C, and $^{29}$Si n.m.r. spectra of methyl β-D-xylopyranoside and the (1→2)-, (1→3)-, and (1→4)-linked methyl β-D-xylopyranosyl-β-D-xylopyranosides have been measured and assigned with the aid of $^{1}$H-$^{1}$H, $^{1}$H-$^{13}$C, and $^{1}$H-$^{29}$Si chemical shift correlated 2D n.m.r. experiments. The sites of glycosylation can be identified by the absence of correlation of the signals for the protons concerned with coupling to the silicon.[44]

Evidence for solvent-induced conformational changes in methyl β-D-maltoside has been obtained from n.m.r. studies of the deuterated molecule. The observed values of three bond coupling constants for C(1)-H(4') and C(4')-H(1) in $D_2O$, deuterated DMSO and dioxan were compared with the calculated values based on conformational equilibria, and the results showed that the solvent plays a major role in determining the conformation at the glycosidic linkage, the differences in intramolecular hydrogen bonding in each solvent being the dominant factor.[45] The determination of long-range $^{13}$C-$^{1}$H coupling in carbohydrates by selective 2D heteronuclear J-resolved n.m.r. spectroscopy has been demonstrated for heteronuclear couplings to H-1 of methyl α-D-glucopyranoside, which are observed for every non-anomeric carbon atom in the pyranose ring, and for the $^{13}$C-$^{1}$H couplings across the glycosidic linkages in methyl β-cellobioside and maltoheptaose.[46] The use of n.O.e. has been described for determining the conformational behaviour of α-cellobiose 1-phosphate, by pre-irradiation of H-1' and observation of the effects in the signals for H-3', H-5', H-5, H-3, H-6, and H-4. Theoretical calculations on methyl α-cellobiose were also reported.[47] The same n.O.e. method was also applied to the conformational analysis of glycosyl-(1→3)-glycosides[48] and glycosyl-

(1→2)-glycosides,[49] and it was shown that each type could be arranged in two groups with characteristic $^{1}$H and $^{13}$C n.m.r. spectra depending on the anomeric configuration.

Sucrose in DMSO-$d_6$ solution is sufficiently conformationally stable, due to intramolecular hydrogen bonding, to manifest secondary isotope multiplet (SIMPLE) n.m.r. for its partially deuterated hydroxy groups. There are two main inter-residue hydrogen bonds in competition: OH-(1')---O(2) and OH-(3')----O(2) in which the former predominates, as well as weaker networks of hydrogen bonds.[50] The preferred rotamers about the C5-C6 bond have been investigated by using (6*R*) and (6*S*)-6-$^{2}H_{1}$-disaccharides. Measurements of three-bond couplings between H-5 and H-6 showed the preferred rotamers for the two anomeric methyl glycosides of β-D-Glu*p*-(1→6)-D-Glu*p*-[6-$^{2}H_{1}$], and for β-D-Gal*p*-(1→6)-β-D-Gal*p*-[6-$^{2}H_{1}$]-OMe and β-D-Gal*p*-(1→6)-β-D-Gal*p*-[6-$^{2}H_{1}$]-OMe.[51] $^{13}$C n.m.r. spectra have been obtained for all isomers of D-Glc*p*-(1→6)-D-Glc*p* and D-Glc*p*-(1→6)-D-Gal*p*, and the differences in shifts obtained with respect to the relevant monosaccharides were used to calculate the $^{13}$C n.m.r. spectra of (1→6)-linked oligo- and polysaccharides. Good agreement was obtained between the calculated and measured spectra of raffinose, stachyose, dextran, and pustulan, although temperature effects were appreciable.[52] Some monosaccharide and oligosaccharide substrate analogues of lysozyme, *vis*, D-GlcNAc, β-D-GlcNAc-(1→4)-D-GlcNAc, and β-D-GlcNAc-(1→4)-β-D-GlcNAc-(1→4)-D-GlcNAc, were investigated by $^{1}$H and $^{13}$C n.m.r. spectroscopy and their conformations deduced. Normal and 2D techniques were applied at 50° to the sugars in deuterium oxide and compared with theoretical spectra generated by conformation simulation methods. Two preferred orientations of the hydroxymethyl group were found for each saccharide, and the analysis was compared with that from *X*-ray crystal data and conformational-energy calculations.[53] The positions of polar substituents and fatty-acid chains in eight lipid A precursors from a 3-deoxy-D-mannooctulosonic acid-deficient mutant of *Salmonella typhimurium* containing β-D-GlcN$H_2$*p*(1→6)-α-D-GlcN$H_2$*p* 1,4'-diphosphate units substituted at N-2 and O-3 of each glucosamine with 3-hydroxytetra-decanoyl side chains, and aminodeoxypentose attached to the 1'-phosphate group, have been elucidated using $^{1}$H, $^{13}$C and $^{31}$P n.m.r. spectroscopy.[54] A 400 MHz $^{1}$H n.m.r. study of cellulose triacetate showed that the spectrum changed between 25° and 50°C. The phenomenon was investigated using *O*-β-D-Glc*p*(1→3)-*O*-β-D-Glc*p*(1→4)-*O*-β-D-Glc*p* undecaacetate, using computer-generated models of methyl 4,6-di-*O*-acetyl-β-D-glucopyranoside to determine the rotamer states of t

5-acetoxymethyl groups.[55] A 2D-DEPT n.m.r. experiment, adjusted for detection of $^{3}J_{CO-CH}$ long-range couplings, has been demonstrated as an alternative approach to sequencing small to medium size oligosaccharides with methyl α-L-Rha*p*(1→4)-α-L-Rha*p*-(1→2)-α-L-Rha*p* repeating units, provided the $T_2$ relaxation times and chemical shifts for the carbon atoms and protons are known.[56]

## 6 N.m.r. of Nuclei Other than $^{1}$H or $^{13}$C

2D Heteronuclear $^{29}$Si-$^{1}$H shift correlated n.m.r. spectra of trimethylsilylated sugar derivatives have permitted unambiguous assignment of the signals for the nuclei in all the TMS groups.[57] Empirical assignment rules for $^{29}$Si have been tested by using $^{29}$Si and $^{13}$C n.m.r. spectra of all pertrimethylsilylated *O*-acetyl- and *O*-benzyl 1,6-anhydro-β-D-glucopyranose derivatives. The rule based on shielding order must either be restricted to the order δ(Si-2) > δ (Si-4), which holds for all compounds studied, or modified for each sub-class of compounds with the same *O*-constituents. The rule founded on Hammett dependence gives correct predictions for Si-3 chemical shifts. When both rules can be applied they yield identical assignments; in other cases they complement each other.[58] An extension of this work has examined the assignment using $^{29}$Si satellites in $^{13}$C n.m.r. spectra measured with selective $^{29}$Si decoupling.[59]

References to the use of 2D $^{29}$Si n.m.r. spectroscopy to determine sites of glycosylation in xylopyranose oligosaccharides[44] and to $^{31}$P n.m.r. spectroscopy in oligosaccharides[54] have been made above.

## References


1 P.Kaliannan, S.Vishveshwara, and V.S.R.Rao, *Int. J. Quantum Chem.*, 1985, **27**, 181 (*Chem. Abstr.*, 1985, **102**, 185 355).

2 Q.Li, S.Zhu, and Y.Zhang, *Fenzi Kexue Yu Huaxue Yanjiu*, 1984, **4**, 419 (*Chem. Abstr.*, 1985, **102**, 20 237).

3 D.A.Pearlman and S.H.Kim, *J. Biomol. Struct. Dyn.*, 1985, **3**, 85 (*Chem. Abstr.*, 1985, **103**, 155 982).

4 E.Diez, A.L.Esteban, F.J.Bermejo, C.Altona, and F.A.A.M.De Leeuw, *J. Mol. Struct.*, 1984, **125**, 49 (*Chem. Abstr.*, 1985, **102**, 131 282).

5 F.A.A.M.De Leeuw, P.N.Van Kampen, C.Altona, E.Diez, and A.L.Esteban, *J. Mol. Struct.*, 1984, **125**, 67 (*Chem. Abstr.*, 1985, **102**, 131 283).

6 V.E.Verovskii, G.M.Lipkind, and N.K.Kochetkov, *Bioorg. Khim.*, 1984, **10**, 1680 (*Chem. Abstr.*, 1985, **102**, 204 193).

7 A.M.Bianucci, G.Catelani, F.Colonna, and L.Monti, *Carbohydr. Res.*, 1985, **140**, 144.

8 D.B.Davies and J.C.Christofides, *Stud. Org. Chem. (Amsterdam)*, 1985, 20 (Nat. Prod. Chem.), 305 (*Chem. Abstr.*, 1985, **103**, 123 810).

9 J.Reuben, *J. Am. Chem. Soc.*, 1985, **107**, 5867.

10 H.Booth and K.A.Khedhair, J. Chem. Soc., Chem. Commun., 1985, 467.
11 P.K.Agarawal, D.C.Jain, R.K.Gupta, and R.S.Thakur, Phytochemistry, 1985, 24, 2479.
12 D.B.Davies, Nucl. Magn. Reson., 1984, 13, 207 (Chem. Abstr., 1985, 102, 20 486).
13 K.Bock and C.Pedersen, Carbohydr. Res., 1985, 145, 135.
14 D.G.Gillies and D.Lewis, J. Chem. Soc., Perkin Trans. II, 1985, 1155.
15 I.M.Vazquez, N.B.D'Accorso, I.M.E.Thiel, and A.M.Schuller, An. Asoc. Quim. Argent., 1984, 72, 583 (Chem. Abstr., 1985, 103, 71 588).
16 W.Ribien, W.Schmid, and E.Zbiral, Carbohydr. Res., 1985, 141, 21.
17 M.L.Sinnott and W.S.S.Wijesundera, Carbohydr. Res., 1985, 136, 357.
18 C.Du Mortier and R.M.De Lederkremer, Carbohydr. Res., 1985, 139, 47.
19 J.Stanek Jr., J.Moravcova, and J.Jary, J. Carbohydr. Chem., 1985, 4, 79.
20 R.Csuk and N.Mueller, Congr. Ampere Magn. Reson. Relat. Phenom., Proc., 22nd, 1984, 345 (Chem. Abstr., 1985, 102, 204 212).
21 A.M.Iribarren and A.B.Pomilio, An. Asoc. Quim. Argent., 1984, 72, 573 (Chem. Abstr., 1985, 103, 54 393).
22 C.Chang, D.J.Ashworth, L.Chern, J.Da Silva Gomes, C.Lee, P.W.Mou, and R. Narayan, Org. Magn. Reson., 1984, 22, 671 (Chem. Abstr., 1985, 102, 113 855).
23 D.B.Davies, P.Rajani, M.MacCoss, and S.S.Danyluk, Magn. Reson. Chem., 1985, 23, 72 (Chem. Abstr., 1985, 102, 221 138).
24 S.G.Zavgorodnii and V.L.Florent'ev, Bioorg. Khim., 1984, 10, 1496 (Chem. Abstr., 1985, 102, 167 072).
25 M.G.Taylor, R.H.Marchessault, S.Perez, P.J.Stephenson, and C.A.Fyfe, Can. J. Chem., 1985, 63, 270.
26 Y.Asakawa, M.Toyota, and L.J.Harrison, Phytochemistry, 1985, 24, 1505.
27 D.J.Nelson and M.J.Lavin, J. Carbohydr. Chem., 1985, 4, 91.
28 L.Szilagyi and Z.Gyorgydeak, Carbohydr. Res., 1985, 143, 21.
29 M.Ranganathan and P.Balaram, Org. Magn. Reson., 1984, 22, 717 (Chem. Abstr., 1985, 102, 113 842).
30 I.A.Toufeili and S.Z.Dziedzic, Aust. J. Chem., 1985, 38, 1425.
31 T.B.Grindley and C.Wickramage, J. Carbohydr. Chem., 1985, 4, 171.
32 J.Harangi, A.Liptak, Z.Szurmai, and P.Nanasi, Carbohydr. Res., 1985, 136, 241.
33 A.Liptak, Z.Szurmai, V.A.Olah, J.Harangi, L.Szabo, and P.Nanasi, Carbohydr. Res., 1985, 138, 1.
34 W.Korytnyk and O.Dodson-Simmons, Prog. Chin. Biol. Res., 1985, 172 (Chem. Abstr., 1985, 102, 185 364).
35 S.Tsukamoto, K.Hayashi, and H.Mitsuhashi, Chem. Pharm. Bull., 1985, 33, 2294.
36 G.I.Birnbaum and D.R.Bundle, Can. J. Chem., 1985, 63, 739.
37 R.Christian, G.Schulz, and F.M.Unger, Tetrahedron Lett., 1985, 26, 3951.
38 D.Davoust, N.Platzer, C.Derappe, M.Lemonnier, B.Ferrari, and A.A.Pavia, Carbohydr. Res., 1985, 143, 233.
39 S.Gasa, M.Nakamara, A.Makita, M.Ikura, and K.Hikichi, Carbohydr. Res., 1985, 137, 244.
40 D.B.Davies, Nucl. Magn. Reson., 1985, 14, 211 (Chem. Abstr., 1985, 103, 208 967).
41 P.H.Bolton, J. Magn. Reson., 1985, 62, 143 (Chem. Abstr., 1985, 103, 47 026).
42 J.Reuben, J. Am. Chem. Soc., 1985, 107, 1747.
43 E.F.Hounsell, D.J.Wright, A.S.R.Donald, and J.Feeney, Biochem. J., 1984, 223, 129 (Chem. Abstr., 1985, 102, 24 954).
44 J.Schraml, E.Petrakova, J.Pelnar, M.Kvicalova, and V.Chvalovsky, J. Carbohydr. Chem., 1985, 4, 393.
45 S.Perez, F.Taravel, and C.Vergelati, Nouv. J. Chim., 1985, 9, 561.
46 M.J.Gidley and S.M.Bociek, J. Chem. Soc., Chem. Commun., 1985, 220.
47 G.M.Lipkind, A.S.Shashkov, and N.K.Kochetkov, Carbohydr. Res., 1985, 141, 191.
48 G.M.Lipkind and N.K.Kochetkov, Bioorg. Khim., 1984, 10, 1229 (Chem. Abstr., 1985, 102, 95 947).
49 G.M.Lipkind and N.K.Kochetkov, Bioorg. Khim., 1984, 10, 1670 (Chem. Abstr., 1985, 102, 204 187).

50 J.C.Christofides and D.B.Davies, J. Chem. Soc., Chem. Commun., 1985, 1533.

51 H.Ohrui, Y.Nishida, M.Watanabe, H.Hori, and H.Meguro, Tetrahedron Lett., 1985, 26, 3251.

52 M.Forsgren, P.-E.Jansson, and L.Kenne, J. Chem. Soc., Perkin Trans. I, 1985, 2383.

53 J.Boyd, R.Porteous, N.Soffe, and M.Delepierre, Carbohydr. Res., 1985, 139, 35.

54 S.M.Strain, I.M.Armitage, L.Anderson, K.Takayama, N.Qureshi, and C.R.H. Raetz, J. Biol. Chem., 1985, 260, 16089.

55 V.S.Rao, F.Sauriol, A.S.Perlin, and M.T.Phan Viet, Can. J. Chem., 1985, 63, 2507.

56 G.Batta and A.Liptak, J. Chem. Soc., Chem. Commun., 1985, 368.

57 E.Liepinsh, I.Sekacis, and E.Lukevics, Magn. Reson. Chem., 1985, 23, 10 (Chem. Abstr., 1985, 102, 185 356).

58 J.Schraml, S.Kucar, J.Zeleny, and V.Chvalovsky, Collect. Czech. Chem. Commun., 1985, 50, 1176.

59 J.Past, J.Puskar, J.Schraml, and E.Lippmaa, Collect. Czech. Chem. Commun., 1985, 50, 2060.

# 22
# Other Physical Methods

## 1 I.r. Spectroscopy

The spectrum of sucrose in aqueous solution has been studied to examine further the potential of F.t.i.r. spectroscopy for determining carbohydrate configurations. While bands characteristic of an α- rather than β-linked D-glucopyranosyl unit were present, no criteria for discerning the stereochemistry of the fructose unit were found.[1] Some bands in the hydroxyl stretching region of crystalline carbohydrates have been identified through a study of the relationships between bond length, hydrogen bond energies and absorption frequency.[2] Intramolecular hydrogen-bonding in the fourteen mono-, di-, and tri-O-methyl ethers of phenyl β-D-glucopyranoside in dilute carbon tetrachloride solution has been probed by i.r. spectroscopy, results being interpreted in terms of hydroxyl group conformations.[3] A near-i.r. reflectance study (750-2500 nm) of carbohydrates has included six mono- and two di-saccharides, and has revealed resonances characteristic of certain structural types. The results have been applied mainly to the analysis of polysaccharides.[4]

## 2 Mass Spectrometry

The fast-atom bombardment (FAB) m.s. of nucleosides, nucleotides, and oligonucleotides has been reviewed,[5] and the FAB mass spectra of the eight major nucleosides of RNA and DNA, pseudouridine, and 2',3'-O-isopropylideneadenosine have been compared with their e.i., c.i., and desorption c.i. spectra.[6] FAB spectra of chlorogenic acid and related quinic acid derivatives [7] and of the antibiotics everninomycin,[8] nebramycin, and lividomycin A[9] have been reported. The unimolecular dissociation (MIKE or mass ion kinetic energy) spectra of solvated and cationized molecular ions generated from underivatized aldohexoses and their methyl glycosides by the FAB technique reveal two kinds of fragment ions whose relative abundances allow anomers and stereoisomers to be distinguished.[10-12]

The use of selected matrices in molecular secondary-ion m.s. for structural characterization of naturally occurring oligosaccharides has been described.[13]

Mono-, di- and tri-saccharides appear to be volatile compounds under the conditions of field desorption (f.d.)-m.s., as evidenced by energy deficit measurements. The electric field has been shown to play an essential role in the desorption of $[M+H]^+$ ions.[14] The optimization of conditions for desorption c.i.($NH_3$)-m.s. has been studied using sucrose.[15] Fourteen flavonoid glycosides, acylated in the sugar moiety by cinnamoyl, coumaroyl, hydroxybenzoyl and galloyl groups, have been examined by f.d.-m.s., the most intense peaks corresponding to molecular, aglycone, and anhydrosugar ions.[16] The results obtained on six flavonoid glycosides and one xanthone glycoside by desorption c.i.-, FAB-, and f.d.-m.s. have been compared. Information on molecular weight, sugar sequence, and aglycone structure could be obtained, with a greater intensity of high-mass ions being generated in the negative-ion mode.[17]

Positive- and negative-ion e.i.-m.s. of a variety of peracetylated aryl β-D-glycosides has been examined.[18] The mass spectrum of levoglucosenone has been re-examined.[19] The usefulness of selected-ion monitoring in the g.c.-m.s. analysis of alditol acetates derived from sugars in complex samples has been demonstrated with the determination of muramic acid and rhamnose in bacterial peptidoglycan-polysaccharide complexes in mammalian tissue.[20] The g.c.-m.s. identification of per-O-trimethylsilylated N-(1-deoxyhexitol-1-yl)amino acids, such as those released from non-enzymically glycosylated mammalian proteins, has been studied, the fragmentation patterns permitting the amino acid moieties to be identified.[21] The c.i.- and e.i.-m.s. of peracetylated methyl 2-amino-2,4-dideoxy- and 3-amino-3,4-dideoxy-DL-pentopyranosides and the corresponding peracetylated alditols have been examined in detail,[22] while e.i.-m.s. of their underivatized N,N-dimethyl analogues has permitted a ready differentiation between the regioisomers.[23]

Peracetylated xylo-oligosaccharides have been studied by e.i. and c.i.($NH_3$) techniques, the latter giving information on the size of the oligomer.[24] Epimeric monosaccharides and their deoxy analogues could be distinguished from their relative ion intensities on trimethylsilylation g.c.-c.i.($NH_3$)-m.s.[25] The c.i.($CH_4$ or i-$C_4H_9$)-

m.s. of methyl (methyl α-D-galactopyranuronate) and its O-methyl ethers has been examined with the aid of deuterated analogues.[26] Partially acetylated methyl β-D-xylopyranosides have been examined by e.i.-, c.i.-, and collisional activation-m.s., and an unambiguous determination of the number and position of acetyl groups achieved. The structure of the ions occurring after elimination of ketene has been studied using MNDO-semiempirical quantum chemical methods.[27] The molecular mass and mass of sugar units present in several triterpenoid glycosides with xylose, glucose, and rhamnose moieties, and their peracetylated derivatives, have been determined by negative-ion c.i.($CH_4$-$CH_2Cl_2$)-m.s.[28]

A number of reports have appeared on the coupling of h.p.l.c. and m.s. systems for the analysis of sugars. The technique of thermospray ionization permits the use of standard h.p.l.c. columns with flows of 1 ml $min^{-1}$ and has been demonstrated for the analysis of adenosine and riboflavin [29] and ribonucleoside anticancer drugs and their impurities.[20,31] Thermospray ionization is soft, so that molecular ions, i.e. $[M+H]^+$ for positive-and $[M-H^-]$ for negative-ionization modes, are the only species produced. Thus, in the analysis of 3-deazauridine, additional data for identification was obtained by high-resolution direct-probe e.i.-m.s.[31]

H.p.l.c.-direct inlet positive-ion c.i.-m.s. analysis of underivatized N,O-acylated sialic acids, 2-deoxy-2,3-didehydro-N-acetylneuraminic acid, and sialyl-α(2 → 3)-lactose can differentiate sialic acid moieties (from the m.s. fragmentation) and the localization of the O-substituents (from h.p.l.c. retention time).[32] Micro-bore fused-silica column h.p.l.c.-m.s. has also been investigated. With conventional amine-bonded silica or reversed-phase columns of 0.22 mm i.d., and a direct capillary inlet, monosaccharides and cardiac glycosides gave interpretable e.i.- or c.i.-mass spectra.[33] With a 0.025 mm i.d. column and a desolvation chamber in which the mobile phase provided for chemical ionization, positive-and negative-ion c.i. spectra of sucrose, adenosine, and a flavonoid disaccharide have been recorded.[34]

## 3 X-ray Crystallography

The crystal structure of dehydro-L-ascorbic acid and the structure of its dimer in solution have been reviewed (in Japanese).[35]

Specific crystal structures have been reported as follows:

Free Sugars and Simple Derivatives Thereof.- D-Glucose 6-(sodium hydrogen phosphate),[36] dipotassium glucose 1-phosphate,[37] trisodium fructose 1,6-diphosphate,[38] 2,3:5,6-di-*O*-isopropylidene-α-D-mannofuranose,[39] and 1,2-*O*-(*S*)-(1-aminomethylethylidene)- and 1,2-*O*-(*R*)-(1-tert-butoxyethylidene)-α-D-glucopyranose.[40]

Glycosides and Derivatives Thereof.- Methyl 2,3-di-*O*-acetyl-4-*O*-trityl-β-D-xylopyranoside,[41] isobutyl 2,3,4-tri-*O*-acetyl-1-thio-β-D-xylopyranoside (an *S*-glycoside),[42] methyl tyveloside (*i.e.* methyl 3,6-dideoxy-α-D-*arabino*-hexopyranoside),[43] five cardiac glycosides with unsubstituted or 3,4-*O*-isopropylidenated β-D-digitoxopyranoside moieties,[44] 6- and 7-*O*-(β-D-glucopyranosides) of esculetin (*i.e.* 6,7-dihydroxycoumarin),[45] the secoiridoid β-D-glucopyranosides gentiopicrin hemihydrate[46] and decentapicrin A[47] (the latter having a *m*-hydroxybenzoyl substituent on *O*-3 of the sugar), 3,3,5-trimethyl-4-(3-oxo-1-butenyl)-4,5-epoxycyclohexyl 2,3,4,6-tetra-*O*-acetyl-β-D-glucopyranoside,[48] and the *C*-glycoside, 3,5-di-*O*-benzoyl-2-deoxy-β-D-*erythro*-pentofuranosylbenzene.[49]

Disaccharides and Derivatives Thereof.- α,α-Trehalose anhydrous,[50] α-lactose monohydrate (a redetermination),[51] 2-acetamido-1,3,6-tri-*O*-acetyl-2-deoxy-4-*O*-(2,3,4,6-tetra-*O*-acetyl-α-L-idopyranosyl)-α-D-glucopyranose,[52] and the (*S*)-orthoester-linked glucosylidene-mannoside (1).[53]

Anhydro-sugars.- Methyl 2,3-anhydro-4-deoxy-α-DL-*ribo*- and -α-DL-*lyxo*-hexopyranosides.[54]

Halogen-, Nitrogen-, and Sulphur-containing Compounds.- The unsaturated 5-*C*-fluoro-uronoside (2),[55] methyl and ethyl 2-azido-2-deoxy-β-D-galactopyranosides,[56] 2-[(2,2-diacetylvinyl)amino]-2-deoxy-α-D-glucopyranose,[57] methyl 4-acetamido-3-*O*-tosyl-2,4,6-tri-

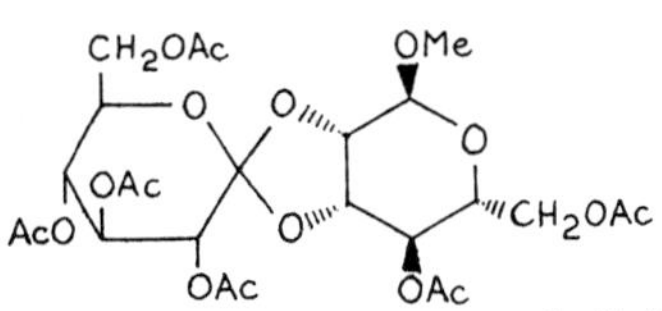

(1)

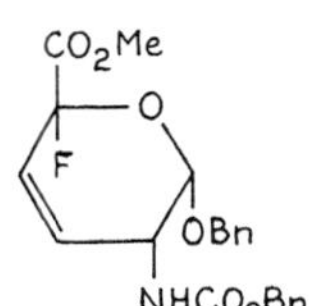

(2)

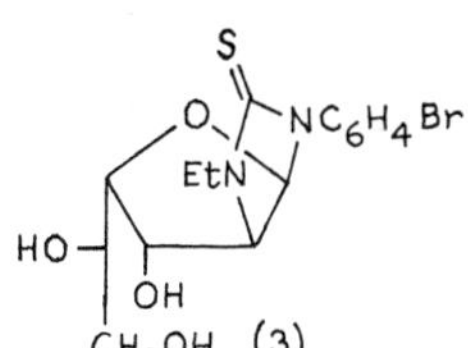

(3)

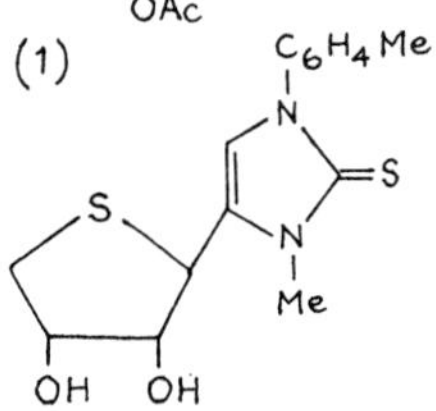

(4)

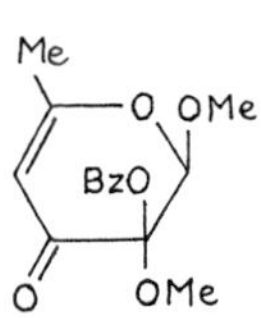

(5)

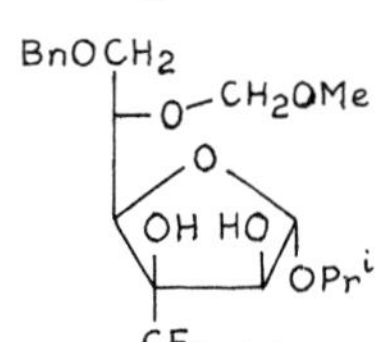

(6)

deoxy-α-DL-<u>arabino</u>-hexopyranoside,[58] 2-acetamido-3-amino-2,3-dideoxy-D-glucofuranurono-6,3-lactam,[59] β-D-galactopyranosylamine,[60] 2,3-<u>O</u>-isopropylidene-4-<u>O</u>-dimethylphosphoryl-5-<u>O</u>-trityl-D-ribononitrile,[61] di-<u>N</u>,<u>N</u>'-(2,3,4,6-tetra-<u>O</u>-acetyl-β-D-glucopyranosyl)-<u>N</u>-phenylhydrazine,[62] and the imidazole-2-thione derivatives (3)[63] and (4).[64]

<u>Unsaturated Compounds</u>.- Phenyl 3,4,6-tri-<u>O</u>-acetyl-2-deoxy-α-D-<u>threo</u>-hex-2-enopyranosylbenzene,[65] 6-<u>C</u>-(2-furyl)-1,2:3,4-di-<u>O</u>-isopropylidene-α-D-<u>glycero</u>-D-<u>galacto</u>-hexopyranose,[66] and the dihydropyran-4-one (5).[67]

<u>Branch-Chain Sugars</u>.- 1,6-Anhydro-2-deoxy-2,4-di-<u>C</u>-methyl-3-<u>O</u>-benzyl-5-<u>O</u>-mesyl-α-L-idofuranose and 1,6-anhydro-2-deoxy-2,4-di-<u>C</u>-methyl-3,4-di-<u>O</u>-benzyl-β-D-glucopyranose,[68] and the 3-<u>C</u>-trifluoromethyl derivative (6).[69]

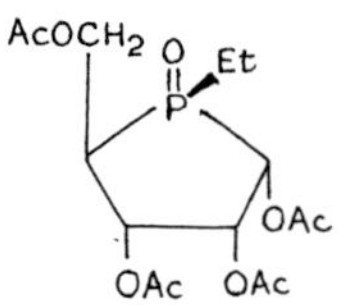
(7)

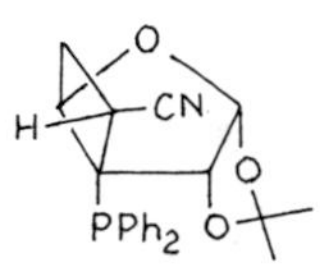
(8)

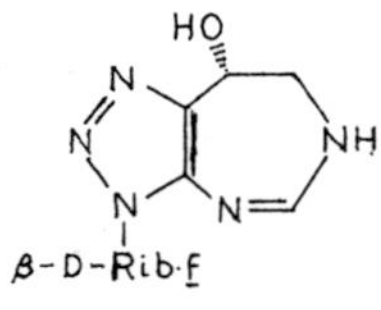
(9)

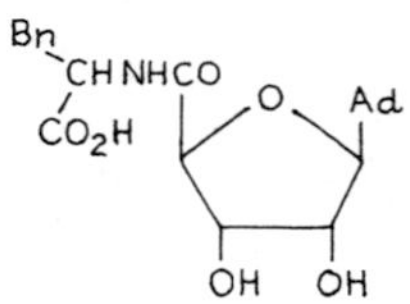
(10)

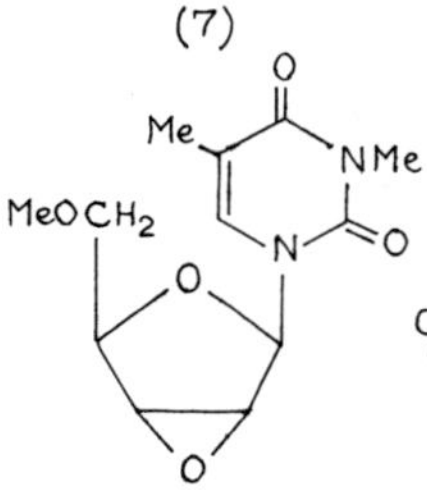
(11)

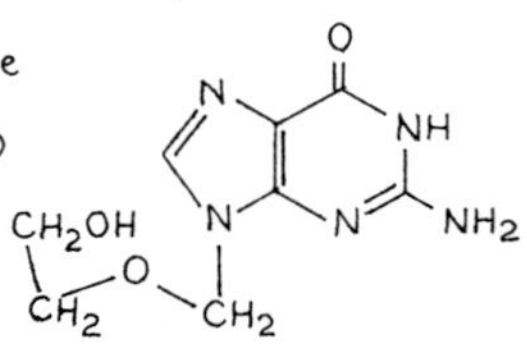
(12)

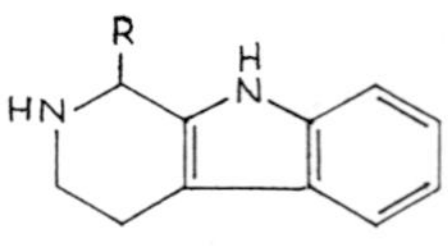
(13)

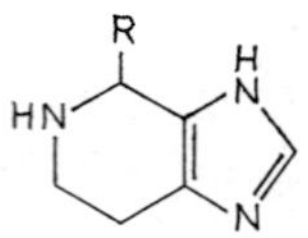
(14)

R = α-D-Ara-<u>f</u>

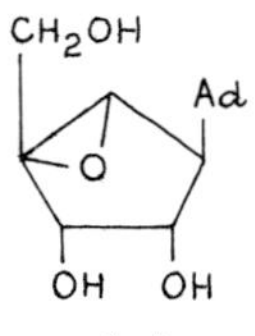
(15)

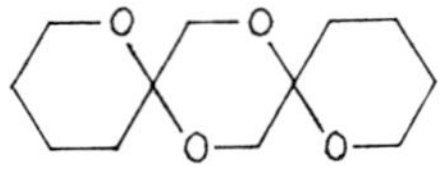
(16)

<u>Sugar Acid Derivatives</u>.- 2-<u>O</u>-Acetyl-3,4-<u>O</u>-(<u>R</u>)-benzylidene-D-ribono-1,5-lactone,[70] <u>N</u>-isopropyl-D-gluconamide and <u>N</u>,<u>N</u>-diethyl-D-gluconamide,[71] calcium D-glucarate (a redetermination),[72] and sodium α-L-guluronate dihydrate.[73]

<u>Inorganic Derivatives</u>.- The ring phosphorus sugar analogue (7),[74] the 5-diphenylphosphino-2-oxabicyclo[3.2.0]heptane-3,4-diol derivative (8),[75] and two copper-nucleotide complexes {Ca[Cu(guanosine

5'-monophosphate)$_2$(ethylenediamine)$(H_2O)_2$].$8H_2O$[76] and [Cu(benzimidazole)$(H_2O)_5$][inosine 5'-monophosphate][77]}.

Nucleosides, Nucleotides, and Derivatives.- Anhydrous xanthosine,[78] 8-methylguanosine,[79] uridine-6-thiocarboxamide and 6-cyanouridine,[80] 3'-*O*-benzoylthymidine,[81] 2',3',5'-tri-*O*-acetylguanosine,[82] 5'-*O*-(L-tryptophanyl)uridine,[83] 3-amino-1-β-D-ribofuranosyl-*s*-triazolo-[5,1-*c*]-*s*-triazole,[84] 2-azacoformycin (9),[85] the 1:1 complex of 3',5'-di-*O*-acetyl-5-bromo-2'-deoxyuridine with its 5-iodo-analogue,[86] the L-phenylalanyl D-ribofuranuronosylamide nucleoside (10),[87] the 2',3'-anhydronucleoside (11),[88] cytidine 5'-*O*-dimethyl phosphate,[89] the seco-nucleoside (without bonding between C-2' and C-3') 1-(β-D-2',3'-secoribofuranosyl)-5,6-dichlorobenzimidazole,[90] the acyclic nucleoside analogue acyclovir (12),[91] the two *C*-α-D-arabinofuranosyl nucleoside analogues (13) and (14),[92] and 5-*C*-(β-D-galactopyranosyl)-1,3-di-*N*-methylbarbituric acid.[93]

Antibiotics.- Tiazofurin N-oxide (*i.e.* 2-β-D-ribofuranosylthiazole-4-carboxamide $N^3$-oxide),[94] neplanocin C (15),[95] and an acetyl derivative of de-desosaminyl mycinamicin (which contains a 4-*O*-acetyl-6-deoxy-2,3-di-*O*-methyl-β-D-allopyranosyl moiety).[96]

Others.- The anhydride (16) formed from acidic treatment of 1,6-dihydroxy-2-hexanone (a 3,4,5-trideoxyketohexose).[97]

## 4 E.s.r. Spectroscopy

Free radicals produced by reaction of xylitol, galactitol, glucitol, and mannitol with hydroxy radicals or by direct γ-radiolysis have been studied by spin-trapping and e.s.r. spectroscopy, the spectra being analyzed on the basis of similar results from glycerol and its $d_8$ analogue. The radical species $-\dot{C}HCHO$, $-CO\dot{C}H_2$, and $-CO\dot{C}H-$ were detected, the latter in the alditols only.[98] Radicals induced by γ-irradiation of α-D-glucose, methyl α-D-glucopyranoside, and maltose in the solid state have been investigated by use of nitroxide spin-trapping. E.s.r. analysis of sugar-nitroxide solutions after h.p.l.c. separation revealed 8-12 radical species, the majority not discernable by the spin-trapping method.[99] The CIDEP (chemically induced dynamic electron polarization) technique has been used to detect the neutral ascorbate radical in photochemical reactions between ascorbic acid 6-palmitate and benzophenone.[100]

## 5 Polarimetry, Circular Dichroism, and Related Studies

The c.d. spectra of methyl β-D-xylopyranoside, methyl α-L-rhamnopyranoside, and 1,6-anhydro-β-D-glucopyranose mono- to tri-acetates have been studied, the results being discussed considering the mechanism of optical rotatory strength. The sign of rotation was influenced by solvent. With the diacetates, the main source of optical rotatory strength was due to the interaction of transitional dipole moments.[101] Temperature-dependent c.d. spectra of α- and β-isomers of 5-substituted-2'-deoxyuridines have been reported.[102]

Nakanishi and co-workers have continued to develop the exiton chirality method. The remarkable additivity relationship in the coupled c.d. of pyranose p-methoxycarbonylbenzyl ethers has permitted the identification of constituent sugars and their glycoside linkages in oligosaccharides.[103] The p-dimethylaminocinnamate group has been shown to be a useful new chromophore for this method (in which diols or β-aminoalcohols are disubstituted, and their absolute configurations determined from their c.d. spectra), as it has a strong, long-wavelength absorption (362 nm), which does not overlap with absorptions in the basic substrate. Its use was exemplified for methyl 6-O-trityl-α-D-galactopyranoside amongst other compounds.[104]

## 6 U.v. Spectroscopy

The absorption and photoluminescence spectra of cytidine, cytosine, and related nucleotides in aqueous ethyleneglycol at 77 K have been measured.[105]

## References


1 P.L.Polavarapu, S.R.Chatterjee, and D.F.Michalska, Carbohydr. Res., 1985, 137, 253.

2 V.P.Panov and R.G.Zhbankov, Dokl. Akad. Nauk SSSR, 1984, 28, 821 (Chem. Abstr., 1985, 102, 95 929).

3 A.Matsumoto and M.Takasuka, Carbohydr. Res., 1985, 137, 13.

4 R.J.Henry, Carbohydr. Res., 1985, 141, 13.

5 D.L.Slowikowski and K.H.Schram, Nucleosides Nucleotides, 1985, 4, 309.

6 D.L.Slowikowski and K.H.Schram, Nucleosides Nucleotides, 1985, 4, 347.

7 A.Sakushima, S.Hisada, S.Nishibe, and H.Brandenberger, Phytochemistry, 1985, 24, 325.

8 A.K.Ganguly, B.N.Pramanik, V.M.Girijavallabhan, O.Sarre, and P.L.Bartner, J. Antibiot., 1985, 38, 808.

9 Y.Fang, Z.Lu, and G.Lin, Huaxue Xuebao, 1985, 43, 406 (Chem. Abstr., 1985, 103, 215 686).

10 G.Puzo, J.-C.Promé, and J.-J.Fournié, Carbohydr. Res., 1985, 140, 131.

11 G.Puzo and J.C.Promé, Spectrosc.: Int. J., 1984, 3, 155 (Chem. Abstr., 1985, 102, 95 928).
12 G.Puzo, J.J.Fournié and J.C.Promé, Anal. Chem., 1985, 57, 892 (Chem. Abstr. 1985, 102, 124 858).
13 M.Suzuki, K.I.Harada, A.Tatematsu, and H.Kambara, Springer Ser. Chem. Phys., 1984, 36, 419 (Chem. Abstr., 1985, 102, 113 830).
14 D.E.C.Rogers and P.J.Derrick, Org. Mass Spectrom., 1984, 19, 490 (Chem. Abstr., 1985, 102, 79 242).
15 N.Takeda, K.Harada, M.Suzuki, and A.Tatematsu, Org. Mass Spectrom., 1985, 20, 236 (Chem. Abstr., 1985, 103, 104 346).
16 G.G.Zapesochnaya, A.N.Stepanov, A.A.Perov, and S.Z.Ivanova, Khim. Prir. Soedin., 1984, 582 (Chem. Abstr., 1985, 102, 149 659).
17 B.Domon and K.Hostettmann, Phytochemistry, 1985, 24, 575.
18 Yu.N.El'kin, V.G.Voinov, and Ya.V.Voznyi, Khim. Prir. Soedin., 1984, 623 (Chem. Abstr., 1985, 102, 79 234).
19 Y.Halpern and J.P.Hoppesch, J. Org. Chem., 1985, 50, 1557.
20 R.S.Whiton, P.Lau, S.L.Morgan, J.Gilbart, and A.Fox, J. Chromatogr., 1985, 347, 109.
21 D.J.Walton, W.A.Szarek, D.B.Maclean, and R.V.Gerard, Carbohydr. Res., 1985, 137, 31.
22 J.H.Banoub, F.Michon, R.Roy, A.Grouiller, and H.Bazin, Carbohydr. Res., 1985, 144, 127.
23 E.Baranowska, A.Banaszek, and A.Zamojski, Pol. J. Chem., 1984, 58, 439.
24 S.Sun, Y.Long, X.Wang, K.Huang, R.Li and D.Lu, Huaxue Xuebao, 1984, 42, 880 (Chem. Abstr., 1985, 102, 113 822).
25 T.Murata, Bunseki Kagaku, 1985, 34, 414 (Chem. Abstr., 1985, 103, 160 774).
26 V.I.Kadentsev, I.A.Trushkina, O.S.Chizhov, V.I.Grishkovets, A.E.Zemlyakov, and V.Ya.Chirva, Bioorg. Khim., 1984, 10, 1242 (Chem. Abstr., 1985, 102, 95 976).
27 V.Kovacik, E.Petrakova, V.Mihalov, I.Tvaroska, and W.Heerma, Biomed. Mass Spectrom., 1985, 12, 49 (Chem. Abstr., 1985, 103, 37 662).
28 G.Fu and Y.Xu, Huaxue Xuebao, 1984, 42, 1157 (Chem. Abstr., 1985, 102, 204 184).
29 J.R.Chapman, J. Chromatogr., 1985, 323, 153.
30 R.D.Voyksner, J.T.Bursey, and J.W.Hines, J. Chromatogr., 1985, 323, 383.
31 R.D.Voyksner, F.P.Williams, and J.W.Hines, J. Chromatogr., 1985, 347, 137.
32 A.K.Shukla, R.Schauer, U.Schade, H.Moll, and E.Th.Rietschel, J. Chromatogr. 1985, 337, 231.
33 H.Alborn and G.Stenhagen, J.Chromatogr., 1985, 323, 47.
34 P.Hirter, H.J.Walther, and P.Dätwyler, J. Chromatogr., 1985, 323, 89.
35 M.Ohmori, Seikatsu Eisei, 1985, 29, 105 (Chem. Abstr., 1985, 102, 209 200).
36 T.Lis, Carbohydr. Res., 1985, 135, 187.
37 N.Narendra and M.A.Viswamitra, Curr. Sci., 1984, 53, 1018 (Chem. Abstr., 1984, 101, 238 417).
38 N.Narendra, T.P.Seshadri, and M.A.Viswamitra, Acta Crystallogr., Sect. C, 1985, 41, 31.
39 B.Sheldrick, W.Mackie and D.Akrigg, Acta Crystallogr., Sect. C, 1985, 41, 431.
40 F.H.Cano, C.Foces-Foces, A.Alemany, M.Bernabe, M.L.Jimeno, and M.Martin-Lomas, Carbohydr. Res., 1985, 139, 65.
41 L.G.Vorontsova, M.O.Dekaprilevich, and O.S.Chizhov, Izv. Akad. Nauk SSSR, Ser. Khim., 1985, 1563 (Chem. Abstr., 1985, 103, 151 272).
42 G.A.Jeffrey, J.R.Ruble, and B.Sepehrnia, Carbohydr. Res., 1985, 144, 197.
43 G.I.Birnbaum and D.R.Bundle, Can. J. Chem., 1985, 63, 739.
44 D.C.Rohrer, M.Kihara, T.Deffo, H.Rathore, K.Ahmed, A.H.L.From, and D.S.Fullerton, J. Am. Chem. Soc., 1984, 106, 8269.
45 K.Ueno, M.Shiraki, M.Sato, and N.Saito, Bull. Chem. Soc. Jpn., 1985, 58, 230.
46 B.Kojic-Prodic, A.L.Spek, W.G.Van Der Sluis, and R.P.Labadie, Acta Crystallogr., Sect. C, 1985, 41, 798.
47 B.Kojic-Prodic, A.L.Spek, W.G.Van Der Sluis and R.P.Labadie, Acta Crystallogr., Sect. C, 1985, 41, 379.

48 F.Pavelcik, K.Havetta, and V.Suchy, Acta Crystallogr., Sect. C, 1985, 41, 1272.
49 J.Gunning, S.Neidle, T.A.Millican, M.A.W.Eaton, G.A.Mock, and J.Mann, Acta Crystallogr., Sect. C, 1985, 41, 720.
50 G.A.Jeffrey and R.Nanni, Carbohydr. Res., 1985, 137, 21.
51 J.H.Noordik, P.T.Beurskens, P.Bennema, R.A.Visser, and R.O.Gould, Z. Kristallogr., 1984, 168, 59 (Chem. Abstr., 1985, 102, 195 586).
52 A.Neuman, J.Becquart, D.Avenel, H.Gillier-Pandraud, and P.Sinaÿ, Carbohydr. Res., 1985, 139, 23.
53 K.Asano, S. Horito, A.Saito, J.Yoshimura, and K.Ueno, Carbohydr. Res., 1985, 138, 1.
54 J.W.Krajewski, P.Gluzinski, Z.Urbaczyk-Lipkowska, A.Banaszek, L.Parkanyi, and A.Kalman, Carbohydr. Res., 1985, 144, 13.
55 J.Kiss, P.C.Wyss, G.Flesch, W.Arnold, K.Noack, and P.Schönholzer, J. Carbohydr. Chem., 1985, 4, 347.
56 N.Darbon, Y.Oddon, E.Guy, B.Ferrari, A.A.Pavia, G.Pepe, and J.P.Reboul, Acta Crystallogr., Sect. C, 1985, 41, 1100.
57 M.J.Dianez, A.Lopez-Castro, and R.Marquez, Acta Crystallogr., Sect. C, 1985, 41, 149.
58 J.T.Welch and B.-M.Svahn, J. Carbohydr. Chem., 1985, 4, 421.
59 N.Sugawara, H.Iwasaki, N.Suzuki, and S.Okuda, Carbohydr. Res., 1985, 142, 195.
60 C.T.Samudzi, J.R.Ruble, G.A.Jeffrey, Carbohydr. Res., 1985, 142, 39.
61 R.Meuwly and A.Vasella, Helv. Chim. Acta, 1985, 68, 997.
62 B.K.Sadybakasov, V.A.Afanas'ev, M.Yu.Antipin, and Yu.T.Struchkov, Zh. Strukt. Khim., 1984, 25, 104 (Chem. Abstr., 1985, 102, 185 378).
63 C.F.Conde, M.Millan, A.Conde, and R.Marquez, Acta Crystallogr., Sect. C, 1985, 41, 277.
64 E.Moreno, S.Perez, A.Lopez-Castro, and R.Marquez, Acta Crystallogr., Sect. C, 1985, 41, 1465.
65 H.Gillier-Pandraud, R.Brahmi, V.Bellosta-Dechavanne, and S.Czernecki, Can. J. Chem., 1985, 63, 491.
66 J.W.Krajewski, P.Gluzinski, Z.Urbanczyk-Lipkowska, A.Zamojski, and P.Luger, Carbohydr. Res., 1985, 139, 55.
67 F.W.Lichtenthaler, S.Nishiyama, P.Köhler, and H.J.Lindner, Carbohydr. Res., 1985, 136, 13.
68 L.G.Vorontsova, M.O.Dekaprilevich, and O.S.Chizhov, Izv. Akad. Nauk SSSR, Ser. Khim., 1985, 1568 (Chem. Abstr., 1985, 103, 151 273).
69 K.Kawada, O.Kitagawa, T.Taguchi, Y.Hanzawa, Y.Kobayashi, and Y.Iitaka, Chem. Pharm. Bull., 1985, 33, 4216.
70 N.Baggett, J.G.Buchanan, M.Y.Fatah, C.H.Lachut, K.J.McCullough, and J.M.Webber, J. Chem. Soc., Chem. Commun., 1985, 1826.
71 N.Darbon-Meyssonnier, Y.Oddon, E.Decoster, A.A.Pavia, G.Pepe, and J.P.Reboul, Acta Crystallogr. Sect. C, 1985, 41, 1324.
72 C.Burden, W.Machie, and B.Sheldrick, Acta Crystallogr., Sect. C, 1985, 41, 693.
73 F.Mo, T.J.Brobak, and I.R.Siddiqui, Carbohydr. Res., 1985, 145, 13.
74 P.Luger, E.Müller, H.Yamamoto, and S.Inokawa, Carbohydr. Res., 1985, 145, 25.
75 S.Saito, T.Kato, Y.Morita, and H.Miyamae, Chem. Pharm. Bull., 1985, 33, 3092.
76 S.Mangani and P.Orioli, J. Chem. Soc., Chem. Commun., 1985, 780.
77 M.D.Poojary, N.S.Begum, H.Manohar, and R.Bau, J. Chem. Soc., Chem. Commun., 1985, 821.
78 B.Lesyng, C.Marck, and W.Saenger, Z. Naturforsch., C: Biosc., 1984, 39, 720 (Chem. Abstr., 1985, 102, 24 975).
79 K.Hamada, I.Honda, S.Fujii, T.Fujiwara, and K.-i.Tomita, Acta Crystallogr., Sect. C, 1985, 41, 1486.
80 V.Cody and T.I.Kalman, Nucleosides Nucleotides, 1985, 4, 587.
81 Y.Yamagata, H.Miyashiro, K.Tomita, T.Ishida, and M.Inoue, Chem. Pharm. Bull., 1985, 33, 455.

82 C.C.Wilson, J.N.Low, and P.Tollin, Acta Crystallogr., Sect. C, 1985, 41, 1123.
83 I.Saito, H.Sugiyama, T.Matsuura, and K.Fukuyama, Tetrahedron Lett., 1985, 26, 4467.
84 S.G.Wood, N.K.Dalley, R.D.George, R.K.Robins, and G.R.Revankar, Nucleosides Nucleotides, 1984, 3, 187 (Chem. Abstr., 1985, 102, 167 100).
85 B.I.Birnbaum, J.R.Brisson, S.H.Krawczyk and L.B.Townsend, J. Am. Chem. Soc., 1985, 107, 7635.
86 C.C.Wilson and P.Tollin, Acta Crystallogr., Sect. C, 1985, 41, 465.
87 F.Takusagawa, H.M.Berman, M.Dabrow, and R.K.Robins, Acta Crystallogr., Sect. C, 1985, 41, 400.
88 M.Marton-Meresz, J.Kuszmann, J.Lango, L.Parkanyi, and A.Kalman, Nucleosides Nucleotides, 1984, 3, 221 (Chem. Abstr., 1985, 102, 204 219).
89 R.G.Brunnan, N.S.Kondo, and M.Sundaralingam, Nucleic Acids Res., 1984, 12, 6813 (Chem. Abstr., 1985, 102, 62 548).
90 G.I.Birnbaum, R.Stolarski, Z.Kazimierczuk, and D.Shugar, Can. J. Chem., 1985, 63, 1215.
91 G.I.Birnbaum, M.Cygler, and D.Shugar, Can. J. Chem., 1984, 62, 2646.
92 I.M.Piper, D.B.MacLean, R.Faggiani, C.J.L.Lock, and W.A.Szarek, Can. J. Chem., 1985, 63, 2915.
93 M.Millan, C.F.Conde, A.Conde, and R.Marquez, Acta Crystallogr., Sect. C, 1985, 41, 274.
94 N.K.Dalley, C.R.Petrie, III, G.R.Revankar, and R.K.Robins, Nucleosides Nucleotides, 1985, 4, 651.
95 K.Kinoshita, S.Yaginuma, M.Hayashi, and K.Nakatsu, Nucleosides Nucleotides, 1985, 4, 661.
96 K.Kinoshita, S.Satoi, M.Hayashi, K.-I.Harada, M.Suzuki, and K.Nakatsu, J. Antibiot., 1985, 38, 522.
97 W.A.Szarek, O.R.Martin, R.J.Rafka, and T.S.Cameron, Can. J. Chem., 1985, 63, 1222.
98 M.Kuwabara, Z.Y.Zhang, O.Imanami, and G.Yoshii, Radiat. Phys. Chem., 1984, 24, 217 (Chem. Abstr., 1985, 102, 204 199).
99 J.P.Agnel, C.Thiery, C.Battesti, P.Vincent, and J.Raffi, Anal. Lett., 1985, 18, 1013 (Chem. Abstr., 1985, 103, 142 289).
100 M.T.Craw, M.C.Depew, and J.K.S.Wan, J. Am. Chem. Soc., 1985, 107, 1084.
101 S.Bystrický, T.Sticzay, S.Kučár, E.Petrakova, and R.Toman, Collect. Czech. Chem. Commun., 1985, 50, 683.
102 Yu.Yu.Volodin, B.S.Kikot, V.M.Lobachev, and Yu.S.Mordashev, Biofizika, 1984, 29, 725 (Chem. Abstr., 1985, 102, 132 404).
103 K.Nakanishi, M.Kuroyanagi, H.Nambu, E.M.Oltz, R.Takeda, G.L.Verdine, and A.Zask, Pure Appl. Chem., 1984, 56, 1031 (Chem. Abstr., 1985, 102, 79 241).
104 G.L.Verdine and K.Nakanishi, J. Chem. Soc., Chem. Commun., **1985**, 1093.
105 V.L.Rappoport and V.M.Bakulev, Lyumin. Anal. Med.-Biol. Issled., 1983 237 (Chem. Abstr., 1985, 102, 95 982).

# 23
# Separatory and Analytical Methods

## 1 Chromatographic Methods

General.- It has been noted that the presence of TRIS [2-amino-2-(hydroxymethyl)propane-1,3-diol] may interfere with the analysis of amino-sugars, as in the examination of enzyme-hydrolyzed glycoproteins in TRIS-buffer. It behaves like an amino-sugar or an amino acid in many systems, *e.g.* paper chromatography, paper electrophoresis, and with ninhydrin reagent. In the analysis of alditol acetate derivatives by g.c., it forms a tetraacetate which elutes with the sugar peaks.[1] Methods for assaying streptomycin and kanamycin antibiotics in biological samples have been covered in a review on antituberculosis drug analyses.[2]

A new procedure developed for characterizing small amounts (0.5-1.0 nmoles) of oligosaccharides involves: i) conversion to u.v.-absorbing *N*-aryl-1-amino-1-deoxy-alditol derivatives by reductive amination with aniline derivatives, and separation by h.p.l.c. on the basis of size and sugar composition; ii) treatment with *exo*-glycosidases to determine sequence and anomeric configuration (the derivatives still being substrates); iii) analysis of the derivatives by FAB-m.s., which is facilitated by the potential positive charge; and iv) analysis by permethylation and capillary g.c.-m.s.[3]

Gas-Liquid Chromatography.- The g.c. analysis of monosaccharide mixtures has been the subject of a review with 222 references.[4]

In connection with the analysis of oligo- and polysaccharides by the permethylation - "ionic hydrogenation" procedure, a series of partially methylated anhydro-D-alditol acetates has been prepared by reductive cleavage of permethyl and partially methylated methyl glycosides and a library of capillary g.c. retention times and e.i.- and c.i.-m.s. data compiled. The usefulness of this data was demonstrated in the analysis of a microgram quantity of a bacterial membrane oligosaccharide.[5]

The substitution of sodium acetate or N-methyl-imidazole for pyridine in the preparation of alditol acetate derivatives for the capillary g.c. analysis of neutral and amino-sugars using selected ion monitoring m.s. detection has been examined. While N-methyl-imidazole effected full acetylation of alditols in the presence of borates and water, it led to a higher detector background than the more tedious use of sodium acetate. Muramic acid produced two alditol acetate derivatives, the lactams (1) and (2), in a reproduceable fashion only in the sodium acetate-catalyzed procedure

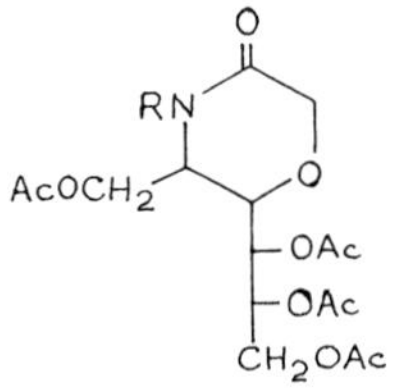

(1) R=H

(2) R=Ac

as exemplified by the g.c.-m.s. analysis of muramic acid and rhamnose in bacterial peptidoglycan-polysaccharide complexes in mammalian tissue.[6] In the analysis of glycoproteins using alditol acetate derivatives, the initial hydrolysis with acid-form resin in the presence of free aqueous acid has been found to cause partial degradation of sugars, despite earlier claims to the contrary.[7]

Mixtures containing nearly all possible partially methylated alditol acetates have been prepared as standards for the methylation analysis of cell-wall polysaccharides. Incomplete Hakamori methylation (i.e. limiting amounts of potassium dimsyl) of neutral monosaccharides with purification of the derivatives on a $C_{18}$ reversed-phase column was followed by standard hydrolysis, reduction, and acetylation. Relative capillary g.c. retention data was reported.[8] By utilizing three capillary columns with different phase polarity, all cases of co-chromatography in the g.c. analysis of partially methylated alditol acetates derived from the neutral sugar units in plant cell walls have been resolved.[9] Retention data for the same sugar derivatives on a low-polarity capillary column have also been detailed.[10] The advantageous use of lithium dimsyl (from n-BuLi) in the Hakamori permethylation procedure is covered in Chapter 5.

A new procedure for the quantitative analysis of the carbohydrate constituents of glycoproteins involves: i) simultaneous action of neuraminidase and neuraminic acid aldolase; ii) hydrolysis (4 M $CF_3CO_2H$, 125°, 1 h); and iii) capillary g.c. analysis of the derived O-methyloxime peracetates. Since a complete separation of all the O-methyloximes and related alditol acetate derivatives was achieved, the carbohydrate chains released from glycoproteins by alkaline borohydride treatment could be conveniently examined.[11]

N-Methyl-bis(trifluoroacetamide) has been recommended as a suitable reagent for trifluoroacetylation of carbohydrates for g.c. analysis, permitting direct injection of the reaction mixture. Its use was demonstrated in the analysis of cyclitols, hexoses, alditols, and oligosaccharides up to DP 6.[12] The capillary g.c. separation of thirteen trifluoroacetylated trisaccharides has been reported, those with reducing sugar moieties giving two peaks. Those with (1 → 6)-links eluted quickly while those comprised of pentose units (e.g. xylotriose) were strongly retarded, the opposite of their behaviour as trimethylsilyl ether derivatives.[13] The 5-mononitrate of 1,4:3,6-dianhydro-D-sorbitol (i.e. isosorbide), an active metabolite of the vasodilator isosorbide 2,5-dinitrate, has been determined in plasma by g.c. analysis of its per-trifluoroacetate.[14]

O-Cinnamoylation has been suggested as a sugar derivatization procedure since the derivatives can be formed quantitatively and are hydrolytically stable at pH 4-8. It was applied to the trace analysis of 3-methylthymidine (at 50 ng) as its 3',5'-diester by g.c. with electron-capture detection.[15]

A new silylating reagent, "trimethylsilyl-N,N-dimethylcarbaminate", has been shown to be superior to the previous best reagent (BSTFA) for derivatizing ribonucleosides for g.c. It can be used as the solvent for derivatization (at 60°C, 0.5 h).[16] The g.c. separation of naturally occurring inositols and their mono-O-methyl ethers following per-trimethylsilylation or per-acetylation has been reported.[17] The products from oxidative and non-oxidative degradation of D-galacturonic acid in alkali have been identified by trimethylsilylation-g.c.-m.s. examination. Thirteen hydroxy-monocarboxylic acids and -2,6-dicarboxylic acids were identified.[18]

The products from periodate oxidation of some mono-O-isopropylidene derivatives of aldoses and alditols, particularly triose and tetrose fragments, have been identified by g.c.-m.s.[19] The optical purity of 2,3-O-isopropylidene-D-glyceraldehyde obtained from D-mannitol has been determined with great accuracy to be >99.4%, by g.c. analysis after conversion to the (R)-α-trifluoro-α-methoxyphenylacetate. The L-enantiomer derived from L-serine was 95% optically pure.[20]

Thin-Layer Chromatography.- A separation of 2-deoxy-2-fluoro-D-mannose and -D-glucose on sodium dihydrogen phosphate-impregnated silica has been developed.[21] Adenosine, its cyclic and non-cyclic nucleotides, and related nucleosides and bases have been separated on silica gel.[22] The t.l.c. and h.p.t.l.c. behaviour of 18 carbohydrates on silica gel has been reported.[23] Impressive separations of homologous series of D-gluco-oligosaccharides (DP 1 to 20-30) obtained from partial hydrolysis of cyclosophoraose (a cyclic 1,2-β-linked glucan), amylose, dextran, and luteose (a 1,6-β-linked glucan) have been achieved on various silica gel layers.[24] The contributions of 2-acetamido-2-deoxy-D-glucose and glucuronic acid moieties to the mobility of hyaluronate oligosaccharides on silica gel have been determined and compared with the results of an earlier h.p.l.c. study (Vol.17, p.234, ref.46). The mobility of a given oligomer could be expressed in terms of the additive contributions of these constituent units, with the exception of glucuronic acid units in the t.l.c. assay.[25]

High-Pressure Liquid Chromatography.- A new europium-impregnated calcium fluoride scintillation detector has been developed to quantitatively detect $^{14}$C-radiolabel in aqueous h.p.l.c. eluant and applied to the determination of labelled free sugars (separated on amine-bonded silica) in a study of $^{14}CO_2$-photosynthetic fixation and metabolism in green plants.[26] A high-stability u.v. light source, based on the u.v. luminescence of crystals stimulated by energetic electrons, has been incorporated into a novel detector, and demonstrated for the detection of nucleosides eluting from a reversed-phase column. A detection limit of 3-30 ng was achieved in this preliminary study.[27]

The h.p.l.c. of carbohydrates has been reviewed.[28] The specific application of h.p.l.c. to the analysis of sucrose, fructose, glucose, and polysaccharides in sugar cane and related commercial extracts has been detailed.[29]

A study of the mechanism by which carbohydrates are retained on silica and amine-bonded silica stationary phases with aqueous eluants has confirmed previous findings that the carbohydrates are partitioned into a stagnant, water-rich phase associated with the stationary phase (c.f. Vol.16, p.253). Differing retentions of sugars could be explained in terms of specific sugar-hydroxyl hydration, e.g. secondary-equatorial hydroxy groups are more strongly hydrated than primary or secondary-axial hydroxy groups.[30]

Reducing-sugars, separated on amine-bonded silica, have been determined by post-column reaction with copper bis(phenanthroline) and amperometric detection (in which the reduced copper complex is re-oxidized). The method was compatible with organic eluants, could detect down to 1 ng of glucose, and was applied without problem to the analysis of sugars in fruit juices and wines. It was claimed to be more sensitive than fluorimetric detection.[31] Reasonably rapid separations of complex mixtures of mono- and di-saccharides have been achieved on a novel, more stable, amino-bonded silica phase, with a selective post-column colorimetric detection.[32] 2-Deoxy-D-glucose (an experimental antiviral agent!) has been determined in topical formulations by reversed-phase h.p.l.c. with u.v. detection at 195 nm.[33]

Separations of mono- and di-saccharides on a novel copper silicate gel [34] and silica gel modified with copper(II)-ammonia complex in the eluant [35] have been reported.

The retention of sugars, sugar alcohols, amino acids, and hydroxyacids on polystyrene-based cation-exchange resins has been examined as a function of the counter-ion. Retention of polyhydroxylic compounds decreased in the order K>Na>Li, due to decreasing retention of "free" water in the resin. While the $Ca^{2+}$-form resin retained even less "free" water, polyols and ribose were strongly retained due to complex formation. It was confirmed that the best separation of hexoses and pentoses is obtained with the $Ca^{2+}$-form resin, while for sugar alcohols and polyols, the $La^{3+}$-form was best.[36] Specific applications of cation-exchange resin columns include the determination of low levels of monosaccharides, alditols, and cyclitols in sheep plasma on a $Ca^{2+}$-form column with u.v. detection at 190 nm, an extension of this involving collection of eluate fractions to monitor the fate of infused radioactive glucose and fructose,[37] and the analysis of sugars, fermentation alcohols, and the illegal additive diethylene glycol in wine and grape-juice on an $H^{+}$-form column, or on $Ca^{2+}$- and $Pb^{2+}$-form resins connected in series using water as eluant.[38]

Two enzyme-based methods and one h.p.l.c. procedure have been assessed for the analysis of fructose and glucose in cane juice and synthetic sucrose solutions.[39]

Micro-bore h.p.l.c. columns (25 or 220 μm i.d. fused silica) with reversed-phase or primary amine-bonded silica packings have been coupled to mass spectrometers and used for the analysis of monosaccharides, cardiac glycosides, and related molecules.[40,41] (See also refs. 84 and 85).

Dansyl hydrazone (i.e. 5-dimethylamino-1-naphthalenesulphonic hydrazone) derivatives have been used for the reversed-phase analysis of neutral sugars in glycoprotein hydrolyzates, with the single peaks per sugar being detected at 254 nm,[42] and of glucose in small volumes of body fluids with fluorimetric detection.[43] Fructose, glucose, and sorbitol in lens tissue have been determined as their per-nitrobenzoate derivatives on silica gel (c.f. Vol.18, p.240, ref.28).[44]

Water-soluble (sugars and aliphatic acids) and lipophilic [furfural and 5-(hydroxymethyl)furfural] organics in wood hydrolyzates have been analyzed in a single run using a column-switching procedure. The latter, long retention components were separated on a short $H^+$-form strong cation-exchange resin column, while the former passed through both this and a longer similar resin column.[45] Typical cellulose and biomass hydrothermolysis products, which contain cello-oligosaccharides, xylose, fructose, arabinose, 1,6-anhydro-β-D-glucose, dihydroxyacetone, 5-(hydroxymethyl)furfural, furfural, and ethanol, could be separated satisfactorily on series-coupled columns of $Ca^{2+}$- or $H^+$-form and $Ag^+$-form polystyrene-based ion-exchange resin columns, using water at 85-95°C as eluant.[46]

Extensive investigations into the separations of oligosaccharides on amine-bonded silica have been reported, the influence of specific structural features on chromatographic properties being probed. From a study of 65 neutral oligosaccharides, it was concluded that the most important features were: i) retention increased with the number of sugar residues, ii) the presence of fucose or 2-acetamido-2-deoxy-D-glucose residues reduced retention, and iii) (1→6)-linkages, even as branching, caused a dramatic decrease in retention.[47] Impressive separations of homologous series of D-gluco-oligosaccharides with 1,2-β-, 1,4-α-, 1,6-α-, or 1,6-β-linkages, and DP 1-35, have been reported.[24] The h.p.l.c. of 20 di- and 13 tri-saccharides has also been examined.[48,49] The separation of oligosaccharides, particularly starch-derived oligoglucosides of DP up to 8, on a reversed-phase silica packing doped with some aminopropyl-bonded groups, and using water as eluant, has been examined. The amine groups catalyzed anomerization so that each oligosaccharide gave a single peak, yet sugars did not appear to be lost due to glycosylamine formation since the column material had a near-neutral pH.[50]

Oligosaccharide moieties of glycoproteins have been converted into 1-deoxy-1-(2-pyridylamino)alditol derivatives by reductive amination and analyzed by h.p.l.c. on reversed-phase[51] and amine-modified silica[52] packings. The efficiency of this reductive amination was investigated with lactose, 2-acetamido-2-deoxy-D-glucose, 2,5-anhydromannose, and with 2-amino-2-deoxy-D-glucose hydrazone (a model for oligosaccharide hydrazones released by hydrazinolysis). In the latter case prior N-acetylation and hydrazone hydrolysis ($H^+$-form resin) permitted efficient reductive amination.[52]

Reversed-phase analyses of bile acid 3-glucuronides [53,54] and of 1-(β-D-glucopyranosyl)phenobarbital, a major urinary metabolite of phenobarbital,[55] have been reported. A procedure developed for assaying glucosinolates in plant material involved absorption onto an anion-exchanger, enzymic desulphation, and elution with water followed by derivatization-g.c. or preferably h.p.l.c. (reversed-phase) analysis of the resulting desulphoglucosinolates.[56]

In connection with the catalytic oxidation ($Pt-O_2$) of glucose-1-phosphate, an ion-pair reversed-phase analysis of glucose- and glucuronic acid-1-phosphates and related hydrolysis and oxidation products has been developed.[57] The same group have used "ion-moderated partitioning" (or IMP) chromatography performed on an $H^+$-form polystyrene-based cation-exchange resin with aqueous trifluoroacetic acid as eluant, for the separation of glucose-1-phosphate, fructose, sucrose, and inorganic phosphate.[58]

Twenty-one sugar acids, lactones, and N-acetylated amino-sugars have also been examined on an $H^+$-form cation-exchanger, with dilute sulphuric acid as eluant. The separations effected by this IMP chromatography (see above) appear to be due to a combination of ion- and size-exclusion mechanisms. A number of useful, rapid (<15 min ) separations were reported, particularly of closely related uronic acids and their lactones, and u.v. detection could be employed. Applications in the analysis of polygalacturonic acid hydrolyzates and for monitoring bacteria for 2-ketogluconic acid production were detailed.[59]

2-Amino-2-deoxy-glucose and -galactose and their corresponding alditols have been assayed down to a limit of 5 pmol, by reversed-phase h.p.l.c. of their N-(4-nitrobenzo-2-oxa-1,3-diazol-7-yl)

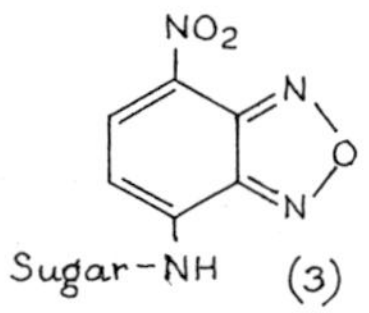

(3)

derivatives (e.g. 3) with fluorescence detection.[60] 3-Deoxy-D-manno-octulosonic acid, N-acetylneuraminic acid, and related derivatives have been determined by h.p.l.c. on a strong anion-exchange resin column [61] and their identification by h.p.l.c.-direct inlet c.i.-m.s. examined.[62]

Ascorbic acid, dehydroascorbic acid, diketogluconic acid, and glucose have been determined down to nanogram levels by h.p.l.c. on a hydrophilic, polyvinylalcohol gel, post-column derivatization with benzamidine, and fluorimetric detection. The method was applied to ascorbic acid in fruit juice.[63] Ascorbic acid, its 2-sulphate, 2-phosphate, 2-O-methyl, and 6-bromo-6-deoxy derivatives, and a related non-enzymic browning product, 5-methyl-3,4-dihydroxy-tetrone, have been determined by reversed-phase h.p.l.c. with electrochemical detection.[64] The sum of ascorbic and dehydro-ascorbic acids in 2 µl samples of human blood has been measured by derivatization with DPQ [6,7-dimethoxy-3-propyl-2(1H)-quinoxalinone], separation on a reversed-phase phenyl column, and fluorimetric detection.[65]

Analysis of aminoglycoside antibiotics by immunoassay and h.p.l.c. methods has been covered in a review on antibiotic monitoring in body fluids.[66] Since N-(2,4-dinitrophenyl) derivatives of amino-sugars are useful in reversed-phase h.p.l.c. analysis, the preparation of the per-N-(2,4-dinitrophenyl) derivative of tobramycin using 1-fluoro-2,4-dinitrobenzene in aqueous acetonitrile has been studied in detail, conditions for its preparation in good yield being elucidated. A sharp pH optimum was observed, since even under moderate alkaline conditions O-substitution became an important side-reaction. Buffers composed of tertiary amines and hydrochloric acid were preferred, because phosphate and phthalate buffers reacted with the derivatization reagent.[67] The following specific assays of antibiotics have been reported: erythromycin and related materials on a polystyrene-based packing,[68] the amino-glycosides kanamycin, streptomycin, dihydrostreptomycin, and particularly neomycin as per N,O-benzoylated derivatives on silica gel,[69] spectinomycin and all known degradation and biosynthesis byproducts as their N-(2-napthalenesulphonate) derivatives on silica gel,[70] aminoglycoside mixture obtained by fermentation (kanamycin, gentamicin, sisomicin, and tobramycin) or synthesis (amikacin) as their N-(2,4,6-trinitrophenyl) derivatives on a reversed-phase column with detection at 350 nm,[71] and spectinomycin in serum by ion-pair reversed-phase h.p.l.c. with detection at

195 nm.[72]

There have been numerous papers dealing with the h.p.l.c. analysis of nucleosides, nucleotides and their constituent bases. Nucleosides and bases found in biological fluids [73] and human urine and plasma[74] have been separated on reversed-phase columns; the latter separation of a very large number of such compounds was achieved in a single run and in a much shorter time than with previously reported methods. The influence of stationary-phase packings varying in alkyl chain length (2, 8, and 18 carbons), pore diameter, and specific surface area, on the separation of twenty nucleosides and bases, has been examined, the $C_{18}$-phase specifically retarding nucleosides relative to bases.[75] Similarly, the effects of changing pH, temperature, flow rate, and the concentration of methanol in the eluant, on the reversed-phase h.p.l.c. of selected biologically important nucleosides have been investigated.[76] An ion-pair reversed-phase method using tetrabutylammonium phosphate in the eluant has been used for the separation of nucleosides and related mono-, cyclo-, and oligo-nucleotides.[77] Reversed-phase analyses of pyrimidine ribo- and deoxyribo-nucleosides and photo-hydrates derived by irradiation (254 nm) of their aqueous solutions,[78] nucleosides and bases released by acidic and enzymic hydrolysis from cerebral DNA of rat foetus,[79] and nucleic acid metabolites from gastrointestinal mucosa and urine of normal and cancer patients [80] have been reported. Pyrimidine nucleoside phosphorylase activity has been assayed by use of an automated reversed-phase analysis of the released nucleosides and bases, using u.v. and radioactive flow detection.[81] The mechanism of retention of nucleotides, nucelosides, and bases on a polyvinyl alcohol gel packing has been examined and is consistent with the solvophobic theory for the latter two classes; this is not the case with octadecylsilica phases, where incomprehensible and undesirable interactions have often been observed.[82] In the separation on silica gel of various deoxynucleoside derivatives used in the synthesis of oligonucleotides, the use of a very polar solvent (e.g. methanol) as a component of the eluant has been found to be effective in reducing peak tailing.[83]

A low-capacity anion-exchange resin phase in a fused-silica microbore column (0.19 mm i.d., i.e. ~1/500th of the cross-sectional area of a conventional column) has given a similar efficiency in the separation of nucleosides, nucleotides, and bases to that achieved on conventional-sized columns.[84] An h.p.l.c.-m.s. system for analysis and unambiguous identification of nucleosides, using a microbore reversed-phase column and volatile buffer systems, has been tested with the detection of pseudouridine in normal urine.[85] A number of other h.p.l.c.-m.s. investigations, some of nucleosides, are reviewed in Chapter 22.

The optimization of the reversed-phase, silica gel adsorption and aqueous ammonia-modified adsorption h.p.l.c. of anticancer 3'-(chloroethylnitrosoureido)-3'-deoxy-nucleosides, their synthetic precursors, and their decomposition products has been examined for both analytical and preparative purposes.[86]

The following analyses of specific nucleosides have been reported by the methodology indicated: adenosine and S-adenosyl derivatives of sulphur aminoacids in rat liver,[87] 5-aza-2'-deoxy-cytidine,[88] 2'-deoxy-5-trifluoromethyluridine,[89] 5'-deoxy-5-fluoro-uridine,[90] 5-bromo-2'-deoxyuridine,[91] and 5-(2-bromo-E-ethenyl)-2'-deoxyuridine[92] by reversed-phase chromatography, 1,4,5,6,8-penta-azaacenaphthylene-2-amino-1,5-dihydro-5-methyl-1-β-D-ribofuranoside and its 5'-monophosphate,[93] 6-azacytidine,[94] and cytosine arabino-side and its nucleotide and metabolites[95] by ion-pair reversed-phase chromatography, 5-fluorocytosine on a polystyrene-based packing with fluorimetric detection,[96] and 3-methylcytosine from DNA treated with carcinogenic methylating agents on a strong cation-exchanger.[97]

The effect of metal ions ($Mg^{2+}$, $Ni^{2+}$, $Zn^{2+}$, and $Cu^{2+}$) in the eluant on the reversed-phase h.p.l.c. retention of nucleosides, nucleotides, and bases has been reported.[98]

Column Chromatography.- Further studies on the preparative separation of glucose and fructose on $Ca^{2+}$-form cation-exchange resins have been reported.[99,100] A study of the partition chromatography of eight gluco-disaccharides, and related disaccharides containing other sugars, on Sephadex G-15 has related retention behaviour to molecular conformation.[101]

Cellodextrins (DP2-8), obtained by either acidic hydrolysis or acetolysis-deacetylation of cellulose and fractionated into components of defined molar mass by size-exclusion chromatography on Bio-Gel P4, have been further fractionated by chromatography of their borate complexes on a borate-form anion-exchange resin. Components of as yet unknown structure were detected, in yields up to 30% for celloheptaose. Such heterogeneity was not revealed by chromatography on $Ag^{+}$- or $Ca^{2+}$-form cation exchangers, or reversed phase columns, and a caution was given over the use of such "impure" cello-oligosaccharides as substrates for studying cellulytic enzyme kinetics.[102]

The separation of enzymatically produced hyaluronic acid oligosaccharides of DP up to 12 has been investigated by gel permeation and anion-exchange chromatography. The best separation and recovery was achieved on DEAE-Sephacel in the formate form.[103]

Nebramycins, closely related aminoglycosides produced by *Streptomyces tenebrarius*, have been separated on a neutral polystyrene-based column, with aqueous sodium hydroxide as eluant and triple-pulse amperometric detection.[104] The separation of ribo- and deoxyribo-nucleosides and bases on Sephadex G10 has been investigated.[105] The naturally occurring form of *S*-adenosylmethionine in soyabean and radish leaves, yeast, and rat liver has been shown to be the (*S*)-sulphonium form by chromatography using an amino acid analyzer that separates the diastereoisomers.[106]

## 2 Electrophoresis

A high-resolution isotachophoresis system, using a 0.25 mm i.d. fused-silica capillary tube and u.v. detection with a photodiode array spectrometer, has been used in the analysis of ascorbic acid in apple juice.[107] The isotachophoretic separation of saccharinic and other organic acids, formed by alkaline degradation of saccharides, has been optimized.[108]

## 3 Other Analytical Methods

The mechanism of the Molisch reaction, the very sensitive colour reaction between aldoses and 1-naphthol in concentrated sulphuric acid, has been elucidated following the isolation of the crystalline

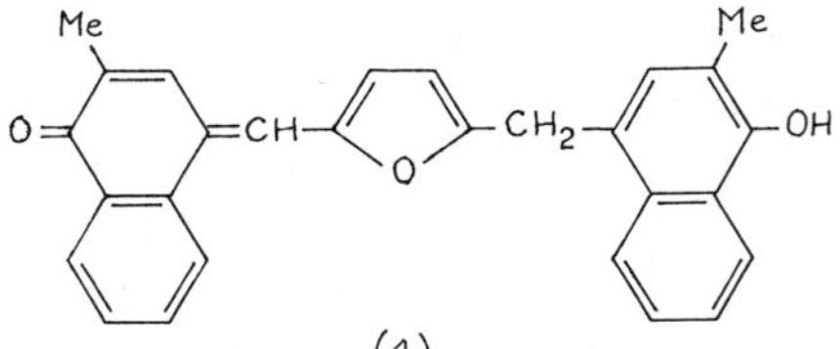

(4)

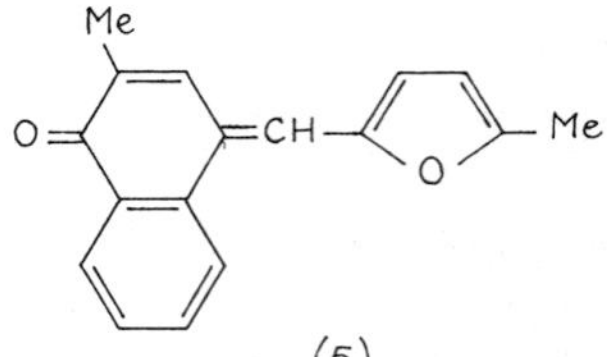

(5)

products (4) and (5) from the related colour reaction of 2-methyl-1-naphthol with glucose and rhamnose, respectively; the 2-methyl substituent prevented subsequent C-2 sulphonation. It was concluded that these sugars were converted to 5-(hydroxymethyl)-

furfural and 5-methylfurfural, respectively, prior to condensation.[109] The colorimetric determination of D-glucose in blood samples by reaction with o-toluidine has been shown to be relatively insensitive to other sugars, and about twice as sensitive to glucose as glucose 1-phosphate.[110]

Glucose in solution has been determined with an experimental enzyme electrode coupled to an iodide-selective electrode. Glucose oxidase was immobilized or entrapped, either alone or with peroxidase, to make the enzyme electrode.[111] A chemically modified lactose electrode has been developed for use in flow injection analysis of lactose in milk. The electrode, made by immobilizing β-galactosidase and glucose oxidase with bovine serum albumin using glutaraldehyde onto a glassy carbon electrode silanized with 3-aminopropyltriethoxysilane, gave a peak current linearly related to lactose concentration in the range $10^{-4}$ to $1.5 \times 10^{-3}$ M.[112]

A titrimetric determination of 2-deoxy-D-ribose has been described. Treatment with excess acidic aqueous periodate gave 1 equiv. of $CH_2(CHO)_2$ and 2 equiv. of iodate. Excess periodate was then masked ($NH_3$ or $NaMoO_4$) and the iodate was titrated, giving results with a 0.5-1% standard deviation.[113]

Total ascorbic and dehydroascorbic acid in human serum has been determined by iodine oxidation followed by reaction of dehydroascorbic acid with 1,2-diamino-4,5-dimethoxy-benzene; the intensely fluorescent product permitted detection down to 5 ng of ascorbic acid with samples as small as 4 µl of serum.[114]

## References


1 H.Krotkiewski, Carbohydr. Res., 1985, 143, 229.

2 M.R.Holdiness, J. Chromatogr., 1985, 340, 321.

3 W.T.Wang, N.C.Le Donne, B.Ackerman,+C.C.Sweeley, Anal. Biochem., 1984, 141, 366 (Chem. Abstr., 1985, 102, 6 980).

4 E.V.Evtushenko and Yu.S.Ovodov, Khim. Prir. Soedin., 1984, 411 (Chem. Abstr. 1985, 102, 79 208).

5 A.Van Langenhove and V.N.Reinhold, Carbohydr. Res., 1985, 143, 1.

6 R.S.Whiton, P.Lau, S.L.Morgan, J.Gilbart, and A.Fox, J. Chromatogr., 1985, 347, 109.

7 C.M.Park, P.E.Reid, and D.A.Applegarth, J. Chromatogr., 1985, 339, 182.

8 E.T.Oakley, D.F.Magin, G.H.Bokelman, and W.S.Ryan, Jr., J. Carbohydr. Chem., 1985, 4, 53.

9 J.A.Lomax, A.H.Gordon, and A.Chesson, Carbohydr. Res., 1985, 138, 177.

10 P.J.Harris, A.Bacic, and A.E.Clarke, J. Chromatogr., 1985, 350, 304.

11 J.-R.Neeser, Carbohydr. Res., 1985, 138, 189.

12 P.Englmaier, Carbohydr. Res., 1985, 144, 177.

13 E.J.-M.Selosse and P.J.Reilly, J. Chromatogr., 1985, 328, 253.

14 A.Marzo and E.Treffner, J. Chromatogr., 1985, 345, 390.

15 J.Adams and R.W.Giese, J. Chromatogr., 1985, 347, 99.

16 E.Csató, P.Keresztes, L.D.Szabó, and D.Knausz, J. Chromatogr., 1985, 348, 280.

17 C.W.Ford, J. Chromatogr., 1985, 333, 167.
18 K.Niemelä and E.Sjöström, Carbohydr. Res., 1985, 144, 93.
19 S.Morgenlie, Carbohydr. Res., 1985, 138, 329.
20 G.Hirth and W.Walther, Helv. Chim. Acta, 1985, 68, 1863.
21 C.J.S.Van Rijn, J.D.M.Herscheid, G.W.M.Visser, and A.Hoekstra, Int. J. Appl. Radiat. Isot., 1985, 36, 111 (Chem. Abstr., 1985, 102, 204 180).
22 M.A.Figueira and J.A.Ribeiro, J. Chromatogr., 1985, 325, 317.
23 K.C.Guven, Acta Pharm. Turc., 1985, 27, 1 (Chem. Abstr., 1985, 102, 167 042).
24 K.Koizumi, T.Utamura, and Y.Okada, J. Chromatogr., 1985, 321, 145.
25 E.Shimada and G.Matsumura, J. Chromatogr., 1985, 328, 73.
26 Y.Nakamura and Y.Koizumi, J. Chromatogr., 1985, 333, 83.
27 K.Jones and D.J.Malcolme-Lawes, J. Chromatogr., 1985, 329, 25.
28 H.Bauer and W.Voelter, High Perform. Liq. Chromatogr. Biochem., 1985, 393 (Chem. Abstr., 1985, 103, 210 070).
29 M.A.Clarke and W.S.C.Tsang, Int. Sugar J., 1984, 86, 215 (Chem. Abstr., 1985, 102, 8 515
30 Z.L.Nikolov and P.J.Reilly, J. Chromatogr., 1985, 325, 287.
31 N.Watanabe, J. Chromatogr., 1985, 330, 333.
32 R.E.A.Escott and A.F.Taylor, HRCCC, J. High Resolut. Chromatogr. Chromatogr. Commun., 1985, 8, 290 (Chem. Abstr., 1985, 103, 215 656).
33 D.E.Hughes, J. Chromatogr., 1985, 331, 183.
34 F.Guyon, A.Foucault, M.Caude, and R.Rosset, Carbohydr. Res., 1985, 140, 135.
35 J.L.Leonard, F.Guyon, and P.Fabiani, Chromatographia, 1984, 18, 600 (Chem. Abstr., 1985, 102, 63 917).
36 H.F.Walton, J. Chromatogr., 1985, 332, 203.
37 J.A.Owens and J.S.Robinson, J. Chromatogr., 1985, 338, 303.
38 G.Bonn, J. Chromatogr., 1985, 350, 381.
39 P.F.Greenfield and G.Geronimos, Int. Sugar J., 1985, 87, 10 (Chem. Abstr., 1985, 102, 151 183).
40 H.Alborn and G.Stenhagen, J. Chromatogr., 1985, 323, 47.
41 P.Hirter, H.J.Walther, and P.Dätwyler, J. Chromatogr., 1985, 323, 89.
42 F.M.Eggert and M.Jones, J. Chromatogr., 1985, 333, 123.
43 D.A.S.Finden, R.R.Fysh, and P.C.White, J. Chromatogr., 1985, 342, 179.
44 P.Lloyd and M.J.C.Crabbe, J. Chromatogr., 1985, 343, 402.
45 D.W.Patrick and W.R.Kracht, J. Chromatogr., 1985, 318, 269.
46 G.Bonn, J. Chromatogr., 1985, 322, 411.
47 W.M.Blanken, M.L.E.Bergh, P.L.Koppen, and D.H.Van den Eijnden, Anal. Biochem., 1985, 145, 322 (Chem. Abstr., 1985, 102, 217 681).
48 Z.L.Nikolov, M.M.Meagher, and P.J.Reilly, J. Chromatogr., 1985, 319, 51.
49 Z.L.Nikolov, M.M.Meagher, and P.J.Reilly, J. Chromatogr., 1985, 321, 393.
50 B.Porsch, J. Chromatogr., 1985, 320, 408.
51 I.Ishii, N.Takahashi, S.Kato, N.Akamatsu, and Y.Kawazoe, J. Chromatogr., 1985, 345, 134.
52 P.W.Tang and J.M.Williams, Carbohydr. Res., 1985, 136, 259.
53 J.Goto, K.Suzaki, M.Ebihara, T.Nambara, and A.Masu, J. Chromatogr., 1985, 345, 241.
54 J.Goto, K.Suzaki, T.Chikai, K.Nagase, and T.Nambara, J. Chromatogr., 1985, 348, 151.
55 V.O.Bhargava, W.H.Soine, and L.K.Garrettson, J. Chromatogr., 1985, 343, 219.
56 R.K.Heaney, Anal. Proc. (London), 1984, 21, 482 (Chem. Abstr., 1985, 102, 92 319).
57 T.T.Tjioe, J.P.van der Weil, R.M.Stikkelman, A.J.J.Straathof, and F.van Rantwijk, J. Chromatogr., 1985, 330, 412.
58 R.M.Stikkelman, T.T.Tjioe, J.P.van der Wiel, and F.van Rantwijk, J. Chromatogr., 1985, 322, 220.
59 K.B.Hicks, P.C.Lim,and M.J.Haas, J. Chromatogr., 1985, 319, 159.
60 T.Imanari, K.Taguchi, S.Tanabe, K.Shinomiya, A.Kunitomo, and H.Suzuki, Chem. Pharm. Bull., 1985, 33, 3057.
61 A.K.Shukla, R.Schauer, F.M.Unger, U.Zähringer, E.T.Rietschel, and H.Brade, Carbohydr. Res., 1985, 140, 1.
62 A.K.Shukla, R.Schauer, U.Schade, H.Moll, and E.Th.Rietschel, J. Chromatogr., 1985, 337, 231.

63 T.Seki, Y.Yamaguchi, K.Noguchi, and Y.Yanagihara, J. Chromatogr., 1985, 332, 283.
64 C.S.Tsao and M.Young, J. Chromatogr., 1985, 330, 408.
65 T.Iwata, M.Yamaguchi, S.Hara, M.Nakamura, and Y.Ohkura, J. Chromatogr., 1985, 344, 351.
66 M.C.Rouan, J. Chromatogr., 1985, 340, 361.
67 D.M.Barends, J.S.Blauw, C.W.Mijnsbergen, C.J.L.R.Govers, and A.Hulshoff, J. Chromatogr., 1985, 322, 321.
68 I.O.Kibwage, E.Roets, J.Hoogmartens, and H.Vanderhaeghe, J. Chromatogr., 1985, 330, 275.
69 T.Harada, M.Iwamori, and Y.Nagai, J. Chromatogr., 1985, 337, 187.
70 K.Tsuji and K.M.Jenkins, J. Chromatogr., 1985, 333, 365.
71 P.Gambardella, R.Punziano, M.Gionti, C.Guadalupi, G.Mancini, and A.Mangia, J. Chromatogr., 1985, 348, 229.
72 N.Kurosawa, S.Kuribayashi, E.Owada, K.Ito, M.Nioka, M.Arakawa, and R.Fukuda, J. Chromatogr., 1985, 343, 379.
73 R.Boulieu, C.Bory, and C.Gonnet, J. Chromatogr., 1985, 339, 380.
74 G.S.Morris and H.A.Simmonds, J. Chromatogr., 1985, 344, 101.
75 A.Rizzi and H.R.M.Lang, J. Chromatogr., 1985, 331, 33.
76 C.Yi, J.L.Fasching, and P.R.Brown, J. Chromatogr., 1985, 339, 75.
77 B.Allinquant, C.Musenger, and E.Schuller, J. Chromatogr., 1985, 326, 281.
78 V.N.Potaman, T.N.Ulmasov, and V.V.Demidov, J. Chromatogr., 1985, 329, 33.
79 S.Iwasaki, H.Tanaka, K.Nakazawa, and M.Arima, J. Chromatogr., 1985, 341, 182.
80 K.Nakano, K.Shindo, T.Yasaka, and H.Yamamoto, J. Chromatogr., 1985, 332, 127; 343, 21.
81 S.A.Siegel, K.L.Woods, and R.N.Dreyer, J. Chromatogr., 1985, 319, 351.
82 H.Wada, J. Chromatogr., 1985, 332, 255.
83 T.Ohkuma and S.Hara, J. Chromatogr., 1985, 323, 227.
84 S.Rokushika, Z.Y.Qiu, and H.Hatano, J. Chromatogr., 1985, 320, 335.
85 F.C.Alderweireldt, E.L.Esmans, and P.Geboes, Nucleosides Nucleotides, 1985, 4, 135.
86 W.F.Brubaker, Jr., and W.H.Prusoff, J. Chromatogr., 1985, 322, 455.
87 D.Fell, L.E.Benjamin, and R.D.Steele, J. Chromatogr., 1985, 345, 150.
88 K.-T.Lin, R.L.Momparler, and G.E.Rivard, J. Chromatogr., 1985, 345, 162.
89 G.Balansard, G.Schwadrohn, E.Vidal, and R.Elias, J. Chromatogr., 1985, 348, 299.
90 L.J.Schaaf, D.G.Ferry, C.T.Hung, D.G.Perrier, and I.R.Edwards, J. Chromatogr., 1985, 342, 303.
91 P.L.Stetson, U.A.Shukla, P.R.Amin, and W.D.Ensminger, J. Chromatogr., 1985, 341, 217.
92 Y.Robinson, N.Gerry, and C.W.Vose, J. Chromatogr., 1985, 338, 219.
93 R.B.Schilcher, J.D.Young, and L.H.Baker, J. Chromatogr., 1985, 337, 55.
94 S.V.Galushko and I.P.Shishkina, J. Chromatogr., 1985, 345, 157.
95 R.L.Schilsky and F.S.Ordway, J. Chromatogr., 1985, 337, 63.
96 C.Lacroix, P.Levert, G.Laine, J.P.Goulle, and A.Gringore, J. Chromatogr., 1985, 345, 436.
97 H.Kawasaki, S.Ninomiya, and H.Yuki, Chem. Pharm. Bull., 1985, 33, 1170.
98 A.S.Cohen and E.Grushka, J. Chromatogr., 1985, 318, 221.
99 T.Sakiyama, K.Nakamura, and T.Yano, Agric. Biol. Chem., 1985, 49, 2619.
100 Yu.E.Kuptsevich, I.D.Stal'naya, O.N.Sosedov, L.A.Nakhapetyan, A.Ya.Pronin, and G.O.Larionov, Prikl. Biokhim. Mikrobiol., 1985, 21, 129 (Chem. Abstr., 1985, 102, 132 374).
101 A.C.Haglund, N.V.B.Marsden, and S.G.Östling, J. Chromatogr., 1985, 318, 57.
102 K.Hamacher, G.Schmid, H.Sahm, and Ch.Wandrey, J. Chromatogr., 1985, 319, 311.
103 P.Nebinger, J. Chromatogr., 1985, 320, 351.
104 J.A.Polta, D.C.Johnson, and K.E.Merkel, J. Chromatogr., 1985, 324, 407.
105 Yu.L.Kaminskii, L.A.Yakovleva, I.F.Ivanova, and L.P.Sosnova, Zh. Obshch. Khim., 1984, 54, 2379 (Chem. Abstr., 1985, 102, 149 690).
106 G.L.Creason, J.T.Madison, and J.F.Thompson, Phytochemistry, 1985, 24, 1151.

107 M.Goto, K.Irino, and D.Ishii, J. Chromatogr., 1985, 346, 167.

108 J.Polonský, L.Repášová, A.Pekarovičová, and M.Košík, J. Chromatogr., 1985, 320, 111.

109 M.Ohta, M.Iwasaki, K.Kouno, and Y.Ueda, Chem. Pharm. Bull., 1985, 33, 2862.

110 S.A.Shcherbakova, N.V.Alekseeva, and E.N.Efimova, Fiziol. Biochim. Osn. Povysh. Prod. S-kh. Zhivotn., 1983, 105 (Chem. Abstr., 1985, 102, 181 798).

111 I.K.Al-Hitti, G.J.Moody, and J.D.R.Thomas, Analyst (London), 1984, 109, 1205 (Chem. Abstr., 1985, 102, 39 253).

112 K.Matsumoto, O.Hamada, H.Ukeda, and Y.Osajima, Agric. Biol. Chem., 1985, 49, 2131.

113 V.Zaharans, Khim.-Farm. Zh., 1984, 18, 1142 (Chem. Abstr., 1985, 102, 95 993).

114 T.Iwata, S.Hara, M.Yamaguchi, M.Nakamura, and Y.Ohkura, Chem. Pharm. Bull., 1985, 33, 3499.

# 24
# Synthesis of Enantiomerically Pure Non-carbohydrate Compounds

Fraser-Reid has reviewed, in characteristic style, some of the recent work in this area by his group.[1]

## 1 Carbocyclic Compounds

An intramolecular Wadsworth-Emmons reaction was used as a key step in the synthesis of (1), an inhibitor of glyoxylase I isolated from a Streptomyces species (Scheme 1), the starting material being derived from D-Mannose.[2] Similar chemistry was used in an improved

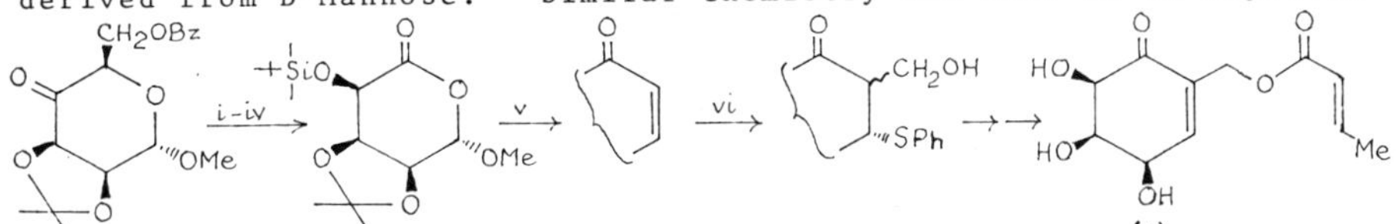


(1)

Reagents: i, $NEt_3$; ii, $NaBH_4$; iii, $O_3$; iv, $Bu^tMe_2SiCl$; v, $\bar{C}H_2P(O)(OMe)_2$; vi, $Me_2AlSPh$, then $CH_2O$

Scheme 1

procedure for the preparation, from D-ribonolactone, of an intermediate previously used (see Vol. 17, p. 248-9) in the synthesis of neplanocin A.[3] D-Lyxose has been converted into methyl shikimate (2; carbons of lyxose numbered) via an intramolecular aldol reaction.[4]

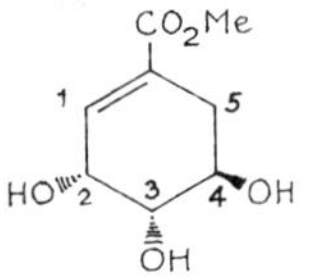


(2)

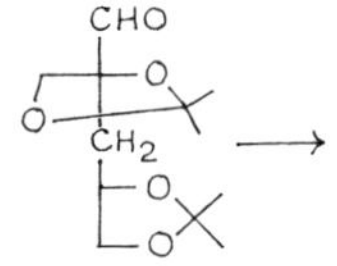


(3)

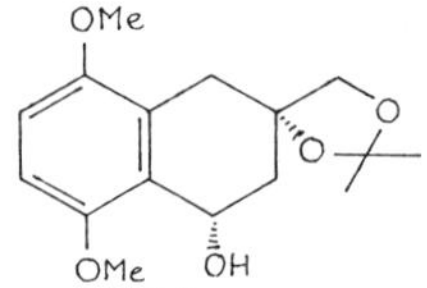


(4)

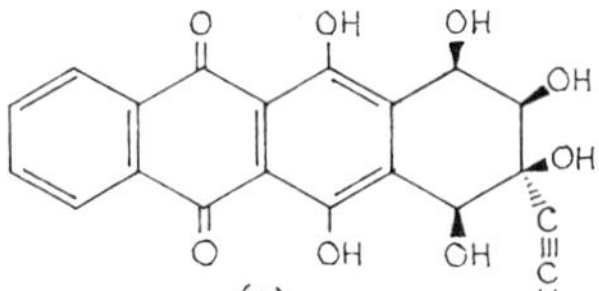


(5)

There has been further activity in the use of sugars as precursors for the A ring of anthracyclinones. The aldehyde (3), previously reported (Vol. 18, p. 248), has been prepared by an improved procedure and converted into (4), a suitable intermediate for an AB + CD approach to the tetracycle.[5] Similar routes to those described earlier (Vol. 18, p. 248) have been used in the synthesis of (5), containing the key tertiary carbinol function, from diisopropylidene glucose via a Marschalk reaction,[6] and earlier work has been improved by the use of DBN or DBU in the cyclizations.[7]

Cycloaddition reactions have been used for the annulation of carbocyclic rings onto carbohydrate templates. The butenolide (6), derived from D-ribonolactone, gave the potential prostacyclin synthon (7) via reaction with butadiene and the chrysanthemic acid

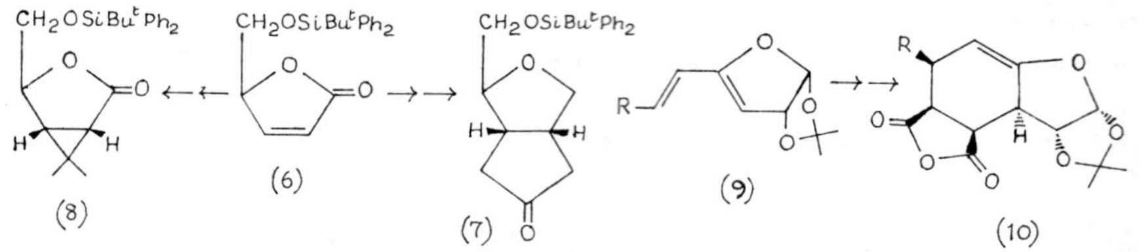

precursor (8) with 2-diazopropane and subsequent photolysis.[8] The dienes (9; R = H or $CH_2OTBDMS$), derived from diacetone glucose, reacted stereospecifically with maleic anhydride to give cyclohexenes (10), from which a range of related compounds were prepared.[9] Reaction of Danishefsky's diene with a glucose-derived enone gave (11), which was transformed into actinobolin (12) (Scheme 2).[10] When nitrocompound (13), derived from diacetone glucose, was treated with

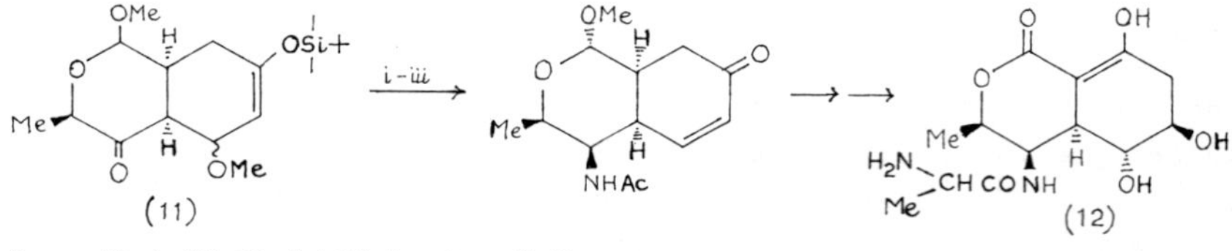

Reagents: i, $NH_2OH$; ii, LAH, then $Ac_2O$; iii, $MnO_2$

Scheme 2

phenyl isocyanate, intramolecular nitrile oxide cyclization gave (14), which was transformed to the oxahydrindene component (15) of the avermectins.[11] A further Diels-Alder reaction is referred to at

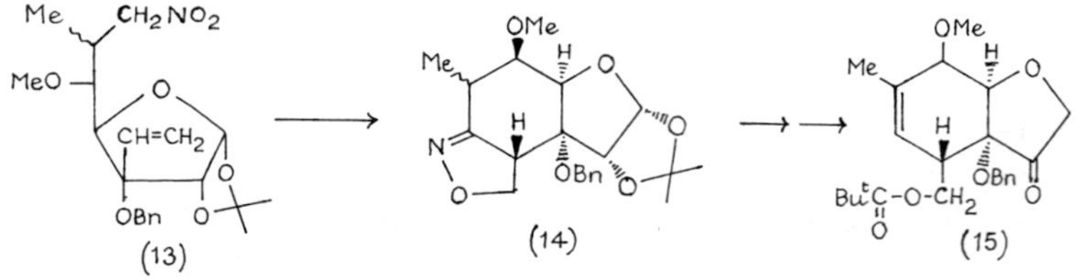

the end of section 5.

The trichothecane skeleton (16) has been constructed from a C-glycoside, the two key formations of carbocyclic rings being outlined in Scheme 3,[12] and progress in this area has been reviewed.[13] A conceptually similar cyclization was used to form the 4-membered

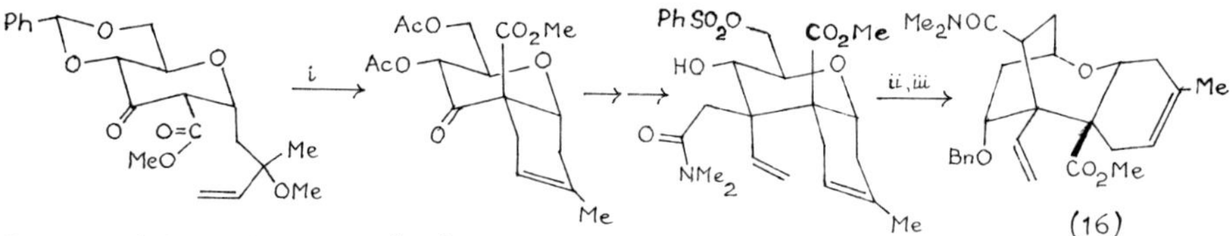

Reagents: i, $SnCl_4$-$Ac_2O$; ii, $KN(TMS)_2$; iii, BnBr

Scheme 3

ring of nitrile (17; R = CN), during a synthesis of the pheromone lineatin (17; R = Me) from D-ribose (sugar carbons numbered).[14] In a synthetic approach to quassimarin, the intramolecular Diels-Alder reaction (18) → (19) proceeded stereoselectively as shown with Lewis acid catalysis; (18) was prepared from glucose, carbons 3 and 4 being incorporated as indicated.[15]

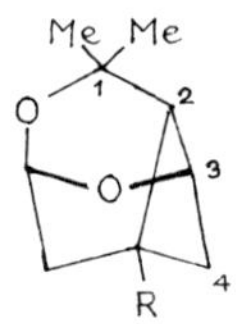

(17)

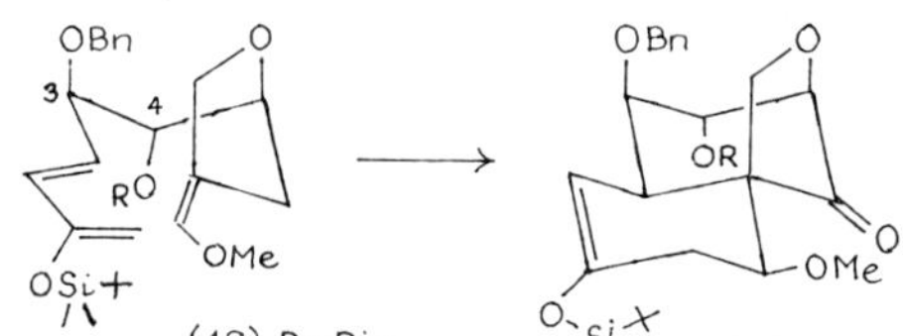

(18) R = Piv (19)

## 2 γ- and δ-Lactones

Full accounts have been given of the synthesis of avenaciolide[16] (see Vol. 9, p. 202) and canadensolide[17] (Vol. 12, p. 209), both from diisopropylidene-glucose. The marine antibiotic (-)-malyngolide (20) was prepared, together with its C-2 epimer, from D-mannose,

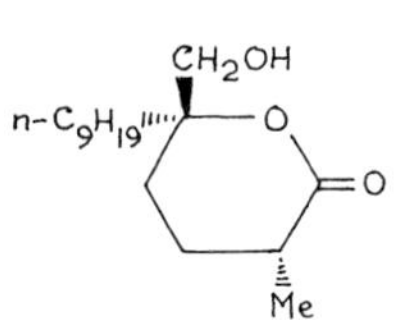

(20)

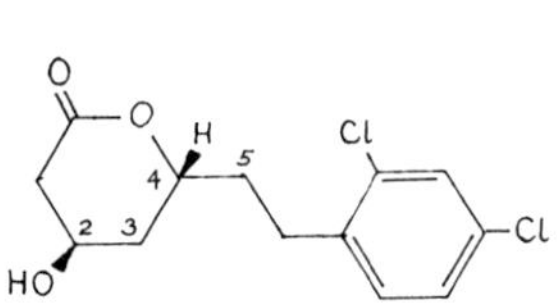

(21)

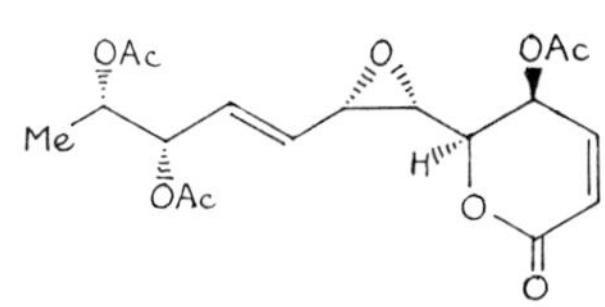

(22)

via 2,3-O-isopropylidene-D-apiose (Vol. 13, p. 126) as intermediate.[18] Simple analogues of mevinolin and compactin such as (21) have been prepared from glucose (sugar numbering on formula),[19] and a synthetic approach to the side chain of olguine (22) has been reported.[20]

The lactone (23), prepared earlier from levoglucosan (see Vol. 17, p. 244), has been converted into (24), a subunit of avermectin $B_{1b}$, by reaction with a chiral alkyne and subsequent partial hydrogena-

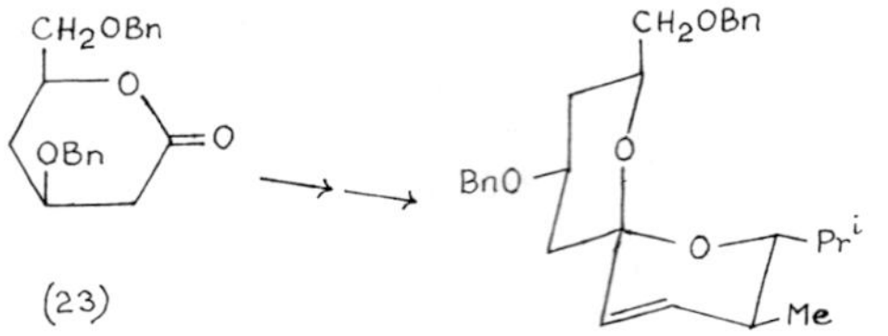


(23) (24)

tion and cyclization.[21] The northern hemisphere of milbemycin has been prepared by an extension of chemistry reported last year (see Vol. 18, p. 252) involving Wittig reaction of a sugar C-1 phosphorane.[22]

'Segment A' (25) of the complex marine natural-product okadaic acid has been prepared by linking together the glucose-derived lactone (26)[23] with the unit (27), which was made either from malic acid or from D-glucal.[24]

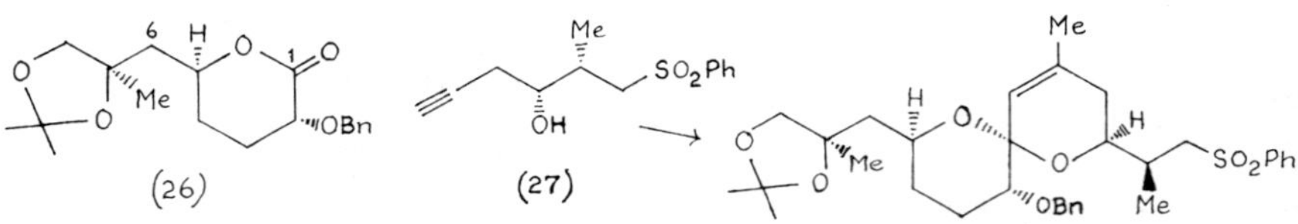


(26) (27) (25)

Both the (R)- and (S)-isomers of spirobi-1,4-dioxan, (28) and (29) respectively, which are dioxa analogues of the olive fruit fly pheromone, have been prepared from fructose derivatives as outlined in Scheme 4.[25] Fructose was also the starting material for an ingenious synthesis, which cannot be adequately summarized here, of the spiroacetal unit (30) related to monensin.[26]

## 4 Other Oxygen Heterocycles

Both the naturally occurring (7R, 8S) isomer of the gypsy moth pheromone disparlure (31) (sugar chain numbering) and its enantiomer have been prepared from D-glucose via a common intermediate.[27] The relative stereochemistry of the fungal toxins AK-I and -II was established by the synthesis of the pair of epoxides (32) from L-

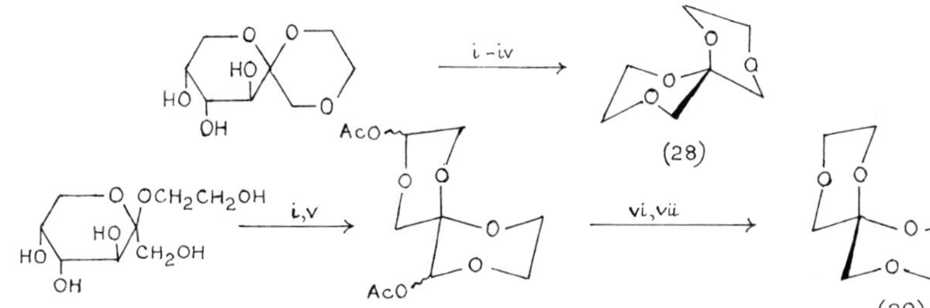

(28)

(29)

Reagents: i, $NaIO_4$; ii, $NaBH_4$; iii, $-C_6H_4-SO_2Cl-Py$; iv, $MeO^-$; v, $Ac_2O$; vi, HBr-HOAc; vii, LAH

Scheme 4

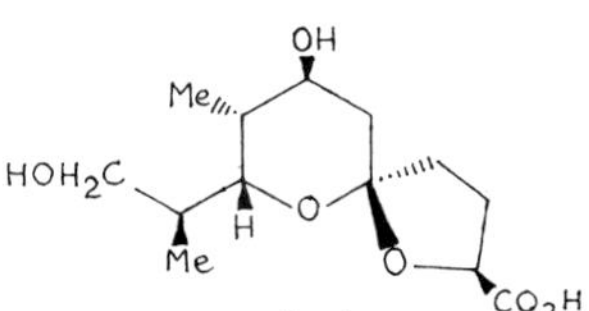

(30)

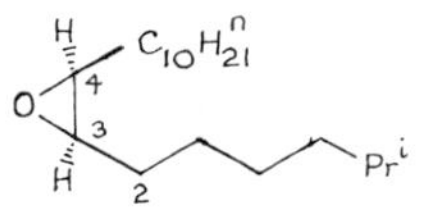

(31)

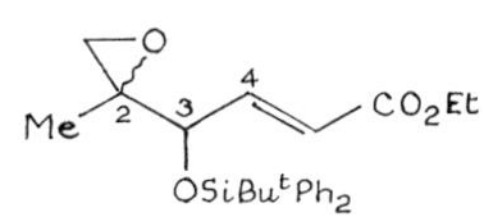

(32)

ascorbic acid (numbering indicated).[28]

α-Methylglucoside has been converted to (S)-(+)-2-methoxy-3,4-dihydro-2H-pyran by elimination of all chirality save that at the anomeric position.[29] The model compound (33) for nogalamycin was

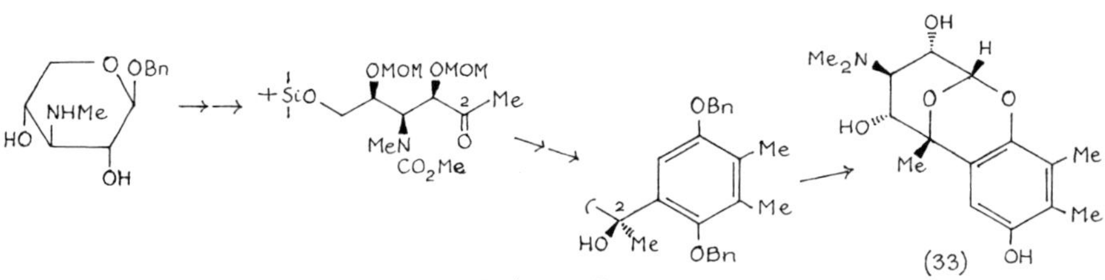

(33)

Scheme 5

synthesized from a gentosamine derivative, itself accessible from D-arabinose, via the key intermediates shown in Scheme 5; the addition of the aryl ring as its organolithium derivative was stereoselective as indicated, which conforms with chelation control.[30] The (-)-isomer of exo-brevicomin (34) has been prepared from the readily available glucose derivative (35) as shown in Scheme 6.[31]

Full details have been reported of the synthesis of pseudomonic acid C from L-lyxose (see Vol. 18, p. 247).[32] The ABC ring structure (36) of the complex marine toxin brevetoxin B has been synthesized from carbohydrate precursors; in the key step, the acid-

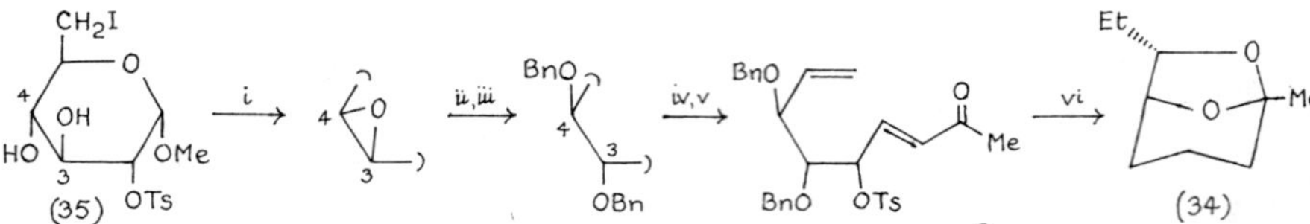

Reagents: i, DIAD-TPP; ii, BnOH-$BF_3$; iii, BnBr-NaH; iv, Zn-$H_2O$/EtOH; v, $Ac\overset{\ominus}{C}HP(O)(OEt)_2$; vi, $H_2$-Pd/C

Scheme 6

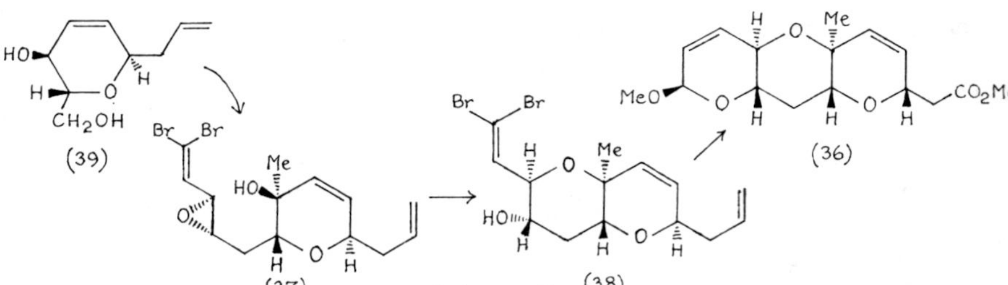

Scheme 7

catalysed cyclization of (37) to (38) (Scheme 7) occurred in the desired 6-endo sense rather than the 5-exo manner due to the presence of the adjacent π-bond, a point investigated with simpler models.[33] The C-glycoside precursor (39) was prepared from triacetyl glucal (see Vol. 16, p. 40).

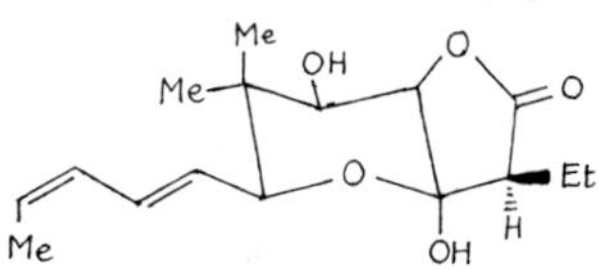

(40)

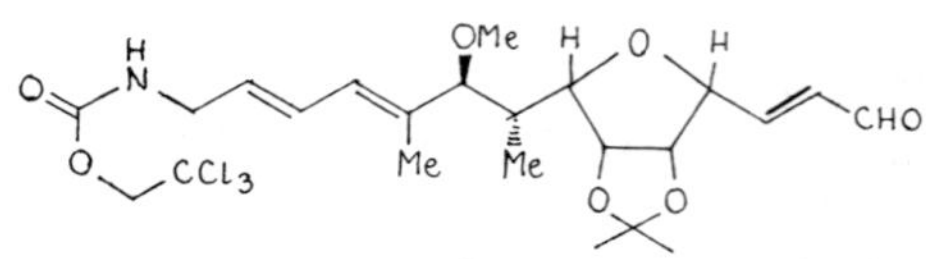

(41)

Goldinonolactone (40) and the chiral tetrahydrofuran (41) were key intermediates in a chiral, but non-carbohydrate-based, synthesis of the antibiotic aurodox;[34] both these intermediates are also accessible from carbohydrates, (40) being made from an L-mannose derivative and (41) from D-mannose via a previously described acetylenic intermediate (see Vol. 8, p. 26).[35] In further superb demonstrations of the capability of the Claisen rearrangement, Ireland and coworkers have prepared the "dimeric" unit (42) of monensin from D-xylose and D-mannose,[36] and then linked it with (43), made from fructose, to give, after free-radical decarboxylation, the 'trimeric' unit (44).[37]

1,4-Anhydro-D-glucitol (45) was used as a precursor for the

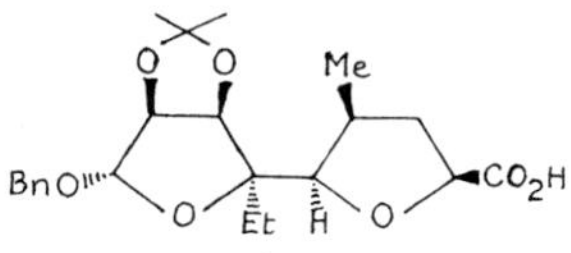

(42)

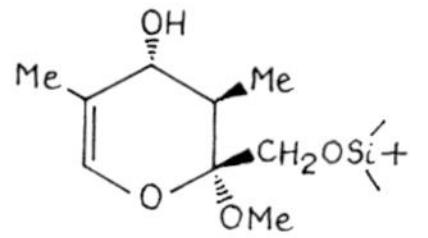

(43)

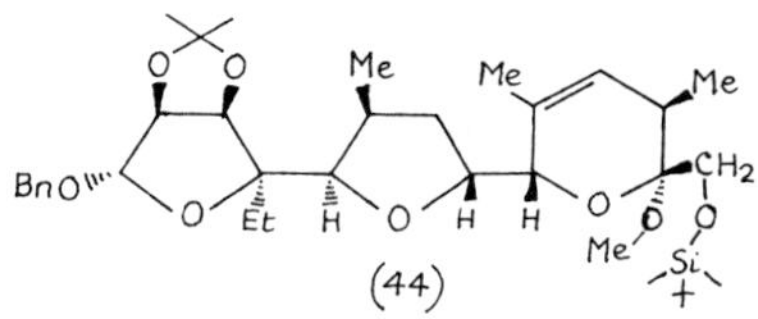

(44)

synthesis of both 11-oxaprostaglandin $F_{2\alpha}$ (46) and the $F_{2\beta}$ isomer (47), as outlined in Scheme 8; mixtures of C-15 isomers were pro-

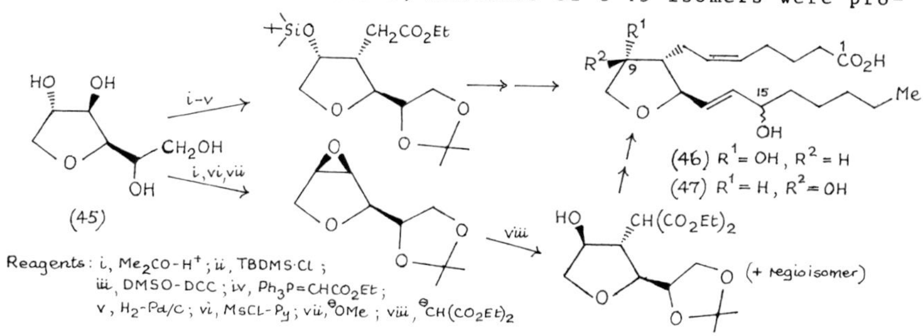

Reagents: i, $Me_2CO-H^+$; ii, TBDMS·Cl; iii, DMSO-DCC; iv, $Ph_3P=CHCO_2Et$; v, $H_2$-Pd/C; vi, MsCl-Py; vii, $^{\ominus}OMe$; viii, $^{\ominus}CH(CO_2Et)_2$

Scheme 8

duced, which could be separated.[38] The isomer of (46) with the $C_1-C_7$ side chain β-oriented has been prepared from 1,4:3,6-dianhydro-D-sorbitol.[39]

Nanaomycin D (48) and its enantiomer kalafungin (49) were both accessible (Scheme 9) from the enone (50), itself derived from L-

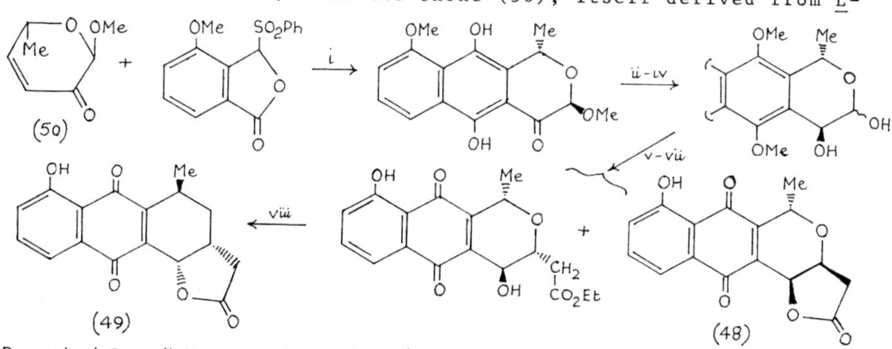

Reagents: i, BuLi; ii, $Me_2SO_4$; iii, $NaBH_4$; iv, $H_3O^+$; v, $Ph_3P=CHCO_2Et$; vi, $Ce(NH_4)_2(NO_3)_6$; vii, $AlCl_3$; viii, $H_2SO_4$, then Δ (Toluene)

Scheme 9

rhamnose.[40,41] The ring system of these antibiotics was prepared in a conceptually similar way by other workers from levoglucosenone (Scheme 10).[42]

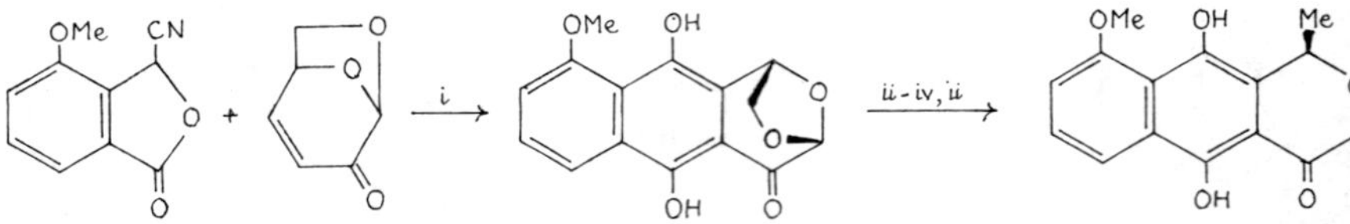

Reagents: i, $MeSO_2CH_2Li$ ; ii, Zn-Cu ; iii, MsCl ; iv, NaI

Scheme 10

## 5 Nitrogen Heterocycles

Oxidative degradation of 5,6-O-isopropylidene-L-ascorbic acid gives the L-threonic acid derivative (51) (Scheme 11), from which was prepared the parent ring system (52) of the monobactams.[43]

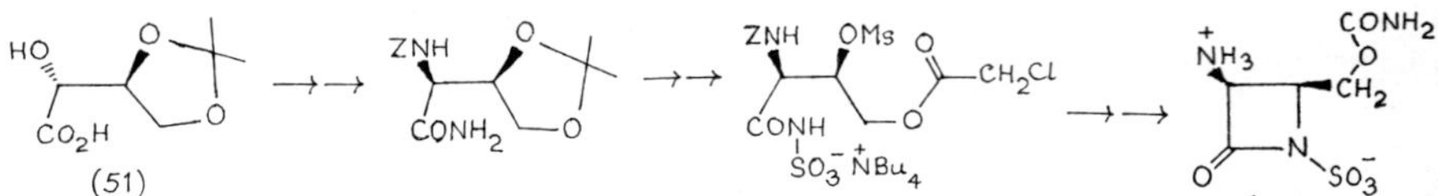

(51) (52)

Scheme 11

A further chiral synthesis of retronecine (53) has been reported. This started from glucose (sugar carbons indicated on formula) and proceeded via an intermediate of type (54); the enantiomer of retronecine was prepared by manipulating the two differentiated hydroxy groups in an alternative way.[44] The mannosidase

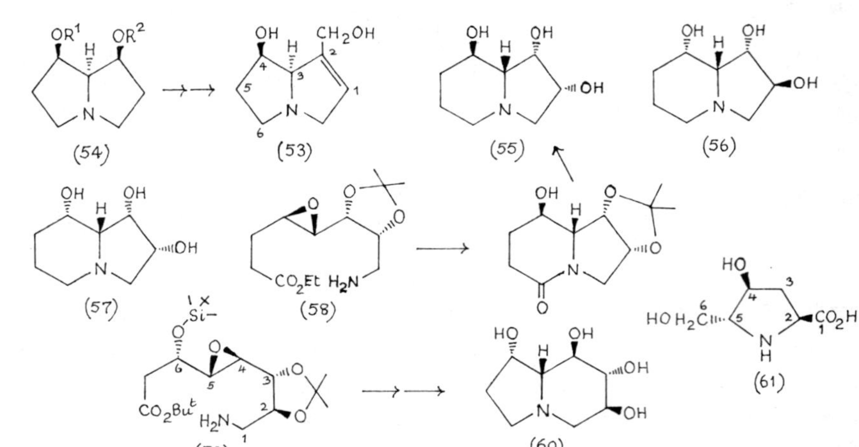

(54) (53) (55) (56) (57) (58) (59) (60) (61)

inhibitor swainsonine (55) continues to attract synthetic attention; a full account has been given[45] of an early synthesis (Vol. 18, p. 253) from D-glucose, and another rather similar approach has been reported.[46] Two stereoisomers of swainsonine, (56) and (57), have been prepared from glucose,[47] and a further synthesis of swainsonine itself involves as a key step the double cyclization (in refluxing ethanol) of the mannose-derived epoxyester (58).[48] A similar cyclization of (59), this time proceeding in the 6-endo mode because of steric restraints, was used in a synthesis of castanospermine (60); D-mannose was used to form (59) (mannose numbers indicated) in a multi-step sequence with no stereocontrol at C-6.[49] Bulgecinine (61), a novel aminoacid found as part of the bulgecin antibiotics, has been prepared from D-glucose (sugar numbering shown) in a 12-step sequence,[50] and a full account has been given of a chiral synthesis of the antibiotic anisomycin from D-ribose (see Vol. 17, p. 248).[51]

The naturally occurring β-D-glucuronidase inhibitor (62) has been prepared using mercuric-ion-induced cyclization (Scheme 12) of

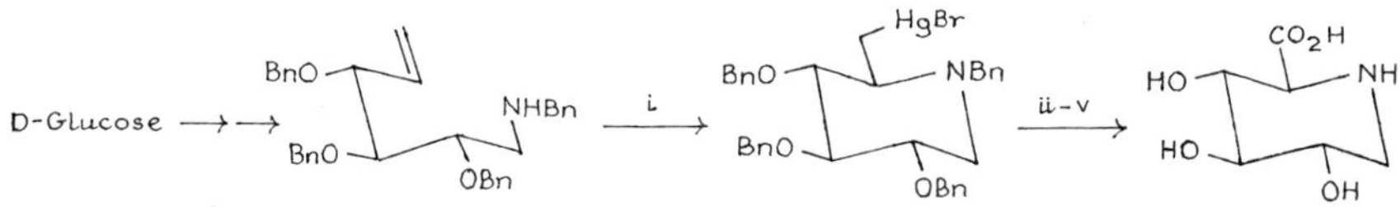

(62)

Reagents: i, $Hg^{2+}$; ii, $NaBH_4$-DMF-$O_2$; iii, DMSO-$(COCl)_2$; iv, $KMnO_4$; v, $H_2$-Pd/C

Scheme 12

an acyclic precursor derived from glucose, followed by reductive oxygenation.[52] (+)-Pseudoconhydrine (63), the alkaloid of common hemlock, has been prepared in chiral form from D-glucosamine; the multi-step procedure is outlined in Scheme 13.[53]

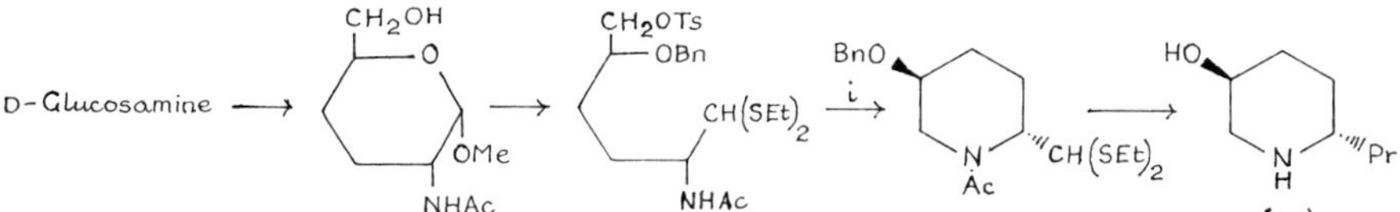

(63)

Reagents: i, NaH-DMF

Scheme 13

The antitumour agent sesbanimide (64) has been a popular target. Extension of the D-xylose derivative (65) at C-1 (Scheme 14) gave the AB ring system (66), and use of L-xylose gave the enantiomer.[54] Alternatively, extension of the related D-xylose derivative (67) at C-5 led ultimately to the AB ring system in the form of (68) (Scheme 15),[55] whilst other workers, following a similar strategy, have used a photochemical route to build up the glutarimide ring (Scheme 16).[56]

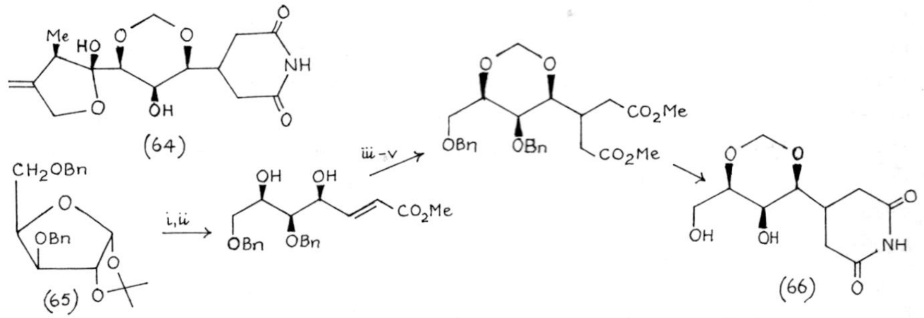

Reagents: i, HCl-HOAc; ii, $Ph_3P{=}CHCO_2Me$ ($C_6H_6$, ref., 30sec); iii, $(MeO)_2CH_2$-TMSOTf; iv, $^{\ominus}CH(CO_2Me)_2$; v, NaCl-$H_2O$-DMSO

Scheme 14

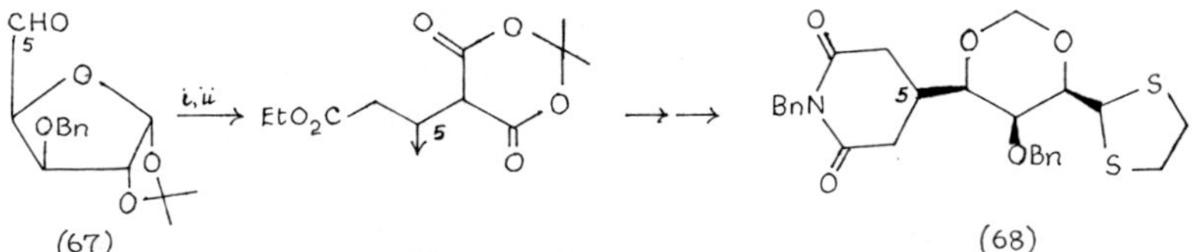

Reagents: i, Meldrum's acid, -HN◯-HOAc; ii, $LiCH_2CO_2Et$

Scheme 15

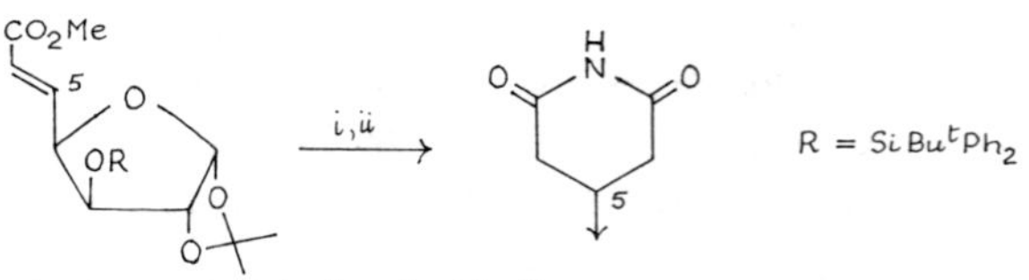

Reagents: i, $ICH_2CONH_2$-$h\nu$; ii, Δ (120°)

Scheme 16

The Diels-Alder adduct (69) of levoglucosenone and butadiene has been employed for the synthesis of the indole alkaloid (-)-alloyohimbine (70) in an overall 12-step process.[57]

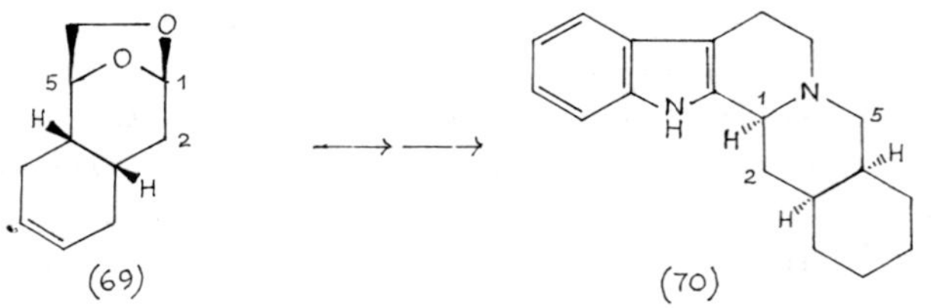

## 6 Acyclic Compounds

The lipoxins are the latest class of arachidonic acid metabolites to attract attention, and in order to determine the stereostructure of lipoxin A the (5S, 6R, 15S) isomer (71) and the corresponding

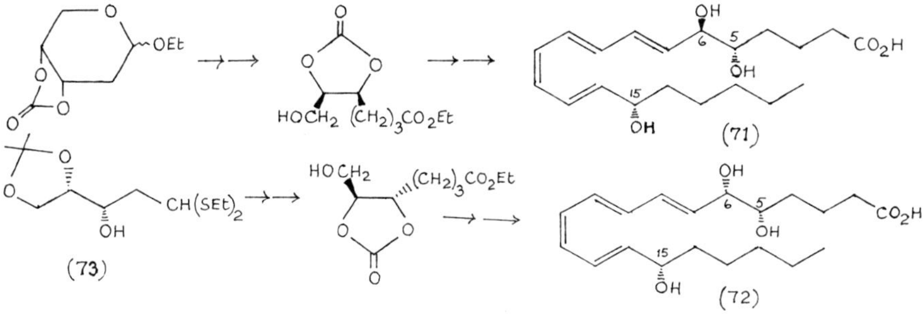

Scheme 17

all-trans system were prepared from 2-deoxy-D-ribose as outlined in Scheme 17; the (5S, 6S, 15S) isomer (72) and its all-trans analogue were also synthesised from L-xylose via the known intermediate (73), and (72) was found to correspond to natural material.[58] The same precursors were used in similar Wittig-type chemistry to prepare the (5S, 14R, 15S) and (5S, 14S, 15S) isomers (74) of lipoxin B and the 8-cis isomers; the natural material was found to be a mixture of the

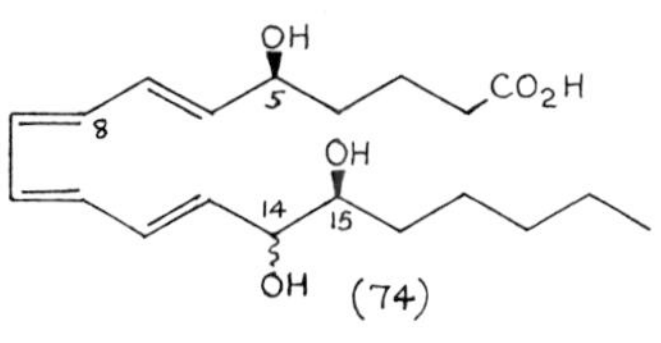

two all-trans isomers (74).[59] The diastereomeric diepoxides (75) and (76) have been prepared from D-mannitol, and opening of the epoxides by organometallic reagents followed by cleavage of the central C-C bond gave enantiomerically pure α-hydroxyaldehydes, suitable for the synthesis of arachidonic acid metabolites; the procedure is exemplified by the formation of (77) from (76) (Scheme 18).[60]

A new acyl peptide fibrinolytic agent, WB-3559D (78), isolated from a flavobacterium species, has been synthesized using the chirality at C-4 of glucose for the R-centre in the product.[61]

A full account has been given of the synthesis (from glucose) of structures corresponding to the C-1 → C-4, C-9 → C-12 and C-7 →

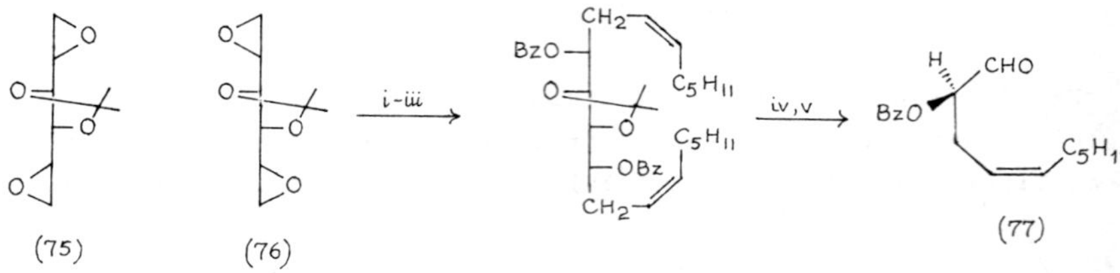

Reagents: i, $LiC{\equiv}C{\cdot}C_5H_{11}$-THF/HMPA ; ii, BzCl ; iii, $H_2$-Lindlar cat.; iv, TFA-$H_2O$ ; v, $Pb(OAc)_4$

Scheme 18

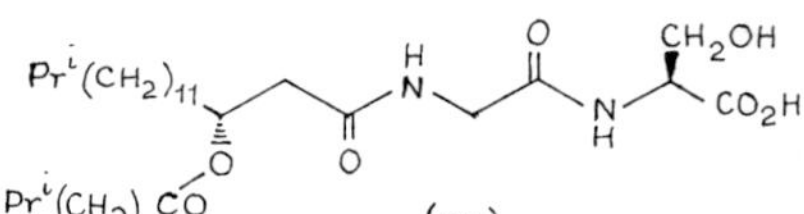

(78)

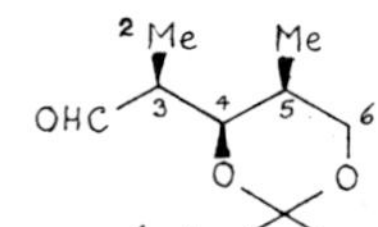

(79)

C-15 fragments of 9-dihydroerythronolide A (see Vol. 17, p. 244-5),[62] and the same workers have also prepared the C-1 → C-5 segment in the form of the aldehyde (79) (sugar chain numbered), again from D-glucose.[63] The segments prepared were found to be identical with compounds prepared by degradation of the antibiotic.[64] The chiral fragment (80) was accessible from the known epoxide (81), itself derived from 1,6-anhydroglucose, as outlined in Scheme 19; an oxida-

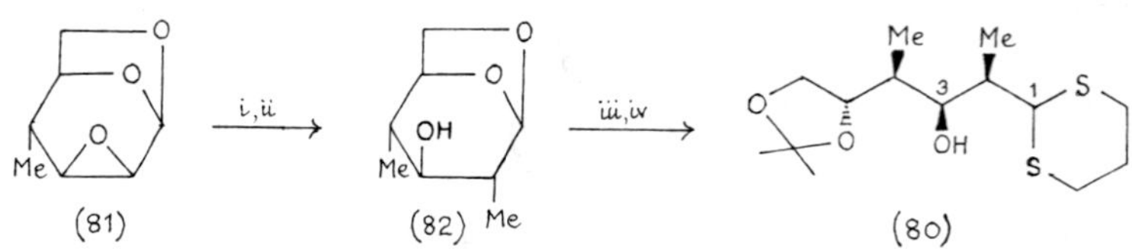

Reagents: i, $LiCH_2SPh$ ; ii, Na-$NH_3$ ; iii, $HS(CH_2)_3SH$-$BF_3$ ; iv, $Me_2C(OMe)_2$- TsOH

Scheme 19

tion-reduction sequence on intermediate (82) led similarly to the C-3 epimer of (80). These fragments contain chiral arrays present in a number of important natural products including tirandamycin, monensin, rifamycin and erythromycin A.[65] Levoglucosan was also the starting material for the preparation of (83), which has been used to synthesize the C-9 → C-13 segment of erythronolide A; a number of stereoisomers of (83) were also made in a stereoselective manner.[66] In connection with an approach to the antibiotic pyridomycin, the chiral acid (84) was prepared by degradation of a branched-chain structure previously prepared from glucose (sugar carbons indicated).[67]

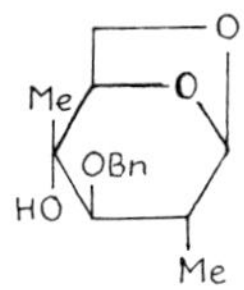
(83)

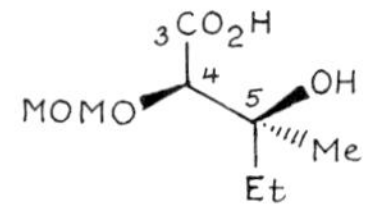
(84)

Acyclic carbon chains bearing methyl groups in a 1,3-, 1,4-, or 1,5-relationship, with or without additional hydroxy functions on the chain, have been chirally prepared from derivatives of S-5-hydroxymethylbutyrolactone, by the use of the concept of a 'replicating chiron'. The starting materials are perhaps easiest obtained from L-glutamic acid, but carbohydrate precursors (D-mannitol, D-ribonolactone) can also be used.[68] The same ideas have been extended to the synthesis of 1,3-polyols.[69]

An interesting fragmentation of the mannose-derived glycal (85) was used in an efficient synthesis of (2S, 3R)-sphingosine (86; R=H) as outlined in Scheme 20. The glycal epimeric at C-3 is available from D-ribonolactone, and this was converted by similar

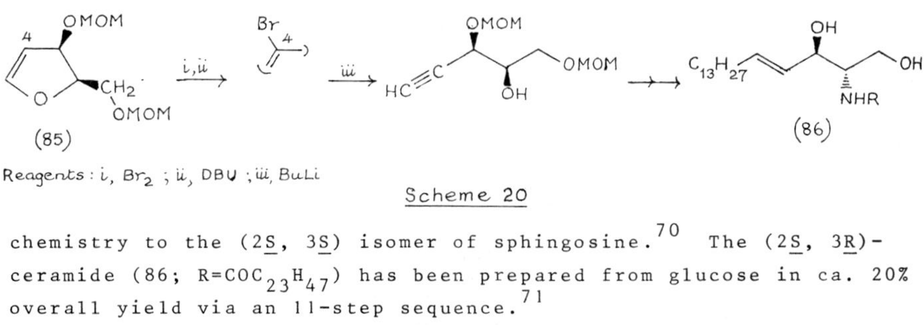
Reagents: i, $Br_2$ ; ii, DBU ; iii, BuLi

Scheme 20

chemistry to the (2S, 3S) isomer of sphingosine.[70] The (2S, 3R)-ceramide (86; $R=COC_{23}H_{47}$) has been prepared from glucose in ca. 20% overall yield via an 11-step sequence.[71]

The (S) isomer of the antiherpetic agent 9-(2,3-dihydroxy-1-propoxymethyl)guanine (87) was prepared by excising the acyclic chain from a D-glucose derivative, the chiral centre of the target

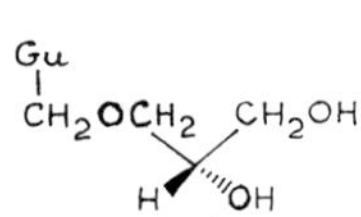
(87)

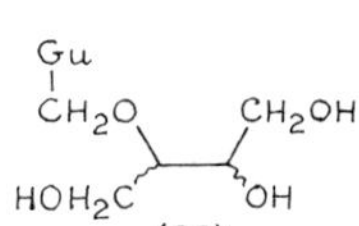
(88)

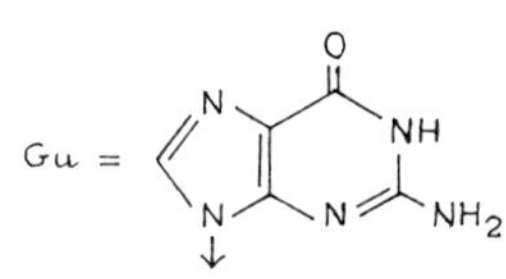

corresponding to C-5 of glucose.[72] In the same area, all four possible diastereoisomers of 9-(1,3,4-trihydroxy-2-butoxymethyl) guanine (88) have been synthesized from D- and L-xylose and D- and L-arabinose.[73]

The stereospecifically deuteriated ribose derivative (89), available from the sugar in five steps, has been used as the starting material in a new synthesis of the (*S*) isomer of 'chiral glycine', as indicated in Scheme 21.[74]

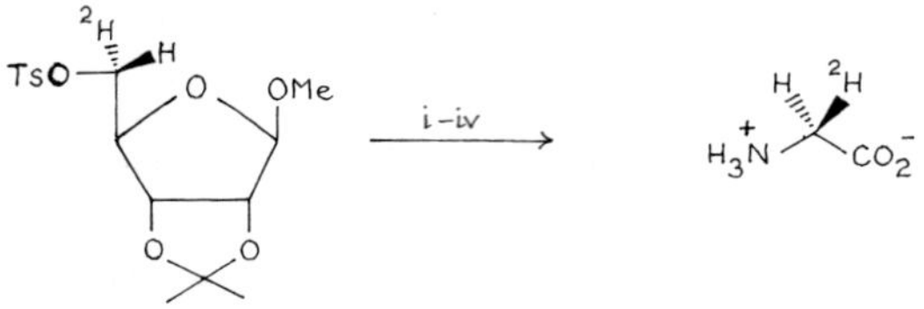

(89)

Reagents: i, $NaN_3$-DMF; ii, HOAc-$H_3O^+$; iii, $KMnO_4$; iv, $H_2$-Pd

Scheme 21

The synthesis of *C*-glycosides required for the synthesis of ambruticin is mentioned in Chapter 3, and the use of a sugar derivative as a chiral auxiliary in the asymmetric synthesis of α-aminophosphonic acids is discussed in Chapter 10.

## References


1 B. Fraser-Reid, *Acc. Chem. Res.*, 1985, *18*, 347.
2 S. Mirza, L.-P. Molleyres, and A. Vasella, *Helv. Chim. Acta*, 1985, *68*, 988.
3 H.-J. Altenbach, W. Holzapfel, G. Smerat, and S.H. Finkler, *Tetrahedron Lett.*, 1985, *26*, 6329.
4 T. Suami, K. Tadano, Y. Ueno and Y. Iimura, *Chem. Lett.*, 1985, 37.
5 J.-C. Florent, A. Genot, and C. Monneret, *Tetrahedron Lett.*, 1985, *26*, 5295.
6 S. Qureshi and G. Shaw, *J. Chem. Soc., Perkin Trans. 1*, 1985, 875.
7 S. Qureshi, G. Shaw, and G.E. Burgess, *J. Chem. Soc., Perkin Trans. I*, 1985, 1557; in Vol. 18, p. 248, the formula of (11) should be replaced by the enantiomer; G. Shaw, personal communication.
8 J. Mann and A. Thomas, *J. Chem. Soc., Chem. Commun.*, 1985, 737.
9 K.M. Sun, R.M. Giuliano, and B. Fraser-Reid, *J. Org. Chem.*, 1985, *50*, 4774.
10 M.A. Rahman and B. Fraser-Reid, *J. Am. Chem. Soc.*, 1985, *107*, 5576.
11 M. Prashad and B. Fraser-Reid, *J. Org. Chem.*, 1985, *50*, 1564.
12 R. Tsang and B. Fraser-Reid, *J. Org. Chem.*, 1985, *50*, 4659.
13 B. Fraser-Reid and R. Tsang, *Stud. Org. Chem. (Amsterdam)*, 1985, 20 (Nat. Prod. Chem.) 197 (*Chem. Abstr.*, 1985, *103*, 142192).
14 A.A. Kandil and K.N. Slessor, *J. Org. Chem.*, 1985, *50*, 5649.
15 R.H. Schlessinger, J.-W. Wong, M.A. Poss, and J.P. Springer, *J. Org. Chem.*, 1985, *50*, 3950.
16 R.C. Anderson and B. Fraser-Reid, *J. Org. Chem.*, 1985, *50*, 4781.
17 R.C. Anderson and B. Fraser-Reid, *J. Org. Chem.*, 1985, *50*, 4786.
18 P.-T. Ho and S. Wong, *Can. J. Chem.*, 1985, *63*, 2221.
19 T.-J. Lee, *Tetrahedron Lett.*, 1985, *26*, 4995.
20 R.M. Rabanal, J. Escudero, M. Martin- Lomas, S. Valverde, A. Perales, and J. Fayos, *Carbohydr. Res.*, 1985, *141*, 49.
21 R. Baker, C.J. Swain, and J.C. Head, *J. Chem. Soc., Chem. Commun.*, 1985, 309.
22 D. Calshaw, P. Grice, S.V. Ley, and G.A. Strange, *Tetrahedron Lett.*, 1985, *26*, 5837.
23 M. Isobe, Y. Ichikawa, and T. Goto, *Tetrahedron Lett.*, 1985, *26*, 5199.

24 M. Isobe, Y. Ichikawa, D. Bai, and T. Goto, Tetrahedron Lett., 1985, 26, 5203.
25 J.Y.C. Chan, L. Hough, and A.C. Richardson, J. Chem. Soc., Perkin Trans. I, 1985, 1457.
26 R.E. Ireland, D. Habich, and D.W. Norbeck, J. Am. Chem. Soc., 1985, 107, 3271.
27 O. Achmatowicz, Jr. and A. Sadownik, J. Carbohydr. Chem., 1985, 4, 435.
28 H. Irie, J. Igarashi, K. Matsumoto, Y. Yanagawa, T. Nakashima, T. Ueno, and H. Fukami, Chem. Pharm. Bull., 1985, 33, 1313.
29 R. Menicagli, C. Malanga, A. Pecunioso, and L. Lardicci, Gazz. Chim. Ital., 1985, 115, 23 (Chem. Abstr., 1985, 103, 37326).
30 M. Kawasaki, F. Matsuda, and S. Terashima, Tetrahedron Lett., 1985, 26, 2693.
31 R.J. Ferrier, P. Schmidt and P.C. Tyler, J. Chem. Soc., Perkin Trans. I, 1985, 301.
32 G.E. Keck, D.F. Kachensky and E.J. Enholm, J. Org. Chem., 1985, 50, 4317.
33 K.C. Nicolaou, M.E. Duggan, C.-K. Hwang, and P.K. Somers, J. Chem. Soc., Chem. Commun., 1985, 1359.
34 R.E. Dolle and K.C. Nicolaou, J. Am. Chem. Soc., 1985, 107, 1691, 1695.
35 R.E. Dolle and K.C. Nicolaou, J. Chem. Soc., Chem. Commun., 1985, 1016.
36 R.E. Ireland and D.W. Norbeck, J. Am. Chem. Soc., 1985, 107, 3279.
37 R.E. Ireland, D.W. Norbeck, G.S. Mandel, and N.S. Mandel, J. Am. Chem. Soc., 1985, 107, 3285.
38 S. Hanessian, Y. Guindon, P. Lavallee, and P. Dextraze, Carbohydr. Res., 1985, 141, 221.
39 J. Thiem and H. Luders, Liebigs Ann. Chem., 1985, 2151.
40 K. Tatsuta, K. Akimoto, M. Annaka, Y. Ohno, and M. Kinoshita, J. Antibiot., 1985, 38, 680.
41 K. Tatsuta, K. Akimoto, M. Annaka, Y. Ohno, and M. Kinoshita, Bull. Chem. Soc. Jpn., 1985, 58, 1699.
42 J.N. Freskos and J.S. Swenton, J. Chem. Soc., Chem. Commun., 1985, 658.
43 C.C. Wei, S. DeBernardo, J.P. Tengi, J. Borgese, and M. Weigele, J. Org. Chem., 1985, 50, 3462.
44 Y. Nishimura, S. Kondo, and H. Umezawa, J. Org. Chem., 1985, 50, 5210.
45 M.H. Ali, L. Hough, and A.C. Richardson, Carbohydr. Res., 1985, 136, 225.
46 T. Suami, K. Tadano, and Y. Iimura, Carbohydr. Res., 1985, 136, 67.
47 N. Yasuda, H. Tsutsumi, and T. Takaya, Chem. Lett., 1985, 31.
48 H. Setoi, H. Takeno, and M. Hashimoto, J. Org. Chem., 1985, 50, 3948.
49 H. Setoi, H. Takeno, and M. Hashimoto, Tetrahedron Lett., 1985, 26, 4617.
50 T. Wakamiya, K. Yamanoi, M. Nishikawa, and T. Shiba, Tetrahedron Lett., 1985, 26, 4759.
51 J.G. Buchanan, K.A. MacLean, R.H. Wightman, and H. Paulsen, J. Chem. Soc., Perkin Trans. I, 1985, 1463.
52 R.C. Bernotas and B. Ganem, Tetrahedron Lett., 1985, 26, 4981.
53 K. Tadano, Y. Iimura, and T. Suami, J. Carbohydr. Chem., 1985, 4, 129.
54 F. Matsuda, M. Kawasaki, and S. Terashima, Tetrahedron Lett., 1985, 26, 4639.
55 M. Shibuya, Heterocycles, 1985, 23, 61.
56 G. Sacripante, C. Tan, and G. Just, Tetrahedron Lett., 1985, 26, 5643.
57 M. Isobe, N. Fukami, and T. Goto, Chem. Lett., 1985, 71.
58 J. Adams, B.J. Fitzsimmons, Y. Girard, Y. Leblanc, J.F. Evans, and J. Rokach, J. Am. Chem. Soc., 1985, 107, 464.
59 Y. Leblanc, B. Fitzsimmons, J. Adams, and J. Rokach, Tetrahedron Lett., 1985, 26, 1399.
60 Y. Le Merrer, A. Dureault, C. Gravier, D. Languin and J.C. Depezay, Tetrahedron Lett., 1985, 26, 319.
61 I. Uchida, K. Yoshida, Y. Kawai, S. Takase, Y. Itoh, H. Tanaka, M. Kohsaka, and H. Imanaka, J. Antibiot., 1985, 38, 1476; Chem. Pharm. Bull. 1985, 33, 424.
62 Y. Oikawa, T. Nishi, and O. Yonemitsu, J. Chem. Soc., Perkin Trans. I, 1985, 1, 7.

63 Y. Oikawa, T. Nishi, and O. Yonemitsu, J. Chem. Soc., Perkin Trans. I, 1985, 19.
64 Y. Oikawa, T. Nishi, and O. Yonemitsu, J. Chem. Soc., Perkin Trans. I, 1985, 27.
65 P.J. Hodges and G. Procter, Tetrahedron Lett., 1985, 26, 4111.
66 A.F. Sviridov, M.S. Ermolenko, D.Y. Yashunsky, A.S. Shashov, and N.K. Kochetkov, Carbohydr. Res., 1985, 136, 101.
67 M. Kinoshita and Y. Mori, Bull. Chem. Soc. Jpn., 1985, 58, 3298.
68 S. Hanessian, P.J. Murray, and S.P. Sahoo, Tetrahedron Lett., 1985, 26, 5623, 5627.
69 S. Hanessian, S.P. Sahoo, and P.J. Murray, Tetrahedron Lett., 1985, 26, 5631.
70 M. Obayashi and M. Schlosser, Chem. Lett., 1985, 1715.
71 K. Koike, Y. Nakamara, and T. Ogawa, Glycoconjugate J., 1984, 1, 107 (Chem. Abstr., 1985, 103, 104777).
72 M. MacCoss, A. Chen, and R.L. Tolman, Tetrahedron Lett., 1985, 26, 1815.
73 M. MacCoss, A. Chen, and R.L. Tolman, Tetrahedron Lett., 1985, 26, 4287.
74 H. Ohrui, T. Misawa, and H. Meguro, J. Org. Chem., 1985, 50, 3007.

# Author Index

In this index the number given in parentheses is the chapter number, which is followed by the relevant reference number(s) within that chapter.